INTERNATIONAL TECHNOLOGICAL UNIVERSITY
This Book is Donated by:
PROF. WAI-KAI CHEN

Date:

LECTURES ON COMPLEX ANALYSIS

ITU Library
Date: ______________

Proceedings of the Symposium on Complex Analysis

LECTURES ON COMPLEX ANALYSIS

21 — 27 May, 1987
Xian, China

Editor
Chi-Tai Chuang
Department of Mathematics
Peking University

World Scientific
Singapore • New Jersey • London • Hong Kong

Published by

World Scientific Publishing Co. Pte. Ltd.
P O Box 128, Farrer Road, Singapore 9128

USA office: World Scientific Publishing Co., Inc.
687 Hartwell Street, Teaneck, NJ 07666, USA

UK office: World Scientific Publishing Co. Pte. Ltd.
P O Box 379, London N12 7JS, England

Library of Congress Cataloging-in-Publication data is available.

COMPLEX ANALYSIS

Copyright © 1988 by World Scientific Publishing Co. Pte. Ltd.
All rights reserved. This book, or parts thereof, may not be reproduced in any form or by any means, electronic or mechanical, including photocopying, recording or any information storage and retrieval system now known or to be invented, without written permission from the Publisher.

ISBN 9971-50-707-2

Printed in Singapore by Utopia Press.

PREFACE

Recently Chinese mathematicians who are interested in the field of the theory and applications of functions of one complex variable organized a symposium on complex analysis which has taken place in the Northwestern University in Xian on May 21-27, 1987. It is supported by the Institute of Mathematics of the Chinese Academy of Sciences, the Peking University, the Northwestern University, the Fudan University, the Beijing University of Technology and the Beijing Institute of Technology. Among the participants in the symposium, besides the Chinese mathematicians, there are seven invited speakers from abroad. They are Carl H. FitzGerald, Irwin Kra, Chung-chun Yang, Carl David Minda and David J. Hallenbeck from the United States, I. Noel Baker from England and Kiyoshi Niino from Japan. We often organize symposiums, but this is the first time that there are speakers from abroad. It turned out that the symposium is a successful one.

At the symposium, lectures on various topics of complex analysis were delivered, such as iteration of entire functions, factorization of meromorphic functions, growth and value distribution of meromorphic functions, harmonic functions, univalent functions, quasiconformal mappings, Riemann surfaces, function spaces, applications of complex functions, special functions, etc. This volume contains twenty-six of these lectures. The editor apologizes for being unable to put all of them in this volume.

Finally, as the editor, I express my gratitude to the contributors of this volume for sending me their manuscripts. I also thank Dr. Phua, Editor-in-Chief, and the editorial staff of the World Scientific Publishing Company for making the publication of this volume possible.

Chi-tai Chuang

CONTENTS

ITERATION OF ENTIRE FUNCTIONS: AN INTRODUCTORY SURVEY

I.N. Baker
Department of Mathematics
Imperial College
London, SW7 2BZ
England

§1. INTRODUCTION

Given a map $f: S \to S$, denote the n-th iterate of f by f^n, $n \in \mathbb{N}$. The study of such situations is of course quite old. Three of the early instances which are still the subject of research are Newton's method of solving equations, Euler's work [17,4,5] on infinite towers of exponentials, and Abel's functional equation $A(f(x)) = A(x)+1$, where f is prescribed and A to be found [1, see also 24].

If $S \subset \mathbb{C}$ and f is analytic in S, a point z_1 in S is a <u>fixed point</u> of f if $f(z_1) = z_1$. The derivative $f'(z_1)$ is the <u>eigenvalue</u> of z_1. More generally z_1 is <u>cyclic with period</u> n if $f^n(z_1) = z_1$, $f^j(z_1) \neq z_1$ for $1 \leq j < n$. Setting $f^j(z_1) = z_{j+1}$, so that $z_{n+1} = z_1$, the set $\{z_1, z_2, \ldots, z_n\}$ is called a <u>cycle</u> (of order n) and $\lambda = (f^n)'(z_i)$, which is the same for all i, is the <u>eigenvalue</u> of the cycle.

The local theory of iteration studies the iterates $f^k(z)$ in the neighbourhood of a fixed point or cycle. The behaviour is rather obvious in the cases (i) $|\lambda| < 1$, when the cycle or fixed point is <u>"attractive"</u>, and (ii) $|\lambda| > 1$ when it is <u>"repelling"</u>.

If $S = \hat{\mathbb{C}}$ or $\mathbb{C}$, the analytic map f is rational or entire, respectively, and f^n is of the same type. We shall, in order to exclude very simple exceptional cases, assume that f is not univalent or constant; that is, f should not be of the form $(az+b)/(cz+d)$, a,b,c,d constant. We may then study the global behaviour of (f^n) which leads to two types of questions:

Q1. For fixed f study the behaviour of $(f^n(z_0))$ as z_0 varies over $\overset{(\wedge)}{\mathbb{C}}$.

Q2. Questions of stability. How does the iteration of f change as we vary f? For instance we may restrict attention to a manifold of functions f, in the simplest cases by allowing a parameter to vary.

The local theory (see e.g. [24] for references) was developed, starting about 1880 by mathematicians such as Koenigs, Leau and Poincaré although some of the more difficult cases when $\lambda = \exp(2\pi i\alpha)$, $\alpha \in \mathbb{R}\backslash\mathbb{Q}$, were not satisfactorily dealt with until Siegel [34] in 1942.

The global theory dates essentially from Julia [22] and Fatou [18], starting in the early 20th century. For entire functions the starting point is Fatou [19]. This theory flourished until about 1930, but met some difficult problems. Some work continued (for example by Kneser [23], Myrberg [32], Baker (references in [4], [7]) but it is only in recent years that a period of much new progress has begun. This progress has come from a mutual interaction between computer experiments (Mandelbrot [28,29]) and new analytic techniques (Sullivan [35-38], Thurston [39], Douady-Hubbard [13]) involving quasi-conformal maps, and topological notions (such as Douady's polynomial-like maps). Ideas from dynamical systems theory and ergodic theory have also contributed much (Ruelle [33], Mañé-Sad-Sullivan [30], Lyubich [25, 26] etc.).

This introductory survey will deal only with the global theory of Fatou and Julia. We shall always assume that f is entire, $f(z) \neq az+b$. Surveys of the general rational case are given by Blanchard [7] (introductory), and, for a more technical and up-to-date account, Lyubich [26].

§2. THE FATOU-JULIA THEORY FOR ENTIRE FUNCTIONS

The main tool is the notion of normal family. Recall the definition: A set of functions g_α analytic in a domain D is normal if every sequence g_{α_n} of functions of the family has a subsequence which is locally uniformly convergent in D. Note that convergence to ∞ is allowed in this definition.

Standing assumption. From now on f is an entire function, not of the form $az+b$, a,b constant.

Definition. $N(f) = \{z;$ there is a neighbourhood D of z such that (f^n) is normal in $D\}$. $N(f)$ is the set of normality and its complement, denoted by $J(f)$ is the Julia set.

It is a basic theorem that $J(f)$ is perfect (and so, in particular, a non-empty closed set). This and the basic properties listed below are proved in Fatou's paper of 1926.

J1. Complete invariance:

$$z \in J(f) \Longleftrightarrow f(z) \in J(f);$$

$$z \in N(f) \Longleftrightarrow f(z) \in N(f).$$

J2. For $p \in \mathbb{N}$: $J(f) = J(f^p)$.

These imply

J3. If N_j are the components of $N(f)$, then $f(N_j) \subset N_k$ for some $k = k(j)$.

J4. {Attractive cycles} $\subset N(f)$
{Repelling cycles} $\subset J(f)$.

(By J2 it is enough to prove this for cycles of order 1, that is, for fixed points).

<u>Definition</u>. The negative orbit of z is

$$o^-(z) = \{(f^n)^{-1}(z),\ n = 0, 1, 2, \ldots\}.$$

<u>Definition</u>. The exceptional set of f is

$$E(f) = \{z;\ o^-(z) \text{ is a finite set}\}.$$

J5. $E(f)$ contains at most two points.

Example. $E(z^2) = \{0, \infty\}$.

J6. Expanding property of $J(f)$:
For any $\beta \in J(f)$, any disc Δ which contains β and for any compact K such that $K \cap E(f) = \phi$, there exists $n_o \in \mathbb{N}$ such that, for all $n > n_o$: $f^n(\Delta) \supset K$.

J1 and J6 imply

J7. Either $J(f)$ is nowhere dense or $J(f) = \mathbb{C}$.

We shall see later that $J(e^z) = \mathbb{C}$.

§3. EXAMPLES

<u>Example 1.</u> The above properties may be verified in the simple case when $f(z) = z^2$. Then $J = \{z: |z|=1\}$ and N has two components $\{z: |z|<1\}$, $\{z: |z|>1\}$.

Note that ∞ is to be considered an attractive fixed point for f: If we take local parameter $Z = 1/z$ we shift ∞ to $Z = 0$ and the map becomes $Z \to Z^2$, with eigenvalue 0 at the fixed point $Z = 0$.

Example 2. For any polynomial f, whose degree by our assumptions, is at least two, there is some $A > 0$ such that

$$|f(z)| > 2|z|, \quad |z| > A,$$

whence $f^n \to \infty$ uniformly in $\{|z|>A\} = D$.

This is a component N_∞ of $N(f)$ which contains D. Since for all branches of f^{-1} we have $f^{-1}(\infty) = \infty$, it follows from J1 that $f^{-1}(N_\infty) \subset N_\infty$. In fact N_∞ is completely invariant and $N_\infty = \{z; f^n(z) \to \infty\}$. The complement of N_∞ consists of $J(f)$ together with any other components of $N(f)$.

Example 3. For any transcendental f the set $J(f)$ is unbounded. If there is a multiply-connected component of f it can be shown to be bounded. A relatively simple example is given by $f(z) = e^{cz}$, $0 < c < 1/e$. Here $N(f)$ is a single simply-connected domain. Its boundary consists of $\bigcup_{n=-\infty}^{\infty} (B+2n\pi i)$, where B is a Cantor set of lines clustering about the segment $[x,\infty)$, and $e^{cx} = x$ (Devaney and Krych [10]). In this case the plane Lebesgue measure of J is zero (Eremenko-Lyubich [15], but the Hausdorff dimension of J is two (McMullen [31]).

§4. THE DIFFERENT TYPES OF NORMAL CONVERGENCE

We interpret Question 1 of §1 in the sense of describing the different limit functions of $(f^n(z))$ in a component N_1 of $N(f)$. From J3 we have

$$N_1 \overset{f}{\to} N_2 \overset{f}{\to} \dots \to N_k \overset{f}{\to} N_{k+1} \to \dots \qquad (4.1)$$

where each N_k is a component of $N(f)$.

The sequence (4.1) is either ultimately periodic, or not. In the first case we have for some p,k that

$$f^p: N_k \to N_k \qquad (4.2)$$

and N_1 is called _preperiodic_. In the second case all N_k are different and N_1 is called a _wandering domain_ (W.D.) of f.

Limits in the preperiodic case (4.2).

We remark first that N_k is simply-connected, except in the case of N_∞ for polynomial f in Example 2 of §3.

There are only four types of behaviour for (f^n) in a preperiodic N_1. This has been known since the days of Fatou and Julia, although it was not known until 1942 that the case of the Siegel disc can in fact occur. We list the cases.

(i) There is an attractive fixed point α of f^p in N_k. Then $f^{np} \to \alpha$ in N_k as $n \to \infty$. The map f^p restricted to N_k is not univalent (since otherwise one has $|(f^p)'(\alpha)| = 1$). Thus N_k contains a singular point of $(f^p|_{N_k})^{-1}$.

The cycle of domains $N_k, N_{k+1}, \dots, N_{k+p-1}$ then contains an attractive cycle of points $\alpha, f(\alpha), \dots, f^{p-1}(\alpha)$, and also $N_k \cup N_{k+1} \cup \dots \cup N_{k+p-1}$ contains a singular point of f^{-1}.

(ii) This may be regarded as a limiting case of (i) when α belongs to the boundary of N_k. In this case the eigenvalue of the fixed point α of f^p is exactly 1. The detailed description of the boundary of N_k near α is somewhat tedious. However, just as in (i) we have $f^{np} \to \alpha$ in N_k and there is again a singular point of f^{-1} in some N_{k+j}, $0 \leq j < p$.

(iii) The case of the Siegel disc. In this case N_k contains a fixed point α of f^p whose eigenvalue is $\lambda = \exp(2\pi i\vartheta)$, $\vartheta \in \mathbb{R}\backslash\mathbb{Q}$. There is a conformal homeomorphism ψ of N_k to the unit disc such that $\psi(\alpha) = 0$, $\psi \circ f^p \circ \psi^{-1}(t) = \lambda t$, $t \in D$. Thus f conjugates f^p to a rotation. In this case it is possible to find integers $n_\nu \to \infty$ such that $f^{pn_\nu} \to \mathrm{Id}$ in N_k. Thus there are non-constant limit functions.

Singular points are involved in the theory of Siegel discs:

$$\partial N_k \subset \overline{\{f^n(c),\ c \text{ singular for } f^{-1},\ n \in \mathbb{N}\}}.$$

A problem which has inspired recent work [20] is: does ∂N_k contain a singular point of f^{-p}?

(iv) N_k is an unbounded domain in which $f^n \to \infty$.

In this case the rate of convergence is in fact bounded by $\log|f^n(z)| = o(n)$ as $n \to \infty$ and there is some path to infinity on which $|f^p(z)| = o(|z|)$ [6]. An example is $e^{-z}+kz$, $k > 1$.

Wandering domains.

The occurrence of these for entire f was first shown in [3] and now many other examples have been given, for example by Herman [21], Sullivan [36], Eremenko and

Lyubich [16].

In the case of a wandering domain N_1 we may have $f^n \to \infty$ arbitrarily fast (Baker [6]) or we may have infinitely many different constant limit functions (Eremenko, Lyubich [16]). Thus the limiting behaviour of f^n in a W.D. may be more complicated than in the preperiodic case and is not yet completely understood.

§5. CASES WHERE W.D. ARE ABSENT

Much of the impetus of the modern revival of iteration theory comes from Sullivan's theorem [35]:

If f *is a rational function then* f *has no W.D.*

Thus polynomials have no W.D.; transcendental f may do so. The proof introduces the subject of quasiconformal mapping into iteration theory. See also §8.

Sullivan's proof can be made to work also for certain classes of entire functions. For example:

If $f(z) = e^{cz}$, c constant, or, more generally, if $f(z) = \int P(z)\exp(Q(z))dz$, P,Q polynomials, then there are no W.D., as shown by Baker and Rippon [5]. This paper gives an accessible and relatively simple account of the method for the classical analyst. These results are included in the more general result of Eremenko and Lyubich [14] that if the singular points of f^{-1} lie over a finite set of points, then f has no W.D.

Relatively "simple" functions such as $z+2\pi i-1+e^{-z}$ have W.D.

§6. ITERATION OF FAMILIES WITH A PARAMETER

We come to Question 2 of §1 and examine only two especially "simple" examples. We look at the cases

(i) $f_c(z) = z^2 + c$,

(ii) $f_c(z) = e^{cz}$,

where c is a complex parameter. These are simple because (a) no W.D. can occur, (b) f_c^{-1} has exactly one (finite) singular point c or 0, respectively, and thus the number of cycles of type 4(i) or (ii) is at most one.

§7. THE QUADRATIC FAMILY z^2+c

For $f_c(z) = z^2+c$ there is exactly one domain, N_∞, of $N(f)$ in which $f^n \to \infty$. This is completely invariant.

If there is any other cycle of domains (4.2) with $N_k \neq N_\infty$ then the orbit $f^n(c)$ must either converge to a cycle of points as in 4(i), 4(ii) or be dense in ∂N_k in case 4(iii). Thus there is at most one cycle of type 4(i) or 4(ii), and an argument of Douady [13] shows that in fact there is at most one cycle of type 4(i), (ii) or (iii). If $f^n(c) \to \infty$ there is no such cycle and $N(f) = N_\infty$.

The Mandelbrot set is defined by

$$M = \{c;\ f_c^n(c) \not\to \infty\}. \tag{7.1}$$

A little analysis shows that this is equivalent to

$$M = \{c;\ |f_c^n(c)| \leqslant 2 \text{ for all } n\}. \tag{7.2}$$

Since $f_c^n(c)$ is an explicit polynomial in c one can plot the sets $\{z:\ |f_c^n(c)| \leqslant 2\}$ and obtain approximations to M using a computer. The well-known pictures of M (e.g. [28], [29]) show a fascinating structure with a main cardioid C_1 touched by smaller approximately circular regions which are touched by yet further regions and so on. M is bounded, lying within $D(0,2)$ and is in fact connected (Douady-Hubbard [11]). There are many deep problems, as yet unsolved. For instance it is not known whether M is locally-connected. An extensive study of M is being made by Douady, Hubbard and their students ([12], [13]).

The importance of M from our present point of view is that it gives a chart in the parameter plane (c-plane) of regions for which f_c shows certain types of iterative behaviour. Thus

$$M = \{c;\ J(f_c) \text{ is connected}\}.$$

For c in the exterior of M $J(f_c)$ is totally disconnected. For c in the cardioid C_1 the behaviour of f_c is much the same as for $c = 0$, $f = z^2$. $N(f)$ has two components N_∞ and N_0. N_0 contains an attractive fixed point α and is a completely invariant component. $f^n \to \alpha$ in N_0, $f^n \to \infty$ in N_∞. N_0 and N_∞ are separated by a Jordan curve which is in fact $J(f_c)$. If $c \neq 0$ this curve has Hausdorff dimension greater than 1.

As c moves from C_1 into the adjacent regions of M, tangent to C_1, $N(f_c)$ becomes more complicated as cycles of higher order appear. One speaks of M as a bifurcation-diagram for the iteration.

We leave the family z^2+c with these few remarks and turn to the next family, whose members are, in contrast, transcendental entire functions.

§8. THE FAMILY $g_c(z) = \exp(cz)$

We recollect from §6 that at most one attractive cycle and no wandering domains can occur. It is also easy to show that §4(iv) cannot occur [5, p.73].

It is thus reasonable to attempt a "Mandelbrot diagram" M_e for e^{cz}. Define $M_e = UD_n$ where $D_c = \{c;\ e^{cz}$ has an attractive n-cycle$\}$. This may be plotted by computer: if there is such a cycle then $g_c^k(0)$ approaches the cycle as $k \to \infty$. This has been done independently by Baker-Rippon [5] and by Devaney [8]. Devaney considers the family of maps $h_c(w) = ce^w$. Since the maps g_c, h_c correspond under the change of variable $w = cz$, they have the same types of cycle and the sets D_n are the same in either case.

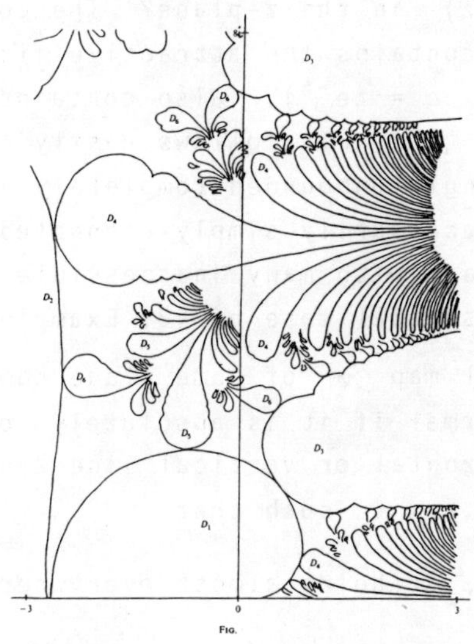

FIG.

A sketch of part of the set is shown in the Figure and properties are described in more detail in [5]. D_1 is a cardioid, D_2 is an unbounded region (almost a half-plane). Any D_p has many D_{pq} tangent to it. Between two adjacent members of a family D_p $(p > 2)$ there is an infinite family of D_{p+1}, all of which go to ∞ in the positive real direction.

There are many problems about the diagram; for instance is $\bigcup_n D_n$ dense in $\mathbb{C}$?

It is easy to determine D_1. For c is in D_1 precisely when $g = e^{cz}$ has an attractive fixed point $w : e^{cw} = w$, $g'(w) = cw$ with $|cw| < 1$. Putting

$t = cw$ gives $t = cw = ce^t$ so that $c = te^{-t}$. Thus

$$D_1 = \{c;\ c = te^{-t} \text{ for some } |t| < 1\}.$$

M_e is a chart in the parameter plane (c-plane) whose regions correspond to types of iterative behaviour for g_c.

In the simplest case, $c \in D_1$. What is then the behaviour of $g_c^n(z)$ in the z-plane? The component N_1 of $N(g_c)$ which contains the attractive fixed point $w(c) = e^t$ (where $c = te^{-t}$), also contains the unique singular point of g_c^{-1}. It follows easily that $N(g_c) = N_1$, a single unbounded completely invariant domain. N_1 is necessarily simply-connected but has a complicated boundary with many inaccessible points, as described in the special case of §3, Example 3.

A topological map φ of some plane domain into $\mathbb{C}$ is called quasiconformal if it is absolutely continuous on almost every horizontal or vertical line and if there is a constant k, $0 \leqslant k < 1$ such that

$$|\varphi_{\bar{z}}| \leqslant k|\varphi_z| \text{ holds almost everywhere in } D.$$

The case $k = 0$ corresponds to conformality.

The importance of quasiconformal maps in the theory appears in the following result. Using ideas of Douady and Sullivan one can show: Given c_1, c_2 in $D_1 \setminus \{0\}$ there exists a quasiconformal homeomorphism $\varphi\colon \mathbb{C} \to \mathbb{C}$, which fixes $0, 1, \infty$ and is such that $g_{c_2} = \varphi \circ g_{c_1} \circ \varphi^{-1}$ in $\mathbb{C}$. Thus the iteration picture for g_{c_1}, g_{c_2} is topologically the same. For example fixed points and cycles of a given order correspond. Further φ depends continuously on c_1, c_2. The map φ cannot of course be taken to be conformal: for example $\varphi(w(c_1)) = w(c_2)$, but the eigenvalues of $w(c_1)$, $w(c_2)$ are different.

It is interesting to consider $c = 1$, $g(z) = e^z$. In this case $g^n(0) \to \infty$ and hence, by the results of §4, there are no periodic components of $N(g)$ is empty and $J(e^z) = \hat{\mathbb{C}}$. There are many other similar examples of this phenomenon.

Can there be a connection between M (for z^2+c) and M_e? Devaney, Goldberg and Hubbard [9] have attempted to interpret the familiar limit

$$\lambda(1 + \frac{z}{n})^n \to \lambda e^z, \quad n \to \infty,$$

in a dynamical sense, to define an analogue M_n of M for $\lambda(1 + \frac{z}{n})^n$ and show that, in a sense, this approaches M_e.

§9. REMARKS AND PROBLEMS

We conclude this short and incomplete survey with a few additional remarks about open problems.

1. There are several ways of constructing wandering domains. However, their theory is by no means complete and the various kinds of possible limiting behaviour not sufficiently understood.
2. There is scope for the study of explicit examples of simple types of f. Thus McMullen [31] finds for example in the family $\sin(cz+d)$ examples where $J(f)$ has positive area but is not the whole plane.
3. The structure of M still holds many problems and that of M_e has barely begun.
4. We have not mentioned the whole area of ergodic problems. For instance one may ask for an invariant measure for the map f, that is a measure μ such that $\mu(A) = \mu(f^{-1}(A))$ in general. Much work has been

done about ergodic problems relating to rational maps and recently some results for transcendental functions also, for example Lyubich's result "e^z is not ergodic" [27].

REFERENCES

1. Abel, N.H.: Détermination d'une fonction au moyen d'une fonction qui ne contient qu'une seule variable. Oeuvres Complètes II, Christiania, 1881, 36-39.

2. Baker, I.N.: The domains of normality of an entire function. Ann. Acad. Sci. Fenn. Ser. A.I. Math., 1, 1975, 277-283.

3. Baker, I.N.: An entire function which has wandering domains. J. Australian Math. Soc. Ser. A, 22, 1976, 173-176.

4. Baker, I.N. and Rippon, P.J.: Convergence of infinite exponentials. Ann. Acad. Sci. Fenn. Ser. A.I. Math. 8, 1983, 179-186.

5. Baker, I.N. and Rippon, P.J.: Iteration of exponential functions. Ann. Acad. Sci. Fenn. Ser. A.I. Math. 9, 1984, 49-77.

6. Baker, I.N.: Infinite limits in the iteration of entire functions (preprint).

7. Blanchard, P.: Complex analytic dynamics on the Riemann sphere. Bull. Amer. Math. Soc. 11, 1984, 85-141.

8. Devaney, R.L.: Julia sets and bifurcation diagrams for exponential maps. Bull. A.M.S. 11, 1984, 167-171.

9. Devaney, R.L., Goldberg, L.R. and Hubbard, J.H.: A dynamical approximation to the exponential map by polynomials (preprint).

10. Devaney, R.L. and Krych, M.: Dynamics of exp z. Ergodic Theory and Dynamical Systems 4, 1984, 35-52.

11. Douady, A. and Hubbard, J.H.: Itération des polynomes quadratiques complexes. C.R. Acad. Sci. Paris, 1982, 294, 123-126.

12. Douady, A. and Hubbard, J.H.: Etude dynamique des polynomes complexes. Publ. Mat. Orsay I, 1984, and II, 1985.

13. Douady, A. and Hubbard, J.H.: On the dynamics of polynomial-like mappings. Ann. Sci. Ecole Normale Sup. 18, 1985, 287-343.

14. Eremenko, A. and Lyubich, M. Yu.: Iteration of entire functions (Russian). Preprint, 1984, No. 6 of the Physico-technical Institute of Low Temperatures, Ukr. S.S.R. Academy of Sciences, Kharkov.

15. Eremenko, A. and Lyubich, M. Yu.: Structural stability in some families of entire functions. Preprint, 1984, No. 29 of the Physico-technical Institute of Low Temperatures, Ukr. S.S.R. Academy of Sciences, Kharkov.

16. Eremenko, A. and Lyubich, M. Yu.: Examples of entire functions with pathological dynamics. To appear in the Bull. London Math. Soc.

17. Euler, L.: De formulis exponentialibus replicatus. Opera omnia, Series Prima XV (1927), 267-297 and Acta Petropolitanae, 1 (1777), 38-60.

18. Fatou, P.: Sur les équations fonctionelles. Bull. Soc. Math. France 47, 1919, 161-271 and 48, 1920, 33-94, 208-314.

19. Fatou, P.: Sur l'itération des fonctions transcendantes entières. Acta Math. 47, 1926, 337-370.

20. Herman, M.R.: Are there critical points on the boundaries of singular domains? Comm. Math. Phys. 99, 1985, 593-612.

21. Herman, M.R.: Exemples de fractions rationelles ayant une orbite dense sur la sphère de Riemann. Bull. Soc. Math. France 112, 1984, 93-142.

22. Julia, G.: Mémoire sur la permutabilité des fractions rationelles. Ann. Sci. Ecole Norm. Sup. (3) 39, 1922, 131-215.

23. Kneser, H.: Reelle analytische Lösungen der Gleichung $\varphi(\varphi(x)) = e^x$ und verwandter Funktionalgleichungen. J.f.d. reine u. angew. Math., 187, 1949, 56-67.

24. Kuczma, M.: Functional equations in a single variable. P.W.N. Warsaw, 1968.

25. Lyubich, M. Yu.: Entropy properties of rational endomorphisms of the Riemann sphere. Ergodic Theory and Dynamical Systems 3, 1983, 351-386.

26. Lyubish, M. Yu.: Dynamics of rational maps: topological picture (Russian). Uspehi Mat. Nauk. 41, 1986, 35-95.

27. Lyubich, M. Yu.: e^z is not ergodic (preprint).

28. Mandelbrot, B.: Fractal aspects of the iteration of $z \to \lambda z(1-z)$ for complex λ, z. Annals N.Y. Acad. Sciences 357, 1980, 249-259; II Physica 7D, 1983, 224-239; III Chaos, fractals and dynamics, 213-224, Lecture Notes in Pure & Applied Mathematics, 8, Dekker, N.Y. 1985; IV Ibid, 225-234; V Ibid, 235-238; VI Ibid, 239-242; VII Ibid, 243-253.

29. Mandelbrot, B.: The fractal geometry of nature. New York: Freeman, 1983.

30. Mañé, R., Sad, P. and Sullivan, D.: On the dynamics of rational maps. Ann. Sci. Ecole Normale Sup. 16, 1983, 193-217.

31. McMullen, C.: Area and Hausdorff dimension of Julia sets of entire functions. Trans. A.M.S. 300, 1987, 329-341.

32. Myrberg, P.J.: Iteration der reellen Polynome zweiten Grades. Ann. Acad. Sci. Fennicae A.I. 256, 1958, 268, 1959 and 336, 1964.

33. Ruelle, D.: Repellers for real analytic maps. Ergodic Theory and Dynamical Systems 2, 1982, 99-107.

34. Siegel, C.L.: Iteration of analytic functions. Ann. of Math. 43, 1942, 607-616.

35. Sullivan, D.: Itération des fonctions analytiques complexes. C.R. Acad. Sci. Paris Sér. A-B, 294, 1982, 301-303.

36. Sullivan, D.: Quasiconformal homeomorphisms and dynamics I. Solution of the Fatou-Julia problem on wandering domains. Ann. Math. 122, 1985, 401-418.

37. Sullivan, D.: Quasiconformal homeomorphisms and dynamics II. Structural stability implies hyperbolicity for Kleinian groups. Acta Math. 155, 1985, 243-260.

38. Sullivan, D.: Quasiconformal homeomorphisms and dynamics III. Topological conjugacy classes of analytic endomorphisms. To appear.

39. Thurston, W.: On the dynamics of iterated rational maps. Preprint 1982.

SOME ASPECTS OF FACTORIZATION THEORY-A SURVEY

Chung-chun Yang
Naval Research Laboratory
Washington, D.C. 20375
USA

1. INTRODUCTION

Rosenbloom [31] investigated the existence and quantitative estimation of the fix-points of a composite function of two entire functions. There he called an entire function F prime if a factorization of the form $F(z) = f(g(z))$, with f and g being entire implies that either f or g is linear. He also asserted without proof that the function $F(z) = e^z+z$ is prime. This function has several special properties, such as it has no fix-point, no multiple zeros, it is periodic mod a polynomial, and its derivative has restricted zeros, etc. Therefore, the primeness of e^z+z can be proven by several different approaches, and the proofs have suggested several directions in the development of the factorization theory. The primeness of e^z+z and its generalizations were subsequently proven by Baker and Gross [4]. Later, Gross [10] extended the study of the primeness of entire functions to meromorphic functions. He defined a meromorphic function $F(z)$ to have a factorization with left factor f and right factor g provided

$$F(z) = f(g(z)), \qquad (1.1)$$

where f is meromorphic and g is entire (g may be meromorphic when f is rational). $F(z)$ is said to be prime(pseudo-prime) if every factorization of form (1.1) implies that either f is bilinear or g is linear (either f is rational or g is a polynomial). F is called E-prime, if F is prime when only entire factors are considered in the factorization of form (1.1).

It seems that given a meromorphic (or entire) function it may be extremely difficult, if not impossible, to determine whether or not it is prime (for instance, it is still unknown [36] whether the function $[z+e^{2z}]/[z+e^{z}]$ is prime or not). The factorization theory has been pursued by numerous authors in various aspects and many interesting results have been obtained. However, for arbitrarily entire or meromorphic functions, no systematic research has actually been developed except for an isolated paper by Ritt [30], in which he completely solved the factorization problem for polynomials. Most of the results that have been derived thus far concern the existence of the fix-points of a composite entire function, the impossibility of factorization (that is the primeness or the pseudo-primeness of certain classes of entire and meromorphic functions), and permutability and unique factorizability of certain classes of entire functions. The tools used in the study of the factorization theory are mainly based on the classical theory for the growth properties of analytic functions and Nevanlinna's value distribution theory [19]. Most methods utilized with the following one or several factors: (1) the growth of the function; (2) zeros distribution; (3) periodicity; (4) the fix-points; (5) the existence of deficient values; and (6) solutions of differential equations.

Through the investigations of Gross and Yang in the U.S.A., Goldberg and Prokopovich in the U.S.S.R., Baker and

Goldstein in England, Steinmetz in Germany, Ozawa, Urabe and Noda in Japan, and Song in China, the theory of factorization has become a new branch in the value-distribution theory of meromorphic functions.

Generally speaking, the research in factorization theory is still far away from its maturity. There are many interesting problems to be studied and resolved. We strongly believe that value-distribution theory can be further perfected by studying factorization theory. Some of the progress in the factorization theory prior to 1980 were summarized by Yang [39] and Gross [12]. In these articles the authors attempted to collect some significant results concerning factorization theory of trancendental functions obtained after 1976 to the present, which are related to the six factors mentioned above. To help the reader to become better prepared in understanding and studying of this subject, we would like to point out two useful observations: 1. According to the definition, it seems inevitable that one must consider meromorphic factors in a factorization in order to determine whether a given entire function is prime or not. For instance, the function $e^{-mz} + e^{nz}$ ($m \neq n$, m,n are two distinct positive integers) is E-prime but not prime. However, by a simple argument Gross [11] proved that any nonperiodic entire function is prime if and only if it is E-prime. Thus, in most cases, we need not be concerned with the meromorphic factors in our discussion of the primality of a given entire function, since most functions that we shall encounter are not periodic. However there do exist some classes of prime entire functions that are periodic, see, e.g. [5, 14, 26]. 2. Let f and g be two entire functions, then $f(g)$ has infinitely many (no) fix-points if and only if $g(f)$ has(has none). Thus, for instance, if one has shown that e^z+z is impossible to be expressed as $f(p)$; f entire, p nonlinear polynomial, then it becomes also impossible for e^z+z to be expressed as $q(g)$; q a nonlinear polynomial, g an entire function. We refer the reader to two books, one by Gross [9] and the other by Chuang and Yang [6] for a better treatment and understanding of this subject and related problems.

2. THE FIX-POINTS OF COMPOSITE ENTIRE FUNCTIONS

As a generalization of the fact that e^z+z is prime Gross [9] asked: If $Q(z)$ is a polynomial $(\not\equiv 0)$ and $\alpha(z)$ is entire $(\not\equiv$constant), is $Q(z)e^{\alpha(z)}+z$ prime? If the answer is "yes", then this would imply that for any two nonlinear entire functions f and g with at least one of them being transcendental, the composition $f(g)$ (and hence $g(f)$) must have infinitely many fix-points. (We shall refer to this statement as the conjecture.) When the order of the composite function $f(g)$ is finite this question has been answered, as a special case, almost simultaneously by Goldstein [7], Prokopovich [29], and Gross and Yang [13] using different approaches.

Theorem 2.1 (Goldstein). Let $p(z)$ be a polynomial of degree $m(\geq 1)$, and let $\phi(z)(\not\equiv$constant) and $\psi(z)$ $(\not\equiv$constant) be entire functions of order $<m$. Suppose that $F(z) = \phi(z) + \psi(z)\exp(p(z)) = f(g(z))$, for some nonlinear entire functions f and g. Then $g(z)$ is a polynomial of degree $k<m$. Moreover, if ρ_f denotes the order of f, then $k\rho_f=m$.

Theorem 2.2 (Prokopovich). Let $F(z) = \psi_1(z) + \psi_2(z)e^{p(z)}$, where $p(z)$ is a nonconstant polynomial and where $\phi_1(z)$ $(\not\equiv$constant) and $\phi_2(z)(\not\equiv 0)$ are two entire functions satisfying

$$T(r, \phi_i) = o(1)T(r,F), \text{ as } r \to \infty, \ i = 1,2.$$

If $F(z)$ can be factorized as $F=f(g)$, where f and g are entire with g being a nonlinear polynomial satisfying $\deg g<\deg p$, then $\phi_1(z)$ and $\phi_2(z)e^{p(z)}$ have g as their common right factor.

Theorem 2.3 (Gross and Yang). Let $p(z)$ be polynomial of degree $m(\geq 1)$, and let $h(z)(\not\equiv 0)$ and $k(z)(\neq\text{constant})$ be two entire functions of order less than m. Then $h(z)e^{p(z)}+k(z)$ is either prime or it can only be factorized as

$$h(z)e^{p(z)}+k(z) = f(L(z)),$$

where $L(z)$ is a nonlinear polynomial of degree n, $f(z) = \mu(z)\exp(cz^d) + \beta(z)$ is an entire function, μ and β are entire functions of order less than m and c is a constant$\neq 0$. Furthermore the following three relationships are satisfied:

(i) $n|m$ (i.e. m/n is an integer)

(ii) $h(z)e^{p(z)} = \mu(L(z))\exp(cL(z)^d)$, $d = m/n$, and

(iii) $k(z) = \beta(L(z))$.

Later Yang [41], and Gross and Osgood [16] have shown that the conjecture is true for certain composite entire functions of infinite order, i.e. when certain restrictions are imposed on the growths of f and g. All their proofs used a beautiful result of Steinmetz's [34], which has become a very powerful tool in solving many kinds of factorization problems.

Theorem 2.4 (Steinmetz). Let $F_o, F_1, \ldots, F_m$ be $m+1$ not identically vanishing entire functions and let $h_o, h_1, \ldots, h_m$ be $m+1$ arbitrarily not identically vanishing meromorphic functions. Let g be a nonconstant entire function and let k be a positive number such that for $i = 0, 1, 2, \ldots m$

$$T(r, h_i) < kT(r, g) \text{ as } r \to \infty , \qquad (2.1)$$

outside possibly a set of r of finite measure.

If F_is and h_is satisfy

$$F_o(g)h_o + F_1(g)h_1 + \ldots\ldots + F_m(g)h_m \equiv 0,$$

then there exists polynomials in z, $p_o, p_1, \ldots p_m$, not all identically zero, such that

$$p_o(z)F_o(z) + p_1(z)F_1(z) + \ldots\ldots + p_m(z)F_m(z) \equiv 0.$$

With this result and an observation that the condition (2.1) in Theorem 2.4 can be relaxed to require that the condition needs only to be held on a sequence of positive numbers $\{r_n\}$, and while on the same sequence,

$$T(r, g') < (1 + o(1))T(r, g).$$

Gross and Osgood obtained the following theorem, which seems to be the best result towards a solution of the conjecture thus far.

Theorem 2.5. Let f and g be two entire functions both of which are nonlinear. Suppose that one of f and g is transcendental. Suppose also that one of f and g is of finite order, while the other, h say, satisfies $\underline{\lim}_{r \to \infty} T(r,h)/r^n = 0$, for some positive number n (This is equivalent to saying that h is of finite lower order). Then $f(g(z)) = z$ has infinitely many zeros.

3. THE CRITERIA OF PSEUDO-PRIMNESS FOR ENTIRE FUNCTIONS

In general, the steps to prove that a given transcendental entire function F is prime are (1) first prove that f is pseudo-prime, (2) prove that F cannot be expressed as $F(z) = f(p(z))$, where f is entire and p is a polynomial of degree no less than two, and (3) prove that $F(z)$ cannot be expressed as $F(z) = q(g(z))$, where q is a polynomial of degree no less than two and g is entire. Now, we introduce some sufficient conditions for determining the pseudo-primeness of a given entire function.

Theorem 3.1 (Goldstein [8]). Let $F(z)$ be a finite order entire function with $\delta(a, F) = 1$ for some complex number a. Then F is pseudo-prime.

(The function $F(z) = \exp(\sin z)$ shows that the restriction on the order of F is a necessary one).

Fuchs and Song raised the question: Does the conclusion remain true under the assumption $\delta(a, F)>0$?

A slight modification of Goldstein's proof leads to the result that if $p(z)$ is a nonconstant polynomial of degree t, and h_1, h_2 are two entire functions of order less than t, then $F(z) = h_1(z)e^{p(z)} + h_2(z)$ is pseudo-prime [15].

Remark. It would be interesting to prove whether or not the condition mentioned about h and h can be relaxed to that

$$T(r, h_i) = o(1)T(r, e^p) \text{ as } r \to \infty, \ i = 1, 2.$$

Theorem 3.2 (Ozawa [27]). Let $F(z)$ be an entire function of finite order whose derivative $F'(z)$ has an infinite number of zeros. Suppose that for any complex number c, the following simultaneous equations

$$F(z) = 0,$$

$$F'(z) = 0,$$

have only a finite number of solutions. Then any factorization of the form $F = f(g)$ with f and g being entire implies that f must be linear whenever g is transcendental (We also call such F a left-prime function).

When no restriction is imposed on the order of $F(z)$, Ozawa also showed that the conclusion remains valid if the function $F(z)$ satisfies $N(r, 1/F) > kT(r, F)$ as $r \to \infty$ for some positive number k.

The prime function $e^z + p(z)$ ($p \neq$ constant) or pseudo-prime function $\cos z$ and other similar forms of functions all satisfy the following type of differential equation

$$a_n(z)w^{(n)}(z) + a_{n-1}(z)w^{(n-1)}(z) + \ldots\ldots + a_o(z) = 0, \tag{3.1}$$

where all the a_is ($i = 0, 1, 2, \ldots.n$) are polynomials. As a simple application of Theorem 2.4, Steinmetz [34] proved the following general result, that particularly implies that any meromorphic solution of equation (3.1) is pseudo-prime.

Theorem 3.3 Let $w(z)$ be a meromorphic function satisfying algebraic differential equation of the form

$$\Omega(z, w, w', \ldots.w^{(n)}) = 0, \tag{3.2}$$

where Ω is a polynomial in $w, w', \ldots.w^{(n)}$ with rational functions as the coefficients. Let $M[w]$ or $M_j[w]$ denote a differential monomial of the form: $a(z)(w)^{k_o}(w')^{k_1}\ldots\ldots(w^{(m)})^{k_m}$; $a(z)$ a rational function$\neq 0$, then $\Omega(z, w, w' \ldots.w^{(n)})$ can be expressed as $\sum_{j=0}^{t} M_j[w]$. Define $\gamma_{M_j} = k_o + k_1 + \ldots. + k_m$ and $\Gamma_{M_j} = k_o + 2k_1 + \ldots. + (m+1)k_m$. Let $\gamma_\Omega = \max_{j=0}^{t} \gamma_{M_j}$ and $\Gamma_\Omega = \max_{j=0}^{t} \Gamma_{M_j}$. If $w(z) = f(g(z))$, where f is transcendental meromorphic and g is entire then f also satisfies a differential equation $\hat{\Omega}(z, w, w', \ldots \ldots)$ of the form in (3.2) with $\gamma_{\hat{\Omega}} \leqslant \gamma_{\Omega}$ and $\Gamma_{\hat{\Omega}} \leqslant \Gamma_{\Omega}$.

The linear differential equation (3.1) is a special case of equation (3.2). We thus have the result that any transcendental meromorphic function satisfying a differential equation of the form (3.1) is pseudo-prime (more precisely, it follows that any non-bilinear meromorphic factor of such a function must be a rational function). Song and Yang [33] obtained some generalizations of this result, that enables one to determine the pseudo-primeness of combinations of several meromorphic functions that are solutions of linear differential equations. Let M denote the family of all meromorphic functions, R the family of all rational functions, and D the class of meromorphic function that satisfy a differential equation of the form as in equation (3.1).

Theorem 3.4. Let $F(z) = \sum_{i=1}^{m} q_i(z)\psi_i(z)$, where $q_j \in R$ and $\psi_j(z) \in D$, $j = 1, 2, \ldots, m$. Then $F(z)$ and all its derivatives $F^{(n)}(z)$, $n = 0, 1, 2, \ldots$ are pseudo-prime.

Recently He and Yang [20] presented a study on the pseudo-primality of the product of two pseudo-prime meromorphic functions, which satisfy a certain linear differential equations with rational functions as the coefficients.

We note that the composite function $e^{\sin z}$ satisfies the nonlinear differential equation: $(w''w)^2 - 2w''w'^2w + w'^4 + (w'w)^2 - w^4 = 0$ (see e.g., Wittich's book: Neuere untersuchungen Uber Eideutige Analytische Funktionen, Springer-Verlag, Berlin, 1955, p.70). It would be interesting to obtain a result analogous to that of Steinmetz's for solutions of nonlinear differential equations.

4. THE DISTRIBUTION OF PRIME FUNCTIONS

It seems to be natural for one to examine the questior of the distribution of the prime functions in the family of all the entire functions. Two such questions were asked:

A. (Gross [9]). Given any entire function F, does there exist a polynomial p such that $F+p$ is prime?

B. (Gross, Osgood, and Yang [17]). Given any entire function F, does there exist an entire function g such that $g(z)F(z)$ is prime?

Noda [28] was able to provide affirmative answers to both of the questions by proving the following result.

Theorem 4.1. Let $F(z)$ be a transcendental entire function. Then the set $\{a | a \in C$ and $F(z) + az$ is not prime$\}$ and the set $\{b | b \in C$ and $(z-b)F(z)$ is not prime$\}$ are at most a countable set.

It does not seem to be a difficult task to find an example where an entire function $F(z)$ and $F(z) + p(z)$ are both not pseudo-prime for some nonconstant polynomial p. It is easy to find an entire function F and a polynomial p ($\neq$constant) such that $F(z)$ is pseudo-prime but $F(z) + p(z)$ is not. When the order of F is restricted to the finite, the previous observations may not be valid. We are inclined to conjecture that if F is a pseudo-prime transcendental entire function F of finite order, then, for any polynomial $p(z)(\neq 0)$, $p(z)F(z)$, and $F(z) + p(z)$ remain pseudo-prime. Recently, Song and Huang [32], as well as Ma [23], noted that $F(z) = \cos z e^{\sin z}$ is prime, but $F^2(z)$ is not even a pseudo-prime function.

Song and Huang [32] also proved that if F is any pseudo-prime entire function, then $F^n(z)$ remains pseudo-prime for any odd positive integer n. It has been conjectured by the present author, that if F is pseudo-prime, then, for any nonconstant polynomial $p(z)$, $F(p)$ remains a pseudo-prime function.

5. PERMUTABILITY, UNIQUE FACTORIZABILITY, AND PRIMALITY

Let $f(z)$ and $g(z)$ be entire functions. We say that f and g are permutable if they satisfy the relation

$$f(g(z)) = g(f(z)) \ , \ \forall \ z \in \text{complex plane.}$$

It is not difficult to prove that if $f(z) = ae^z+c$ $(ab\neq 0)$ and $g(z)$ is any entire function that is permutable with f, then $g(z)$ is the n-th iterate $f_n(z)$ of f for some positive integer $n(f_n(z) = f_{n-1}(f(z)),\ n>1)$ [1]. Baker [2, 3] showed if f is a nonlinear entire function, then the set of entire functions that are permutable with f is countably infinite. It is easy to prove that a nonlinear polynomial and a transcendental entire function are not permutable [1, 18]. In general, it is difficult to determine whether two transcendental entire functions are permutable or not. We are now mentioning some such results.

Theorem 5.1 (Yang and Urabe [42]). Let H denote the class of entire functions h that satisfy $\overline{\lim_{r\to\infty}} \log\log M(r, h)/\log\log r < \infty$. Let $f(z) = \alpha(z)e^{p(z)}$, where $\alpha(z) \in H$ $(\alpha \not\equiv 0)$ and $p(z)$ is a nonconstant polynomial. Let $g(z)$ be any transcendental entire function of finite order that is permutable with f. Then $g(z) = af(z)$ for some constant a that is a root of unity. Moreover, if a is the primitive s-th root of unity ($s \geqslant 2$), then $f(z) = z\phi(z^s)$, where $\phi(z)$ is a transcendental entire function.

Theorem 5.2 (Kobayashi [21]). Let $f(z)$ be $z + ce^{az}$, where a and c are constants with $ac \neq 0$. Let $g(z)$ be a nonconstant entire function of finite order that is permutable with f. Then either $g(z) = f(z) + d$, or $g(z) = z + d$, where d is a constant satisfying $\exp(ad) = 1$.

Many of the techniques used in the proofs of the above theorems are applicable to the study of unique factorizability of certain classes of entire functions. Clearly, if $F(z) = f(g(z))$ is a factorization of F, then so is $F(z) = f_1(g_1(z))$, where $f_1(w) = f(a+w/c)$, $g_1(z) = c(g(z) -a)$, and a and $c \neq 0$ are constants. To avoid this kind of ambiguity we have defined equivalent relations among all the factorizations. Let a nonconstant entire function $F(z)$ have two factorizations $f_1 of_2 of_3 \ldots of_m(z)$ and $g_1 og_2 og_3 o \ldots og_n(z)$ into nonlinear entire factors. If $m=n$ and if there exist suitable linear polynomials T_j ($j = 1, 2, \ldots, n-1$) such that the relations

$$f_1(z) = g_1 oT_1^{-1}(z),\ f_2(z) = T_1 og_2 oT_2^{-1}(z),\ \ldots,\ f_n(z) = T_{n-1} og_n(z) \tag{5.1}$$

hold simultaneously, then the two factorizations are called equivalent (in entire sense). If any two factorizations of $F(z)$ into nonlinear, prime, entire factors are equivalent to each other, then F is called uniquely factorizable. We don't know whether there exists a transcendental entire function (with all its factors transcendental entire functions) that is not uniquely factorizable. Prime functions are uniquely factorizable. Nontrivial examples of entire function that are uniquely factorizable seem to be somewhat rare. Ritt [30] proved that polynomials are uniquely factorizable. More precisely, it means that any two factorizations of a polynomial into prime polynomials contain the same number of polynomials; the degrees of the polynomials in one factorization are the same as those in the other except perhaps for the order in which they occur, and, furthermore, if a polynomial has two nonequivalent factorizations, one can transform from either to a factorization equivalent to the other by suitably converting adjacent pairs of prime polynomials. As one can see an analogous discussion of the factorization for rational or transcendental meromorphic function will be a much more complicated one. (In this case, we shall need to modify the equivalence of two factorizations of a meromorphic function accordingly by allowing in (5.1) the T_js to be bilinear, and f_j, g_j to be meromorphic). Three interesting questions were raised in [12] that thus far remain to be solved: (1) Does there exist a meromorphic but not entire function, that cannot be factored into prime factors? (2) Does there exist a meromorphic function that has no prime factors at all? (3) Does there exist a meromorphic function with a nondenumerable number of non-equivalent factorizations? The uniqueness theory of factorization for rational functions has not been proven. Several classes of uniquely factorizable transcendental entire functions have been constructed (see, e.g. [24]). The entire function $F(z) = z^2 e^{2z}$ is first such an example [9, p. 133]. The following general result is due to Ozawa [25].

Theorem 5.3. Let $f(z)$ be a prime entire function of finite order. Assume that $f(z)$ has infinitely many zeros, and $f(z) = c$, $f'(z) = 0$ have only finitely many common roots for every constant c. Suppose that p is a prime number, then $F(z) = f(z^p)$ is uniquely factorizable.

Other types of such functions are built by considering compositions of various kinds of known prime or special types of functions. Ozawa [27] first proved that for any positive integer n the function $F(z) = z+e_n(z)(e_n(z) = e_{n-1}(e(z)); e_0(z)\equiv z)$ is prime. Later on Yang and Gross conjectured that for any periodic entire function $H(z)$, the function $F(z) = z+e^{H(z)}$ is prime. This is still an open question. Several results related to this problem have been obtained. For instance, the right and left factors of such functions and their generalized forms were characterized by Yang [40], Koont[22], and Urabe [35]. For a nonzero constant b, following Koont [22], we define

$J(b) = \{F(z) = H(z)+cz;\ H$ is entire, periodic with period b and c is a nonzero constant$\}$;

$L(b) = \{F(z) = H_1(z)+ze^{H_2(z)};\ H_1$ and e^{H_2} are entire, periodic with period $b\}$.

Note that $J(b) \subset L(b)$. In this regard, for instance, let $f(z)$ be $z+e^z$ and let $g(z)$ be ze^z, then Urabe [33] showed $f(g(z))$ is uniquely factorizable, whereas Kobayashi [21] showed $g(f(z))$ is uniquely factorizable. The above two classes of functions have yielded numerous types of prime functions and functions of uniquely factorizable. About the latter, most results were obtained by Urabe in a series of papers. In [35] by deriving first that if $F \in J(b)$ and $F = f(g)$ for some nonentire functions f and g, then f and g both belong to $J(b)$. Noting that $F(z) = (z+e^z)o(z+e^z)$ can be written as $z+H(z)$, where $H(z) = e^z(1+\exp(e^z))$ is periodic with period $2\pi i$, Urable was able to show that F is uniquely factorizable. More generally,

Theorem 5.4 [35]. Let

$$F(z) = (z + H_1(z)).\ \exp(H_2(z)), \qquad (5.2)$$

where entire functions $H_1(z)$ and $\exp(H_2(z))$ have period $2\pi i$. If the order of H_1 is finite, then $F(z)$ is uniquely factorizable. Further if there exists an entire function $H_3(z)$ satisfying the identical relation $H_2(z) = H_3(z+H_1(z))$, then $F(z) = (z.\exp(H_3(z))o((z+H_1(z))$ is the only factorization up to equivalent factorizations. Thus, if there is no such identical relation, then F is prime.

Theorem 5.5 (Urabe [37]). Let $F(z)$ be as defined by (5.2) but no restriction on the order of H_1. If $F(z) = f(g(z))$, f and g being nonlinear, then, two functions $z+H_1(z)$ and $H_2(z)$ must have $g(z)$ as their common right factors, that is,

$$z+H_1(z) = K_1(g(z)) \text{ and } H_2(z) = K_2(g(z))$$

for some entire functions K_j $(j = 1,2)$.

As remarked in [37], F is prime, if the two functions $z+H_1(z)$ and $H_2(z)$ have no common nonlinear entire right factor. Further, if $z+H_1(z)$ is prime, then $F(z)$ is uniquely factored as

$$F(z) = (z.\exp[K_2(K_1^{-1}(z))]\circ(z+H_1(z)).$$

Relating to the problem on the uniqueness of the factorization, Urabe [37] raised the following question: When f and g are entire functions and satisfy the identity $f(f(z)) = g(g(z))$, then what can be said about

f and g? Let $f(z) = z - \sin z$ and $g(z) = -(z - \sin z) + 2k\pi i$ with an integer $k(\neq 0)$, then $f(f(z)) = g(g(z))$. However, by posing certain specific conditions on f, Urabe was able to derive the following interesting result.

Theorem 5.6 [37]. Let f be an entire function of the form:

$$f(z) = z + h(e^z) ,$$

where $h(z)$ is a nonconstant entire function satisfying $\exp[k.h(0)] \neq 1$ for any natural number k. Suppose that g is entire and satisfies the following identical relation

$$f(f(z)) = g(g(z)). \qquad (5.3)$$

Then it is necessary that $g(z) \equiv f(z)$.

Therefore, if $f(z) = z + e_m(z)$, $(m \geq 1)$, then the only entire function g that can satisfy the equation (5.3) is f itself. Furthermore, when $m = 1$, it is known that the composite function $(z+e^z)o(z+e^z)$ is uniquely factorizable. In general, when m and k are two distinct positive integers, the uniqueness of the factorization of the composite function $(z+e_m(z))o(z+e_k(z))$ is still an open question. Recently, Yang and Urabe [38] settled the case when $m = 1$, $k = 2$.

Generally, as we mentioned before, it is very difficult to determine whether a given meromorphic but not entire function is prime. Few prime meromorphic functions have been exhibited. The following one is due to Urabe.

Theorem 5.7 [37]. Let

$$F(z) = \frac{P(z) + Q(e^z)}{S(z) + R(z)\, H(z)}$$

where $P(\not\equiv \text{constant})$, $Q(\not\equiv \text{constant})$, $R(\not\equiv 0)$ and S are polynomials such that, for any positive integer k and complex number c, the function $e^{-kz}.[Q(e^z) + c]$ is nonconstant, and $H(z)$ is a period entire function (with period $2\pi i$) with its order and lower order satisfying $1 < \underline{\rho}(H) \leq \rho(H) < 2$. Assume that the numerator and denominator have no common zeros. Then $F(z)$ is prime.

The proof of this Theorem relies on the results on primeness in divisor sense as defined in [36]. An entire function $F(z)(\not\equiv \text{constant})$, with zeros, is said to be prime in divisor sense (pseudo-prime in divisor sense; right-prime in divisor sense; left-prime in divisor sense), if, for any identical relation of the form

$$F(z) = f(g(z))e^{A(z)}, \tag{5.4}$$

where f, $g(\not\equiv \text{constant})$, and A are entire functions, we can deduce the following assertion: f has just one simple zero or g is a linear polynomial (f has only a finite number of zeros or g is a polynomial; g is a linear polynomial whenever f has an infinite number of zeros; f has just one simple zero whenever g is transcendental, respectively). As is shown in [36], the function $z+e^z$ is also prime in divisor sense. Also it has been shown in [36] that in the class of entire functions G, each of which has at least one zero, the subclass that consists of prime functions and the subclass that consists of prime functions in divisor sense are not identically the same. Moreover, Urabe [36] showed that if $F(z, w)$ is an entire function of z and w defined by $F(z,w) = (z+w) - (e^z+e^w)$, and $F(z,w) = f(g(z,w)).e^{A(z,w)}$, where $f(z)$, $g(z,w)$, and $A(z,w)$ are entire functions with f has at least two zeros, then we must have $g(z,w) = a+bz+cw$ a linear function in variables z and w. Under such a setting $F(z,w)$ is called prime in divisor sense. This also suggests that another possible direction in the investigations of factorization theory is the quest of the proper definitions of the primeness and pseudo-primeness of entire or meromorphic functions of several complex variables and some necessary tools in studying them.

REFERENCES

1. I.N. Baker, Zusammensetzungen ganger Functionen, Math. Zeit., 69, (1958), 121-163.

2. ----------, Permutable entire functions, Math. Zeit., 79, (1962), 243-249.

3. ----------, Repulsive fixpoints of entire functions, Math. Zeit., 104(1968), 252-256.

4. I.N. Baker and F. Gross, Further results on factorization of entire functions, in Entire functions and related parts of analysis (Proc. Symp. Pure Math., La Jolla, California, 1966), Amer. Math. Soc., 1968, 30-35.

5. I.N. Baker and C.C. Yang, An infinite order periodic entire function which is prime, Lecture notes in mathematics, Springer-verlag, 599 (Complex Analysis, Kentucky Conference), (1976), 7-10.

6. C.T. Chuang and C.C. Yang, Theory of fix-points and factorization of meromorphic functions (in Chinese), to be published in Mathematical Monograph Series, Peking University.

7. R. Goldstein, On factorisation of certain entire functions, II, Proc. London Math. Soc., vol. 22, (1971), 483-506.

8. ----------, On factorisation of certain entire functions, J. London Math. Soc., 2, (1970), 221-224.

9. F. Gross, Factorization of Meromorphic Functions, U.S. government printing office, Washington, D.C., 1972.

10. ----------, On factorization of meromorphic functions, Trans. Amer. Math. Soc., vol. 131 (1968), 215-222.

11. ----------, Factorization of entire functions which are periodic mod g, Indian Jour. of Pure and Applied Math., 2, no. 3, (1971).

12. ----------, Factorization of meromorphic functions and some open questions, Complex Analysis, Lecture notes, Springer-Verlag, 599, (1976).

13. F. Gross and C.C. Yang, Further results on prime entire functions, Trans. Amer. Math. Soc., vol. 142, (1974), 347-355.

14. ----------, On prime periodic entire functions, Math. Zeit., 174 (1980), 43-48.

15. ----------, On pseudo-prime entire functions, Tohoku Math. J. vol. 26, no. 1, March (1974), 65-71.

16. F. Gross and C.F. Osgood, On fixed points of composite entire functions, J. London Math. Soc., (2), 28 (1983), 57-61.

17. F. Gross, C.C. Yang and C. Osgood, Primeable entire functions, Nagoya Math. J. 51, (1973), 123-130.

18. V. Ganapathy Iyer, On the permutable integral functions, J. London Math. Soc., 34(1959), 141-143.

19. W.K. Hayman, Meromorphic Functions, Oxford Mathematical Monographs, Clarendon Press, Oxford, 1964.

20. Y.Z. He and C.C. Yang, On pseudo-primality of the product of some pseudo-prime meromorphic functions, to appear in the Proceedings of the special session on "Analysis of one complex variable" (ed. by C. Yang), A.M.S. Summer Meeting, Wyoming, 1985), which is to be published by World Scientific Publishing Co., Singapore.

21. T. Kobayashi, Permutability and unique factorizability of certain entire functions, Kodai Math. J. 13, (1980), 8-25.

22. S. Koont, On factorization in certain classes of entire functions, Math. Research Report, no. 76-5, University of Maryland, (1976), 19 pp.

23. L.Z. Ma, On prime entire functions, to appear in the Proceedings as mentioned above ref. [20].

24. M. Ozawa, On uniquely factorizable of entire functions, Kodai Math. Sem. Report, 27, (1977), 342-360.

25. ----------, On uniquely factorizable of meromorphic functions, Kodai Math.J. (1978), 339-353.

26. ----------, On the existence of prime periodic entire functions, Kodai Math. Sem. Report, 29, (1978), 308-321.

27. ----------, On certain criteria for the left-primeness of entire functions, Kodai Math. Sem. Rep., vol. 26, (1975), 186-192.

28. Y. Noda, On factorization of entire functions, Kodai Math. J., 4 (1981), 480-494.

29. G.S. Prokopovich, On superposition of some entire functions, Ukrainskii Matematicheskii Zhurnal, (English transl.), vol. 26, no. 2, March-April, (1974), 188-195.

30. J.F. Ritt, Prime and composite polynomials, Trans. Amer. Math. Soc., 3, (1922), 51-66.

31. P.C. Rosenbloom, The fix-points of entire functions, Medd Lunds Univ. Mat. Sem. Suppl. Bd. M. Riesz (1952), 186-192.

32. G.D. Song and J. Huang, On pseudo-primality of the n-th power of prime entire functions, Kodai Math. J., vol. 10 (1987), 42-48.

33. G.D. Song and C.C. Yang, On pseudo-primality of the combination of meromorphic functions satisfying linear differential equations, in Value distribution theory and its applications (edited by C.C. Yang), Contemporary Math. Series, vol. 25, American Math. Soc., Providence, R.I. 1986.

34. N. Steinmetz, Uber die factoriserbaren Losungen gewohnlicher Differentialgleichungen, Math. Zeit., 170 (1980), 168-180.

35. H. Urabe, Uniqueness of the factorization under composition of certain entire functions, J. Math. Kyoto Univ., 18 (1978), 95-120.

36. ----------, Primeness in divisor sense for entire functions, J. Math. Kyoto Univ., 24 (1984), 127-140.

37. ----------, On factorization of certain entire and meromorphic functions, Journal of Math. Kyoto Univ., vol. 26, no. 2, 1986, 177-190.

38. H. Urabe and C.C. Yang, in preparation.

39. C.C. Yang, Progress in factorization theory of entire and meromorphic functions, in Lecture Notes in Pure and Applied Math., vol. 78, edited by C.C. Yang, Marcel Dekker Inc., 1982.

40. ----------, On the factorization of entire functions, Ill. J. of Math., vol. 21, no. 4, (1977), 898-905.

41. ----------, Further results on the fix-points of composite transcendental entire functions, Jour. Math. Analysis and Applications, vol. 90, no. 1 (1982), 259-269.

42. C.C. Yang and H. Urabe, On permutability of certain entire functions, J. London Math. Soc., (2), 14 (1976), 153-159.

ON THE GROWTH OF A MEROMORPHIC FUNCTION AND OF ITS DIFFERENTIAL POLYNOMIALS

Chi-Tai Chuang
Peking University
P.R. China

In what follows, by meromorphic functions we always mean functions which are meromorphic in the complex plane $\mathbb{C}$. For a meromorphic function $f(z)$ we shall use the notations $m(r, f)$, $N(r, f)$ and $T(r, f)$ introduced by Nevanlinna [1].

1. PRELIMINARY NOTIONS

In 1951 I proved the following theorem [2]:

Theorem 1. If $f(z)$ is a transcendental meromorphic function, then there exists a number $r_o > 0$ such that for $\lambda > 1$ and $r > r_o$, we have

$$T(r, f') < A_o(\lambda)\, T(\lambda r, f) \qquad (1)$$

$$T(r, f) < B_o(\lambda)\, T(\lambda r, f')\ , \qquad (2)$$

where $A_o(\lambda)$ and $B_o(\lambda)$ are two positive functions of λ defined for $\lambda > 1$ and f' is the derivative of f.

This theorem implies that f and f' have the same order and the same lower order and other consequences.

The main purpose of the present work is to extend the

Theorem 1 to the case where f' is replaced by a more general differential polynomial of f and to give some applications. For this it is convenient to give first some definitions.

Definition 1. Let $f(z)$ and $g(z)$ be two meromorphic functions. We say that $f(z)$ and $g(z)$ have nearly the same growth, if there exists a number $r_0 > 0$ such that for $\lambda > 1$ and $r > r_0$, we have

$$T(r, g) < A(\lambda)\, T(\lambda r, f) \tag{3}$$

$$T(r, f) < B(\lambda)\, T(\lambda r, g), \tag{4}$$

where $A(\lambda)$ and $B(\lambda)$ are two positive functions of λ defined for $\lambda > 1$.

Lemma 1. If three meromorphic functions $f(z)$, $g(z)$, $h(z)$ are such that $f(z)$ and $g(z)$ as well as $g(z)$ and $h(z)$ have nearly the same growth, then $f(z)$ and $h(z)$ have nearly the same growth.

Proof. Let $r_0 > 0$ be a number such that (3) and (4) hold for $\lambda > 1$ and $r > r_0$. On the other hand, let $r_0' > 0$ be a number such that for $\lambda > 1$ and $r > r_0'$, we have

$$T(r, h) < A_1(\lambda)\, T(\lambda r, g) \tag{5}$$

$$T(r, g) < B_1(\lambda)\, T(\lambda r, h), \tag{6}$$

where $A_1(\lambda)$ and $B_1(\lambda)$ are two positive functions of λ defined for $\lambda > 1$. Let $r_0'' = \max(r_0, r_0')$ and $\lambda > 1$, $r > r_0''$. Then by (5) and (3) we have

$$T(r, h) < A_1(\lambda) T(\lambda r, g) < A_1(\lambda) A(\lambda) T(\lambda^2 r, f)\,, \tag{7}$$

and by (4) and (6), we have

$$T(r, f) < B(\lambda)T(\lambda r, g) < B(\lambda)B_1(\lambda)T(\lambda^2 r, h). \tag{8}$$

In (7) and (8), replacing λ by $\lambda^{1/2}$, we get

$$T(r, h) < A_1(\lambda^{1/2})A(\lambda^{1/2})T(\lambda r, f) \tag{9}$$

$$T(r, f) < B(\lambda^{1/2})B_1(\lambda^{1/2})T(\lambda r, h) . \tag{10}$$

Since (9) and (10) hold for $\lambda > 1$, $r > r_0''$, Lemma 1 is proved.

Definition 2. Let $\Omega(f)$ be an operator such that to each meromorphic function f corresponds a meromorphic function $\Omega(f)$. We say that the operator $\Omega(f)$ is growth preserving, if for each transcendental meromorphic function f, the two meromorphic functions f and $\Omega(f)$ have nearly the same growth.

Lemma 2. If $\Phi(f)$ and $\Psi(f)$ are two growth preserving operators, then the composite operator $\Omega(f) = \Psi\{\Phi(f)\}$ is also growth preserving.

Proof. Note first, by Definition 1, if two meromorphic functions f and g have nearly the same growth and if one of them, say f, is transcendental, then the other g is also transcendental.

Now let f be a transcendental meromorphic function. By Definition 2, the two meromorphic functions f and $\Phi(f)$ have nearly the same growth, and so $\Phi(f)$ is transcendental. Next again by Definition 2, the two meromorphic functions $\Phi(f)$ and $\Psi\{\Phi(f)\}$ have nearly the same growth. Finally by Lemma 1, f and $\Psi\{\Phi(f)\}$ have nearly the same growth. Hence the composite operator $\Omega(f) = \Psi\{\Phi(f)\}$ is growth preserving.

Examples

1° Consider the operator $\omega_1(f) = f'$ which is the derivative of f. By Theorem 1, $\omega_1(f)$ is growth preserving. Hence by Lemma 2, the operators

$$\omega_1(f) = f'$$

$$\omega_2(f) = \omega_1\{\omega_1(f)\} = f''$$

$$\omega_3(f) = \omega_1\{\omega_2(f)\} = f'''$$

. . . .

$$\omega_p(f) = \omega_1\{\omega_{p-1}(f)\} = f^{(p)}$$

are all growth preserving.

2° Consider a positive integer q and the operator $\pi_q(f) = f^q$. It is well known that for any meromorphic function f we have the formula:

$$T(r, f^q) = qT(r, f)$$

which evidently implies that the operator $\pi_q(f)$ is growth preserving.

Starting from the operators $\omega_p(f)$ and $\pi_q(f)$, we deduce by composition many growth preserving operators such as:

$$\Omega_{q,p}(f) = \pi_q\{\omega_p(f)\} = \{f^{(p)}\}^q$$

$$\Omega_{p,q}(f) = \omega_p\{\pi_q(f)\} = \{f^q\}^{(p)}$$

$$\Omega_{p_1,q,p}(f) = \omega_{p_1}\{\Omega_{q,p}(f)\} = (\{f^{(p)}\}^q)^{(p_1)}$$

$$\Omega_{q_1,p,q}(f) = \pi_{q_1}\{\Omega_{p,q}(f)\} = (\{f^q\}^{(p)})^{q_1}.$$

Note that no matter how far we carry on this process, we always get a growth preserving operator $\Omega(f)$ which is a homogeneous polynomial of f and its successive derivatives $f^{(j)}$ $(j = 1, 2, \ldots, n)$ up to a certain order n, with positive integer coefficients. These homogeneous differential polynomials have application in a joint work by the author and Ma, which appears in this book.

2. LINEAR DIFFERENTIAL POLYNOMIALS

In the paper [3] of 1964, I consider a system of linearly independent meromorphic functions $\psi_k(z)$ $(k = 1, 2, \ldots, p;\ p \geq 1)$ and a meromorphic function $f(z)$, and define

$$L(f) = \frac{(-1)^p}{A_o} \Delta\,(f, \psi_1, \psi_2, \cdots, \psi_p)$$

$$= f^{(p)} + \frac{A_1}{A_o} f^{(p-1)} + \cdots + \frac{A_{p-1}}{A_o} f' + \frac{A_p}{A_o} f \qquad (11)$$

where $A_o = \Delta(\psi_1, \psi_2, \ldots, \psi_p)$ is the Wronskian determinant of $\psi_k(z)$ $(k = 1, 2, \ldots, p)$ and $\Delta(f, \psi_1, \psi_2, \ldots, \psi_p)$ the Wronskian determinant of $f(z), \psi_k(z)$ $(k = 1, 2, \ldots, p)$. Next in the paper [4], I compared the growth of $f(z)$ with that of $F(z) = L(f)$. For this, I need to express inversely $f(z)$ in terms of $F(z)$, and I got the formula:

$$f(z) = \sum_{k=1}^{p} C_k(z)\ \psi_k(z) \qquad (12)$$

where $C_k(z)$ $(k = 1, 2, \ldots, p)$ are meromorphic functions satisfying the identities

$$C'_k = (-1)^{p+k} \frac{\Phi_k}{A_o} F \quad (k = 1, 2, \ldots, p) \qquad (13)$$

and Φ_k is the determinant obtained by omitting the pth row and the kth column of the matrix

$$\begin{pmatrix} \psi_1 & \psi_2 & \cdots & \psi_p \\ \psi_1' & \psi_2' & \cdots & \psi_p' \\ \cdots & \cdots & \cdots & \cdots \\ \psi_1^{(p-1)} & \psi_2^{(p-1)} & \cdots & \psi_p^{(p-1)} \end{pmatrix}.$$

The proof of the formula (12) is closely related to the problem of finding a solution of the differential equation

$$L(w) = F$$

where w is the unknown function and F is the meromorphic function $F = L(f)$ defined by (11). I solved this problem by a classical method of Lagrange, called the method of the variation of constants. Here the main difficulty is to show that there exist meromorphic functions C_k $(k = 1, 2, \ldots, p)$ satisfying the identities (13). In appearance, this is not quite difficult, because we need only to develop the meromorphic function on the right of (13) into a series by Mittag-Leffler theorem and then integrate term by term. However the general term of this series is of the form

$$\sum_{j=1}^{m} \frac{b_j}{(z - z_o)^j} - P(z)$$

where z_o is a pole of the meromorphic function on the right of (13) and $P(z)$ is a polynomial. Integration of the term $b_1/(z - z_o)$ yields $b_1 \log(z - z_o)$ which is a multiple valued function. So in order that we can get by integration a meromorphic function C_k satisfying the identity (13), it is necessary that the residues corresponding to all the poles of the meromorphic function on the right of (13) must be all equal to zero. It is interesting that this is really true.

Basing upon Theorem 1 and the formula (12), I proved the following theorem.

Theorem 2. Let $f(z)$ be a transcendental meromorphic function and $\psi_k(z)$ $(k = 1, 2, \ldots, p;\ p \geq 1)$ be p linearly independent meromorphic functions such that there is a number $\mu > 1$ for which we have

$$T(\mu r, \psi_k) = o\{T(r, f)\} \quad (k = 1, 2, \ldots, p).$$

Then $f(z)$ and the meromorphic function $F(z) = L(f)$ defined by (11) have nearly the same growth. A fortiori, $f(z)$ and $F(z) = L(f)$ have the same order and the same lower order.

This theorem has applications to value-distribution of meromorphic functions. In order to simplify the statement of the results, it is convenient to introduce first some notations.

Let $f(z)$ be a transcendental meromorphic function. We denote by $\sigma(f)$ the set of meromorphic functions $\psi(z)$ such that there is a number $\mu = \mu_\psi > 1$ for which

$$T(\mu r, \psi) = o\{T(r, f)\}$$

and set $\hat{\sigma}(f) = \sigma(f) \cup (\infty)$. If $\phi \in \hat{\sigma}(f)$, we define

$$\delta(\phi, f) = \varliminf_{r\to\infty} \frac{m(r, \frac{1}{f-\phi})}{T(r,f)} = 1 - \varlimsup_{r\to\infty} \frac{N(r, \frac{1}{f-\phi})}{T(r,f)}$$

For the particular case $\phi = \infty$, $m(r, 1/(f-\phi))$ and $N(r, 1/(f-\phi))$ should be replaced by $m(r, f)$ and $N(r, f)$ respectively. We have always $0 \leq \delta(\phi, f) \leq 1$.

The following theorem is proved.

Theorem 3. Let $f(z)$ be a transcendental meromorphic function of finite order. Suppose that there is an element $\phi_o \in \hat{\sigma}(f)$ such that

$$N(\mu r, \frac{1}{f - \phi_o}) = o\{T(r, f)\} \qquad (14)$$

for a number $\mu > 1$. Then for any finite number of distinct elements ϕ_j $(j = 1, 2, \ldots, q;\ q \geq 2)$ of the set $\hat{\sigma}(f)$, we have

$$\sum_{j=1}^{q} \delta(\phi_j, f) \leq 2 - \overline{\lim_{r\to\infty}} \frac{N(r, \Lambda) + N(r, \frac{1}{\Lambda})}{T(r, \Lambda)}$$

where $\Lambda = \Lambda(z)$ is a transcendental meromorphic function such that $f(z)$ and $\Lambda(z)$ have the same order and the same lower order.

By a classical theorem of Nevanlinna [1], if $f(z)$ is a transcendental meromorphic function of finite non-integral order ρ, then

$$\overline{\lim_{r\to\infty}} \frac{N(r,f) + N(r, \frac{1}{f})}{T(r, f)} \geq k(\rho)$$

where $k(\rho) > 0$ is a number depending only on ρ. Hence Theorem 3 implies the following corollary:

Corollary 1. Let $f(z)$ be a transcendental meromorphic function of finite non-integral order ρ. If there is an element $\phi_o \in \hat{\sigma}(f)$ satisfying the condition (14), then for any finite number of distinct elements ϕ_j $(j = 1, 2, \ldots q;\ q \geq 2)$ of the set $\hat{\sigma}(f)$, we have

$$\sum_{j=1}^{q} \delta(\phi_j, f) \leq 2 - k(\rho).$$

Finally note that in (11) if we keep ψ_k (k = 1, 2, ..., p) fixed and let f vary, then L(f) is a linear operator. Theorem 2 yields immediately the following consequence.

Corollary 2. Let $\psi_k(z)$ (k = 1, 2, ..., p; p ⩾ 1) be a system of linearly independent rational functions. Then the corresponding operator L(f) defined by (11) is growth preserving.

3. DIFFERENTIAL POLYNOMIALS OF GENERAL FORM

Consider a transcendental meromorphic function f(z) and its first p derivatives $f^{(j)}(z)$ (j = 1, 2, ..., p). A differential polynomial of f(z) is a polynomial of $f^{(j)}(z)$ (j = 0, 1, 2, ..., p; $f^{(0)}(z) = f(z)$) of the form

$$P(z) = P(f,f',\ldots,f^{(p)}) = \sum_{k=1}^{n} a_k(z) \prod_{j=0}^{p} \{f^{(j)}(z)\}^{S_{kj}} \tag{15}$$

where $a_k(z)$ (k = 1, 2, ..., n) are meromorphic functions such that

$$T(r, a_k) = o\{T(r, f)\} \quad (k = 1, 2, \ldots, n) \tag{16}$$

and where S_{kj} (k = 1, 2, ..., n; j = 0, 1, 2, ..., p) are non-negative integers. We define

$$\bar{d}(P) = \max_{1 \leqslant k \leqslant n} \left(\sum_{j=0}^{p} S_{kj} \right), \quad \underline{d}(P) = \min_{1 \leqslant k \leqslant n} \left(\sum_{j=0}^{p} S_{kj} \right). \tag{17}$$

If $\bar{d}(P) = \underline{d}(P) = d$, then the differential polynomial (15) is said to be homogeneous of degree d. If $\bar{d}(P) > \underline{d}(P)$, then $P(z) = P(f, f', \ldots, f^{(p)})$ can be put in the form

$$P(z) = \sum_{i=1}^{q} P_i(z) \tag{18}$$

where $P_i(z) = P_i(f, f', \ldots, f^{(p)})$ is a homogeneous differential polynomial of degree d_i with

$$\underline{d}(P) = d_1 < d_2 < \ldots < d_q = \bar{d}(P) . \tag{19}$$

Our main result is the following theorem:

Theorem 4. Let $f(z)$ be a transcendental meromorphic function and $P(z) = P(f, f', \ldots, f^{(p)})$ the differential polynomial of $f(z)$ defined by (15). Suppose that $\underline{d}(P) > 0$ and $P(z) \not\equiv 0$. Then we have

$$T(r, f) \leqslant \frac{1}{\underline{d}(P)} T(r, P) + N(r, \frac{1}{f}) - \frac{1}{\underline{d}(P)} N(r, \frac{1}{P})$$

$$+ o\{T(r, f)\} + S(r) \tag{20}$$

where $S(r)$ satisfies for $1 \leqslant r < R$ the inequality

$$S(r) < A \overset{+}{\log} T(R, f) + B \log R + C \overset{+}{\log} \frac{1}{R-r} + D \tag{21}$$

A, B, C, D being positive constants.

For the proof of this theorem, we need several lemmas.

Lemma 3. Let $f(z)$ be a transcendental meromorphic function and $P(z) = P(f, f', \ldots, f^{(p)})$ the differential polynomial of $f(z)$ defined by (15). Suppose that $\underline{d}(P) > 0$. Then we have

$$m(r, \frac{P}{f^{\bar{d}(P)}}) \leqslant \{\bar{d}(P) - \underline{d}(P)\}m(r, \frac{1}{f}) + o\{T(r,f)\} + S(r) \tag{22}$$

where $S(r)$ satisfies the inequality (21).

Proof. Distinguish two cases:

1) $\overline{d}(P) = \underline{d}(P) = d$. Setting

$$\phi_k(z) = a_k(z) \prod_{j=0}^{p} \{f^{(j)}(z)\}^{S_{kj}},$$

we have

$$m(r, \frac{P}{f^d}) \leqslant \sum_{k=1}^{n} m(r, \frac{\phi_k}{f^d}) + \log n$$

$$m(r, \frac{\phi_k}{f^d}) = m\{r, a_k \prod_{j=0}^{p} (\frac{f^{(j)}}{f})^{S_{kj}}\}$$

$$\leqslant m(r, a_k) + \sum_{j=0}^{p} m\,(r, (\frac{f^{(j)}}{f})^{S_{kj}})$$

$$= m(r, a_k) + \sum_{j=0}^{p} S_{kj}\; m(r, \frac{f^{(j)}}{f})$$

$(k = 1, 2, \ldots, n)$.

By the condition (16), we have

$$m(r, a_k) = o\{T(r, f)\}.$$

On the other hand, it is known [1] [3] that for $1 \leqslant r < R$ we have

$$m(r, \frac{f^{(j)}}{f}) < \alpha_1 \overset{+}{\log} T(R,f) + \alpha_2 \log R + \alpha_3 \overset{+}{\log} \frac{1}{R-r} + \alpha_4$$

where α_i $(i = 1, 2, 3, 4)$ are positive constants. It is then clear that (22) holds in this first case.

2) $\overline{d}(P) > \underline{d}(P)$. In this case, making use of the formula (18), we have obtained in the work [5] the inequality

$$m(r, \frac{P}{f^{d_q}}) \leqslant \sum_{i=1}^{q} m(r, \frac{P_i}{f^{d_i}}) + (d_q - d_1)\, m(r, \frac{1}{f}) + (q - 1)\log 2. \tag{23}$$

Since we have shown that (22) holds in the first case, we see by (23) that (22) also holds in the second case.

Lemma 4. Let $f(z)$ and $g(z)$ be two meromorphic functions non identically equal to zero, and let their Laurent expansions in the neighbourhood of the point $z = 0$ be respectively

$$f(z) = az^{\mu} + a'z^{\mu+1} + \ldots (a \neq 0), \quad g(z) = bz^{\nu} + b'z^{\nu+1} + \ldots (b \neq 0).$$

Then we have the inequality

$$T(r, f) \leqslant T(r, g) + N(r, \frac{1}{f}) - N(r, \frac{1}{g}) + m(r, \frac{g}{f}) + \log\left|\frac{a}{b}\right| . \tag{24}$$

The proof of this lemma is given in the work [5].

Proof of Theorem 4. Apply Lemma 4 to the two meromorphic functions $\{f(z)\}^{\bar{d}(P)}$ and $P(z)$, we get

$$T(r, f^{\bar{d}(P)}) \leqslant T(r, P) + N(r, \frac{1}{f^{\bar{d}(P)}}) - N(r, \frac{1}{P}) + m(r, \frac{P}{f^{\bar{d}(P)}}) + c \tag{25}$$

where c is a constant. Making use of the relations

$$T(r, f^{\bar{d}(P)}) = \bar{d}(P)T(r, f), \quad N(r, \frac{1}{f^{\bar{d}(P)}}) = \bar{d}(P)N(r, \frac{1}{f})$$

and the inequality (22), we find from (25),

$$\bar{d}(P)\{T(r, f) - N(r, \frac{1}{f})\} \leq T(r, P) - N(r, \frac{1}{P})$$

$$+ \{\bar{d}(P) - \underline{d}(P)\}\{T(r, f) - N(r, \frac{1}{f})\}$$

$$+ o\{T(r, f)\} + S(r)$$

which yields the inequality (20).

An interesting problem is to eliminate the term $N(r, 1/f)$ in (20). Here we consider only the particular case where the differential polynomial of $f(z)$ has the form

$$Q(z) = f^h P(f, f', \ldots, f^{(p)}) \tag{26}$$

where $h \geq 1$ is a positive integer and $P(f, f', \ldots, f^{(p)})$ is defined by (15). We are going to prove the following theorem:

Theorem 5. Let $f(z)$ be a transcendental meromorphic function and $Q(z)$ the differential polynomial of $f(z)$ defined by (26). Suppose that $P(z) \not\equiv 0$. Then we have

$$T(r, f) \leq \frac{1}{h} T(r, Q) - \frac{\underline{d}(P)}{h} m(r, \frac{1}{f}) + o\{T(r, f)\} + S(r) \tag{27}$$

where $S(r)$ satisfies for $1 \leq r < R$ the inequality (21).

Proof. First by Theorem 4 we have

$$T(r, f) \leq \frac{1}{\underline{d}(P) + h} T(r, Q) + N(r, \frac{1}{f}) - \frac{1}{\underline{d}(P) + h} N(r, \frac{1}{Q})$$

$$+ o\{T(r, f)\} + S(r). \tag{28}$$

Evidently it is sufficient to show that

$$N(r, \frac{1}{Q}) \geq h\, N(r, \frac{1}{f}) - o\{T(r, f)\}\,. \tag{29}$$

To prove this inequality, it is convenient to make use of the symbols $\omega(g, z_0)$ and $\Omega(g, z_0)$ introduced by me in my paper [3] and defined as follows: For any meromorphic function $g(z)$ and any point z_0, $\omega(g, z_0)$ is equal to the order of z_0 or equal to 0 according to z_0 is a pole of $g(z)$ or not. On the other hand, for any meromorphic function $g(z) \not\equiv 0$ and any point z_0, $\Omega(g, z_0)$ is equal to the order of z_0 or equal to 0 according to z_0 is a zero of $g(z)$ or not.

With these notations, we are going to show that for any zero z_0 of the function $f(z)$, we have

$$\Omega(Q, z_0) \geqslant h\, \Omega(f, z_0) - \sum_{k=1}^{n} \omega(a_k, z_0) \tag{30}$$

where $a_k(z)$ $(k = 1, 2, \ldots, n)$ are the coefficients of the sum (15). If z_0 is not a pole of $P(f, f', \ldots, f^{(p)})$, (30) is obvious. Now assume that z_0 is a pole of $P(f, f', \ldots, f^{(p)})$. Then it is easy to see that

$$\Omega(Q, z_0) \geqslant h\, \Omega(f, z_0) - \omega(p, z_0)$$

and

$$\omega(P, z_0) \leqslant \max_{1 \leqslant k \leqslant n} \omega(a_k, z_0) \leqslant \sum_{k=1}^{n} \omega(a_k, z_0) .$$

Hence (30) is also true.

Now consider a circle $|z| \leqslant r$ and make the summation of each side of (30) with respect to all the zeros z_0 of $f(z)$ in the circle $|z| \leqslant r$. We get

$$n(r, \frac{1}{Q}) \geqslant h\, n(r, \frac{1}{f}) - \sum_{k=1}^{n} n(r, a_k)$$

from which we deduce (29) in making use of the condition (16).

Theorem 6. Let $f(z)$ be a transcendental meromorphic function and $P(f, f', \ldots, f^{(p)})$ be defined by (15) such that $P(z) \not\equiv 0$. Let $h \geqslant 1$ be a positive integer. Then $f(z)$ and $Q(z) = f^h P(f, f', \ldots, f^{(p)})$ have nearly the same growth.

For the proof of this theorem, we base upon Theorem 5 and the following lemma which is known [7], [6].

Lemma 5. Let $U(r)$ be a non-negative and non-decreasing function in an interval $r_0 < r < \rho$ $(r_0 \geqslant 0)$. Let a and b be two positive numbers such that $b \geqslant 2a$ and $b \geqslant 8a^2$. If for $r_0 < r < R < \rho$ we have

$$U(r) < a \overset{+}{\log} U(R) + a \log \frac{R}{R-r} + b,$$

then for $r_0 < r < R < \rho$ we have

$$U(r) < 2a \log \frac{R}{R-r} + 2b.$$

Now let us prove Theorem 6. First we are going to show that we can find a number $r_0 > 0$ such that for $\lambda > 1$ and $r > r_0$, we have

$$T(r, f) < B(\lambda)\, T(\lambda r, Q) \tag{31}$$

where $B(\lambda)$ is a positive function of λ defined for $\lambda > 1$. By Theorem 5, we have

$$T(r, f) \leqslant \frac{1}{h} T(r, Q) + o\{T(r, f)\} + S(r)$$

and then by (21), we have for $1 \leqslant r < R$,

$$T(r,f) < \frac{1}{h} T(r,Q) + o\{T(r,f)\} + K(\overset{+}{\log} T(R,f) + \log R$$

$$+ \overset{+}{\log} \frac{1}{R-r} + 1)$$

where $K > 1$ is a constant. Take a number $r_1 > 1$ such that for $r > r_1$ we have

$$o\{T(r,f)\} < \frac{1}{2} T(r,f), \quad \log r > 16K ,$$

and let ρ be a number such that $r_1 < \rho$. Then for $r_1 < r < R < \rho$ we have

$$T(r,f) < 2T(r,Q) + 2K(\overset{+}{\log} T(R,f) + \log R + \overset{+}{\log}\frac{1}{R-r} + 1)$$

$$< a \overset{+}{\log} T(R,f) + a \log \frac{R}{R-r} + b,$$

with

$$a = 2K, \quad b = 2T(\rho, Q) + 2K(\log \rho + 1).$$

Since the conditions $b \geqslant 2a$, $b \geqslant 8a^2$ are satisfied, hence by Lemma 5, we have, for $r_1 < r < R < \rho$,

$$T(r, f) < 2a \log \frac{R}{R-r} + 2b .$$

Keeping r fixed and let $R \to \rho$, we get

$$T(r, f) \leqslant 4K \log \frac{\rho}{\rho - r} + 4T(\rho, Q) + 4K(\log \rho + 1)$$

for $r_1 < r < \rho$. So for $\lambda > 1$ and $r > r_1$, we have, in setting $\rho = \lambda r$,

$$T(r, f) \leqslant 4K \log \frac{\lambda}{\lambda - 1} + 4T(\lambda r, Q) + 4K(\log(\lambda r) + 1).$$

This inequality shows that $Q(z)$ is also a transcendental meromorphic function. Then it is easy to see that we can find a number $r_0 > r_1$ such that for $\lambda > 1$, $r > r_0$, we have the inequality (31), provided that there we take

$$B(\lambda) = 4K \log \frac{\lambda}{\lambda - 1} + 4(K+1) .$$

To complete the proof of Theorem 6, we have to show that inversely we can find a number $r_0' > 0$ such that for $\lambda > 1$ and $r > r_0'$, we have

$$T(r, Q) < A(\lambda) T(\lambda r, f) \tag{32}$$

where $A(\lambda)$ is a positive function of λ defined for $\lambda > 1$. This is easy to show. In fact, for $r \geqslant 1$, we have

$$T(r, Q) \leqslant h T(r, f) + T(r, P)$$

$$T(r, P) \leqslant \sum_{k=1}^{n} T(r, \phi_k) + \log n$$

where

$$\phi_k(z) = a_k(z) \prod_{j=0}^{p} \{f^{(j)}(z)\}^{S_{kj}}$$

and hence

$$T(r, \phi_k) \leqslant T(r, a_k) + \sum_{j=0}^{p} S_{kj} T(r, f^{(j)}).$$

So we get

$$T(r, Q) \leqslant o\{T(r, f) + \sum_{j=0}^{p} \sigma_j T(r, f^{(j)}) \tag{33}$$

where σ_j $(j = 0, 1, 2, \ldots, p)$ are non-negative integers. By the examples given in connection with the Definition 2, we see that (33) implies (32), provided that the function $A(\lambda)$ is suitably chosen.

In view of Theorem 6, it is useful to give some examples of differential polynomials $P(f, f', \ldots, f^{(p)})$ which are non identically equal to zero. In the following three examples, it is always assumed that $f(z)$ is a transcendental meromorphic function.

1) Let $a(z) \not\equiv 0$ be a meromorphic function such that

$T(r, a) = o\{T(r, f)\}$.

Then evidently the differential polynomial

$$P_1(f, f', \ldots, f^{(p)}) = a(z) \prod_{j=0}^{p} \{f^{(j)}(z)\}^{S_j} \not\equiv 0$$

where $S_j (j = 0, 1, 2, \ldots, p)$ are non-negative integers with $\sum_{j=0}^{p} S_j > 0$.

2) Let $P(f, f', \ldots, f^{(p)})$ be defined by (15) and $b(z) \not\equiv 0$ be a meromorphic function such that

$T(\lambda_0 r, b) = o\{T(r, f)\}$

for a number $\lambda_0 > 1$. Then the differential polynomial

$$P_2(f,f', \ldots,f^{(p)}) = f^h P(f,f', \ldots,f^{(p)}) + b(z) \not\equiv 0$$

where $h \geqslant 1$ is a positive integer.

To see this, distinguish two cases. If $P \equiv 0$, this is evident. If $P \not\equiv 0$, then by Theorem 6, $f(z)$ and $Q(z) = f^h P(f, f', \ldots, f^{(p)})$ have nearly the same growth, hence

$T(r, f) < B(\lambda_0)\, T(\lambda_0 r, Q)$.

So it is impossible that $P_2 \equiv 0$.

3) Let $\alpha_i(z)$ $(i = 1, 2, \ldots, h)$ and $\beta_j(z)$ $(j = 1, 2, \ldots, k)$ be two systems of meromorphic functions satisfying the following conditions:

1° Each of the systems $\alpha_i(z)$ $(i = 1, 2, \ldots, h)$ and $\beta_j(z)$ $(j = 1, 2, \ldots, k)$ is linearly independent.

2° Set $p = h + k - 1$. We have

$$T(r,\alpha_i^{(n)}) = o\{T(r,f)\} \; (n = 0,1,2,\ldots,p; \; i = 1,2,\ldots,h)$$
$$T(r,\beta_j^{(n)}) = o\{T(r,f)\} \; (n = 0,1,2,\ldots,p; \; j = 1,2,\ldots,k). \tag{34}$$

Then the differential polynomial

$$P_3(f,f',\ldots,f^{(p)}) = W(\alpha_1,\alpha_2,\ldots,\alpha_h,\beta_1 f,\beta_2 f,\ldots,\beta_k f) \not\equiv 0 \tag{35}$$

where $W(\alpha_1, \alpha_2, \ldots, \alpha_h, \beta_1 f, \beta_2 f, \ldots, \beta_k f)$ denotes the Wronskian of the functions $\alpha_i(z)(i = 1, 2, \ldots, h), \beta_j(z)f(z)$ $(j = 1, 2, \ldots, k)$.

To prove this, it is sufficient to show that the system of functions $\alpha_i(z)$ $(i = 1, 2, \ldots, h)$, $\beta_j(z)f(z)$ $(j = 1, 2, \ldots, k)$ is linearly independent. In fact, if it is not so, then there are constants C_i $(i = 1, 2, \ldots, h)$, $C_j'(j = 1, 2, \ldots, k)$ not all equal to zero, such that

$$\sum_{i=1}^{h} C_i \alpha_i(z) + \sum_{j=1}^{k} C_j' \beta_j(z) f(z) \equiv 0 .$$

C_j' $(j = 1, 2, \ldots, k)$ cannot be all equal to zero, because $\alpha_i(z)$ $(i = 1, 2, \ldots, h)$ are linearly independent. Since $\beta_j(z)$ $(j = 1, 2, \ldots, k)$ are linearly independent, we have

$$\sum_{j=1}^{k} C_j' \beta_j(z) \not\equiv 0.$$

So we may write

$$f(z) = \frac{-\sum_{i=1}^{h} C_i\alpha_i(z)}{\sum_{j=1}^{k} C_j{}'\beta_j(z)} \tag{36}$$

(34) and (36) yield $T(r, f) \leq o\{T(r, f)\}$ and we get a contradiction.

A particular case is that $k = 1$, $\beta_1 = 1$, then the Wronskian in (35) is $W(\alpha_1, \alpha_2, \ldots, \alpha_h, f)$.

In view of the above examples, it is easy to find an operator in the form of a differential polynomial $\mathscr{P}(f, f' \ldots, f^{(p)})$ whose coefficients $\alpha_k(z)$ $(k = 1, 2, \ldots, n)$ are rational functions such that for any transcendental meromorphic function $f(z)$ the corresponding function $\mathscr{P}(f, f', \ldots, f^{(p)}) \not\equiv 0$. Then by Theorem 6, the operator

$$\Omega(f) = f^h \mathscr{P}(f, f', \ldots, f^{(p)}) \tag{37}$$

is growth preserving, where $h \geq 1$ is a positive integer. Consider another growth preserving operator $\omega(f)$. Then the composite operator

$$\Omega\{\omega(f)\} = \omega(f)^h \mathscr{P}(\omega(f), \omega(f)', \ldots, \omega(f)^{(p)})$$

is also growth preserving. For instance, μ being a positive integer, the composite operator

$$\Omega(f^{(\mu)}) = (f^{(\mu)})^h \mathscr{P}(f^{(\mu)}, f^{(\mu+1)}, \ldots, f^{(\mu+p)}) \tag{38}$$

is growth preserving. A simple consequence of this result is that, if $R(z) \not\equiv 0$ is a rational function, then an operator of the form

$$M(f) = R(z) \prod_{j=0}^{p} (f^{(j)})^{S_j}, \tag{39}$$

where S_j $(j = 0, 1, 2, \ldots, p)$ are non-negative integers with $\sum_{j=0}^{p} S_j > 0$, is growth preserving. In fact, if $S_0 > 0$, $M(f)$ is of the form (37). In general, if $S_j = 0$ $(j = 0, 1, \ldots, \mu-1)$ and $S_\mu \neq 0$, $M(f)$ has the form (38). The operator (39) has also application in the joint work by the author and Ma which appears in this book.

Now let us return to Theorem 5 and deduce from it other consequences.

Corollary 3. Let $f(z)$ be a transcendental meromorphic function of finite order. Let $Q(z)$ be defined by (26) and assume that $P(z) \not\equiv 0$. Then we have

$$\varliminf_{r\to\infty} \frac{T(r,Q)}{T(r,f)} \geq h + \underline{d}(P)\, \delta(0, f) , \tag{40}$$

$$\varlimsup_{r\to\infty} \frac{T(r,Q)}{T(r,f)} \geq h + \underline{d}(P)\, \Delta(0,f) \tag{41}$$

where

$$\delta(0,f) = \varliminf_{r\to\infty} \frac{m(r,\frac{1}{f})}{T(r,f)} = 1 - \varlimsup_{r\to\infty} \frac{N(r,\frac{1}{f})}{T(r,f)} , \tag{42}$$

$$\Delta(0,f) = \varlimsup_{r\to\infty} \frac{m(r,\frac{1}{f})}{T(r,f)} = 1 - \varliminf_{r\to\infty} \frac{N(r,\frac{1}{f})}{T(r,f)} . \tag{43}$$

Proof. Since $f(z)$ is of finite order, taking $R = 2r$ in (21), we see that the function $S(r)$ in (27) satisfies the condition

$$S(r) = O(\log r) .$$

From (27), we have

$$h + \underline{d}(P) \frac{m(r,\frac{1}{f})}{T(r,f)} \leq \frac{T(r,Q)}{T(r,f)} + \frac{h(o\{T(r,f)\} + S(r))}{T(r,f)} . \tag{44}$$

Then we get (40) by taking the lower limit of each side of (44). Also (41) is obtained by taking the upper limit of each side of (44).

When the order of $f(z)$ is infinite, then by taking in (21)

$$R = r + \frac{1}{T(r, f)},$$

we see that the function $S(r)$ in (27) satisfies the condition

$$S(r) = O(\log T(r, f) + \log r),$$

provided that $r \notin \sigma(f)$ which is the set of the values of r satisfying simultaneously the inequalities

$$T(r,f) > 0, \quad T\{r + \frac{1}{T(r,f)}, f\} \geq 2T(r,f). \tag{45}$$

It is well known that this set $\sigma(f)$ has finite measure. So we deduce from (44) the following Corollary:

Corollary 4. Let $f(z)$ be a transcendental meromorphic function of infinite order. Let $Q(z)$ be defined by (26) and assume that $P(z) \not\equiv 0$. Then we have

$$\varliminf_{\substack{r\to\infty \\ r\notin\sigma}} \frac{T(r,Q)}{T(r,f)} \geq h + \underline{d}(P)\delta'(0,f) \tag{46}$$

$$\varlimsup_{\substack{r\to\infty \\ r\notin\sigma}} \frac{T(r,Q)}{T(r,f)} \geq h + \underline{d}(P)\ \Delta'(0,f) \tag{47}$$

where

$$\delta'(0,f) = \varliminf_{\substack{r\to\infty \\ r\notin\sigma}} \frac{m(r, \frac{1}{f})}{T(r,f)} = 1 - \varlimsup_{\substack{r\to\infty \\ r\notin\sigma}} \frac{N(r, \frac{1}{f})}{T(r,f)}, \tag{48}$$

$$\Delta'(0,f) = \overline{\lim_{\substack{r\to\infty \\ r\notin\sigma}}} \frac{m(r, \frac{1}{f})}{T(r,f)} = 1 - \underline{\lim_{\substack{r\to\infty \\ r\notin\sigma}}} \frac{N(r, \frac{1}{f})}{T(r,f)} \tag{49}$$

where $\sigma = \sigma(f)$ is defined by (45).

Noting that in (46),

$$\underline{\lim_{\substack{r\to\infty \\ r\notin\sigma}}} \frac{T(r,Q)}{T(r,f)} \leqslant \overline{\lim_{r\to\infty}} \frac{T(r,Q)}{T(r,f)},$$

$$\delta'(0,f) \geqslant \delta(0,f),$$

we have, a fortiori,

$$\overline{\lim_{r\to\infty}} \frac{T(r,Q)}{T(r,f)} \geqslant h + \underline{d}(P)\ \delta(0,f), \tag{50}$$

where $\delta(0, f)$ is defined by (42). In view of (41), the inequality (50) also holds when $f(z)$ is of finite order.

We are going to generalize the inequality (50) and prove the following Corollary:

Corollary 5. Let $f(z)$ be a transcendental meromorphic function and $\phi(z)$ a meromorphic function such that

$$T(r,\phi^{(j)}) = o\{T(r,f)\} \ (j = 0, 1, 2, \ldots, p). \tag{51}$$

Set

$$Q_1(z) = (f-\phi)^h\, P(f, f', \ldots, f^{(p)}) \tag{52}$$

where $h \geqslant 1$ is a positive integer and $P(z) = P(f, f', \ldots, f^{(p)})$ is defined by (15). If $P(z) \not\equiv 0$, then we have

$$\overline{\lim_{r\to\infty}} \frac{T(r,Q_1)}{T(r,f)} \geq h + \underline{d}(P_*)\ \delta(\phi,f) \tag{53}$$

where

$$\delta(\phi,f) = \underline{\lim_{r\to\infty}} \frac{m(r, \frac{1}{1-\phi})}{T(r,f)} = 1 - \overline{\lim_{r\to\infty}} \frac{N(r, \frac{1}{f-\phi})}{T(r,f)} \tag{54}$$

and P_* denotes the differential polynomial

$$P_*(g,g',\ldots,g^{(p)}) = P(g+\phi,g'+\phi',\ldots,g^{(p)}+\phi^{(p)}) \tag{55}$$

of g.

Proof. Set $g(z) = f(z) - \phi(z)$. Then $g(z)$ is also a trancendental meromorphic function and we have

$$T(r,g) = T(r,f) + o\{T(r,f)\}, \tag{56}$$

$$P(z) = P(f,f',\ldots,f^{(p)}) = P(g+\phi,g'+\phi',\ldots,g^{(p)}+\phi^{(p)})$$

$$= P_*(g,\ g',\ \ldots,\ g^{(p)})$$

where

$$P_*(z) = P_*(g,\ g',\ \ldots,\ g^{(p)})$$

is a differential polynomial of $g(z)$. We also have

$$Q_1(z) = g^h P_*(g,\ g',\ \ldots,\ g^{(p)}).$$

Since $P(z) \not\equiv 0$, hence $P_*(z) \not\equiv 0$. Consequently we can apply (50) to $g(z)$ and $Q_1(z)$. So we have

$$\overline{\lim_{r\to\infty}} \frac{T(r,Q_1)}{T(r,g)} \geq h + \underline{d}(P_*)\ \delta(0,\ g). \tag{57}$$

From (56) we see easily that

$$\overline{\lim_{r\to\infty}} \frac{T(r,Q_1)}{T(r,g)} = \overline{\lim_{r\to\infty}} \frac{T(r,Q_1)}{T(r,f)}, \quad \delta(0, g) = \delta(\phi, f)$$

and we get (53).

Finally we give an application of Corollary 5 to differential equations. Consider a differential equation of the form

$$P(w, w', \ldots, w^{(p)}) \equiv \sum_{k=1}^{m} a_k(z) \prod_{j=0}^{p} \{w^{(j)}\}^{S_{kj}} = 0 \qquad (58)$$

where $a_k(z)$ $(k = 1, 2, \ldots, n)$ are meromorphic functions, S_{kj} $(k = 1, 2, \ldots, n;\ j = 0, 1, 2, \ldots, p)$ are non-negative integers and w is the unknown function. Suppose that we can find a trancendental meromorphic function $f(z)$ satisfying the following conditions:

1^{0} $T(r, a_k) = o\{T(r, f)\}$ $(k = 1, 2, \ldots, n)$.

2^{0} There exist a meromorphic function $\phi(z)$ and a positive integer $h \geqslant 1$ such that we have (51) and that, $Q_1(z)$ being defined by (52), we have

$$\overline{\lim_{r\to\infty}} \frac{T(r, Q_1)}{T(r, f)} < h + \underline{d}(P_*)\ \delta(\phi, f)$$

where P_* is defined by (55).

Then by Corollary 5 we have necessarily

$$P(f, f', \ldots, f^{(p)}) \equiv 0 .$$

In other words, the function $w = f(z)$ is a solution of the differential equation (58).

Similar applications to differential equations can also be obtained basing upon Corollaries 3 and 4.

REFERENCES

1. Nevanlinna, R., Le théorème de Picard-Borel et la theorie des fonctions méromorphes, Paris (1929).

2. Chuang Chi-tai, Sur la comparaison de la croissance d'une fonction méromorphe et de celle de sa dérivée (Bull. Sciences Math., 75 (1951), 1-20).

3. Chuang Chi-tai, Une généralisation d'une inégalité de Nevanlinna (Scientia Sinica, 13 (1964), 887-895).

4. Chuang Chi-tai, On the inversion of a linear differential polynomial (Contemporary Mathematics, 48 (1985), 1-20).

5. Chuang Chi-tai, On differential polynomials (Analysis of one complex variable, Singapore (1987)).

6. Chuang Chi-tai, Singular directions of meromorphic functions (in Chinese), Beijing (1982).

7. Milloux H., Les fonctions méromorphes et leurs dérivées. Extensions d'un théorème de M.R. Nevanlinna. Applications, Actualités Scient. et Ind., Paris (1940).

ON THE DISTRIBUTION OF VALUES OF RANDOM DIRICHLET SERIES (I)*

Yu Jia-Rong and Sun Dao-Chun
Wuhan University
P.R. China

Consider a random Dirichlet series

$$\sum_{n=1}^{+\infty} b_n Z_n(\omega) e^{-\lambda_n s} \tag{1}$$

where $\{b_n\}$ is a sequence of complex numbers, $\{Z_n(\omega)\}$ is a sequence of complex random variables defined on a probability space $(\Omega, \mathscr{A}, \mathscr{P})$ $(\omega \in \Omega)$,

$$0 \leqslant \lambda_1 < \lambda_2 < \dots < \lambda_n \uparrow +\infty \tag{2}$$

$s = \sigma + it$, σ and $t \in \mathbb{R}$. Let $\sigma_c(\omega)$ denote the abscissa of convergence of (1).

When $\{Z_n(\omega)\}$ is a Rademacher or Steinhaus sequence [1] and $\sigma_c(\omega) = +\infty$ a.s. (almost surely), (1) represents a random entire function $f(s\ ;\ \omega)$ and we have studied the distribution of its values in terms of the order (R) (Ritt order) [10]. In the case $\sigma_c(\omega) = 0$ a.s., (1) represents

---

* Partially supported by the Science Foundation of People's Republic of China

a random analytic function $f(s\ ;\ \omega)$ in the right half plane and we have also studied the same problem for this random function in terms of the order (R) of a function analytic in the right half plane [11] [12]. We have discovered that the results for a Rademacher or Steinhaus sequence are also valid for a Gause sequence [13].

In this paper we introduce N sequences of random variables of which the Rademacher, Steinhaus and Gauss sequences are special cases. When $\{Z_n(\omega)\}$ in (1) is an N sequence and $\sigma_c(\omega) = -\infty$ or 0. a.s., we study the distribution of values of the random function represented by (1) in terms of the (p, q) - order (R) [2] (p and q are integers and $p \geqslant q \geqslant 0$), an extension of the order (R), and obtain some results generalizing and improving our previous results for the infinite order (R). The corresponding results for the finite order (R) will be given in another paper.

I. DEFINITIONS AND LEMMAS

Let $(\Omega, \mathscr{A}, \mathscr{P})$ be a probability space, where Ω is a set $\neq \emptyset$, $\mathscr{A}$ consists of sub-sets of Ω and it is closed under the operations of complementation and of taking countable unions and $\mathscr{P}$ is a probability on $(\Omega, \mathscr{A})$. Let $\mathscr{B}$ consist of all Borel sets on the complex plane $\mathbb{C}$ and $Z : \Omega \to \mathbb{C}$. If $\forall\, B \in \mathscr{B}, Z^{-1}(B) \in \mathscr{A}$, then $Z(\omega)$ is called a complex random variable defined on $(\Omega, \mathscr{A}, \mathscr{P})$. We define a real random variable in a similar way. Let

$$\mathscr{B}_Z = \{B \mid B \subset \mathbb{C} \text{ and } Z^{-1}(B) \in \mathscr{A}\}$$

and

$$\mu_Z(B) = \mathscr{P}(Z^{-1}(B)) \qquad \forall B \in \mathscr{B}_Z .$$

Evidently $\mathscr{B} \subset \mathscr{B}_Z$ and $(\mathbb{C}, \mathscr{B}_Z, \mu_Z)$ is also a probability space determined by the random variable $Z(\omega)$. If $(\mathbb{C}, \mathscr{B}_Z, \mu_Z)$ and $(\mathbb{C}, \mathscr{B}_{-Z}, \mu_{-Z})$ are the same probability space, we say that $Z(\omega)$ is a <u>symmetric random variable</u>.

If the sequence of random variables $\{Z_n(\omega)\}$ satisfies

$$\mathscr{B}_Z = \mathscr{B}_{Z_n} \text{ and } \mu_Z = \mu_{Z_n}, \quad \forall n \in \mathbb{N} = \{1,2,3,\ldots\}, \tag{3}$$

where Z is a complex or real random variable, we say that $\{Z_n(\omega)\}$ is a sequence of <u>equally distributed random variables</u>. Evidently all the $Z_n(\omega)$ have the same finite or infinite variance.

Let $\{Z_n(\omega)\}$ be a sequence of independent, symmetric and equally distributed complex or real random variables of finite variance. Suppose that there exists $k_o \in \mathbb{N}$ such that

$$\int_{|Z| \leqslant 1} |Z|^{-1/k_o} \mu(d\, Z) < +\infty, \tag{4}$$

where μ is the common measure μ_{Z_n} in (3). Then $\{Z_n(\omega)\}$ is called an <u>N-sequence of random variables</u>.

For example, the Rademacher and Steinhaus sequences and the Gauss sequence of complex or real random variables[1] are all N-sequences. We can verify this by the the following two lemmas.

<u>Lemma 1.</u> Let $Z(\omega)$ be a complex or real random variable. If there exists $\delta > o$ such that

$$\mathscr{P}(\{\omega \mid |z(\omega)| \leqslant \delta\}) = 0,$$

then (4) is verified,

Proof. We have, for $k_o \in \mathbb{N}$,

$$\int_{|Z|\leqslant 1} |Z|^{-\frac{1}{k_o}} \mu(dZ) \leqslant \delta^{-1/k_o} < +\infty .$$

Lemma 2. Let $Z(\omega)$ be a complex or real random variable of continuous type with a bounded probability density p. Then (4) is verified.

Proof. Let K be a finite bound of p and take $k_o = 4$ in (4). For example, in the complex case,

$$\int_{|Z|\leqslant 1} |Z|^{-1/4} \mu(dZ) = \int_{x^2+y^2\leqslant 1} (x^2 + y^2)^{-1/4} p(x,y)\, dx\, dy$$

$$\leqslant \int_0^1 r^{-1/2} K\, 2\pi dr < +\infty .$$

Lemma 3. Suppose that $\{Z_n(\omega)\}$ be an N-sequence defined on $(\Omega, \mathscr{A}, \mathscr{P})$. Then for $\omega\in\Omega$, $\exists N(\omega) > 0$ a.s. such that $\forall\, n > N(\omega)$,

---

1) For the Rademacher sequence $\{Z_n\}$,

$\mu_Z(Z_n = \pm 1) = \frac{1}{2}$.

For the Steinhaus sequence $\{Z_n\}$, $Z_n = e^{2\pi i \theta_n}$, where $\theta_n(\omega)$ is uniformly distributed on $[0, 1]$. For the Gauss sequence in the real case,

$$Z_n \in \mathbb{R} \text{ and } \mu_Z(Z_n \in B) = \frac{1}{\sqrt{2\pi}} \int_B e^{-x^2/2} dx$$

for any Borel set B on $\mathbb{R}$ and in the complex case,

$$\mu_Z(Z_n \in B) = \frac{1}{2\pi} \int_B e^{-|z|^2/2} dx\, dy, \quad \forall\, B \in \mathscr{B},\ z = x + iy$$

$$n^{-k_o} \leq |Z_n(\omega)| \leq n^{k_o} \qquad (5)$$

<u>Proof</u>. By (4), $\mathscr{P}(Z(\omega) = 0) = 0$, where $Z(\omega)$ is the random variable in (3). We have

$$+\infty > \int_{|Z|\leq 1} |Z|^{-1/k_o} \mu(dZ)$$

$$\geq \sum_{m=1}^{+\infty} \int_{(m+1)^{-k_o} \leq |Z| < m^{-k_o}} |Z|^{-1/k_o} \mu(dZ)$$

$$\geq \sum_{m=1}^{+\infty} (m^{-k_o})^{-1/k_o} \mu((m+1)^{-k_o} \leq |Z| < m^{-k_o})$$

$$= \sum_{n=1}^{+\infty} \sum_{m=n}^{+\infty} \mu((m+1)^{-k_o} \leq |Z| < m^{-k_o})$$

$$= \sum_{n=1}^{+\infty} \mu(|Z| < n^{-k_o})$$

$$= \sum_{n=1}^{+\infty} \mu(|Z_n| < n^{-k_o}) .$$

Hence by Borel-Cantelli lemma,

$$\mathscr{P}\Big(\bigcap_{j=1}^{+\infty} \bigcup_{n=j}^{+\infty} \{\omega \,|\, |Z_n(\omega)| < n^{-k_o}\}\Big) = 0$$

and consequently for $\omega \in \Omega$, $\exists N(\omega) > 0$ a.s. such that $\forall\ n > N(w)$,

$$|Z_n(\omega)| \geq 1/n^{k_o}$$

Since the variances are finite, we have

$$+\infty > \int_\Omega |Z(\omega)|^2 \;\mathscr{P}(d\omega) \geq \int_{|Z|\geq 1} |Z|\,\mu(dZ)$$

$$\geq \sum_{m=1}^{+\infty} m\,\mu(m < |Z| \leq m+1) = \sum_{n=1}^{+\infty} \mu(|Z| > n)$$

$$\geq \sum_{n=1}^{+\infty} \mu(|Z_n| > n^{k_o}).$$

Then by Borel-Cantelli lemma we obtain another half of the conclusion.

<u>Lemma 4</u>. Suppose that $\{Z_n(\omega)\}$ is a sequence of independent symmetric random vairables of finite variances σ_n^2 defined on the probability space $(\Omega, \mathscr{A}, \mathscr{P})$[2]), then $\forall\, H \in \mathscr{A}$, $\exists N = N(H) \in \mathbb{N}$ such that for any sequence $\{b_n\}$ of complex numbers and for any $N'' > N$,

$$\int_H \Big|\sum_{n=N}^{N''} b_n Z_n(\omega)\Big|^2 \;\mathscr{P}(d\omega) \geq \frac{1}{2} \inf_{n\geq 1}$$

$$\Big\{\int_H \frac{|Z_n(\omega)|^2}{\sigma_n^2} \mathscr{P}(d\omega)\Big\} \sum_{n=N}^{N''} |b_n|^2 \;\sigma_n^2 \quad . \tag{6}$$

<u>Proof</u>. Let

$$A = \inf_{n\geq 1} \Big\{\int_H \frac{|Z_n(\omega)|^2}{\sigma_n^2} \;\mathscr{P}(d\omega)\Big\}.$$

---

2) This sequence is not necessarily an N-sequence or a sequence of equally distributed random variables.

When $\sigma_n = 0$, we put

$$\int_H \frac{|Z_n(\omega)|^2}{\sigma_n^2} \mathscr{P}(d\omega) = +\infty$$

as a convention. We can assume that $A > 0$. $\forall N'$ and $N'' \in \mathbb{N}$ and $N' < N''$, we have

$$\int_H \Big| \sum_{n=N'}^{N''} b_n Z_n(\omega)\Big|^2 \mathscr{P}(d\omega)$$

$$= \sum_{n=N'}^{N''} |b_n|^2 \int_H |Z_n(\omega)|^2 \mathscr{P}(d\omega) +$$

$$\sum_{\substack{N' \leqslant k, j \leqslant N'' \\ k \neq j}} b_k \overline{b_j} \int_H Z_k(\omega) \overline{Z_j(\omega)} \mathscr{P}(d\omega).$$

If $\sigma_{n_0} = 0$, $Z_{n_0}(\omega) = 0$ a.s. Hence in the above formula we can take off the terms with such indices n_0 and suppose that all $\sigma_n > 0$. Then we have

$$\int_H \Big| \sum_{n=N'}^{N''} b_n Z_n(\omega)\Big|^2 \mathscr{P}(d\omega) \geqslant \Big(\sum_{n=N'}^{N''} |b_n|^2 \sigma_n^2 \Big)$$

$$\inf_{n \geqslant 1} \Big\{ \int_H \frac{|Z_n(\omega)|^2}{\sigma_n^2} \mathscr{P}(d\omega) - Y, \qquad (7)$$

where

$$Y = \sum_{\substack{N' \leqslant k, j \leqslant N'' \\ k \neq j}} \Big(\sigma_k \sigma_j |b_k b_j| . \Big| \int_H \frac{Z_k(\omega) \overline{Z_j(\omega)}}{\sigma_k \sigma_j}$$

$$\mathscr{P}(d\omega) \Big| \Big).$$

Evidently

$$Y \leqslant \Big(\sum_{\substack{N' \leqslant k, j \leqslant N'' \\ k \neq j}} (|b_k|^2 |b_j|^2 \sigma_k^{\,2} \sigma_j^{\,2} \Big)^{1/2}$$

$$\times \Big(\sum_{N' \leqslant k, j \leqslant N''} \Big| \int_H \frac{Z_k(\omega)\overline{Z_j(\omega)}}{\sigma_k \ \sigma_j} \ \mathscr{P}(d\omega) \Big|^2 \Big)^{1/2}$$

$$\leqslant \Big(\sum_{n=N'}^{N''} |b_n|^2 \ \sigma_n^{\,2} \Big) \Big(\sum_{\substack{N' \leqslant k, j \leqslant N'' \\ k \neq j}}$$

$$\Big| \int_H \frac{Z_k(\omega)\overline{Z_j(\omega)}}{\sigma_k \ \sigma_j} \ \mathscr{P}(d\omega) \Big|^2 \Big)^{1/2}. \tag{8}$$

Since $\{\frac{Z_k(\omega) \ \overline{Z_j(\omega)}}{\sigma_k \ \sigma_j}\}_{k>j}$ and $\{\frac{Z_n(\omega) \ \overline{Z_j(\omega)}}{\sigma_k \ \sigma_j}\}_{k<j}$ are two orthonormal systems and $\{\int_H \frac{Z_k(\omega) \ \overline{Z_j(\omega)}}{\sigma_k \ \sigma_j} \ \mathscr{P}(d\omega)\}_{k<j}$ can be considered as the Fourier coefficients of the function $\chi = \{\begin{matrix} 1 & (\omega \in H) \\ 0 & (\omega \bar{\in} H) \end{matrix}$, with respect to the two systems, we have

$$\sum_{1 \leqslant k, j < +\infty} \Big| \int_H \frac{Z_k(\omega) \ \overline{Z_j(\omega)}}{\sigma_k \ \sigma_j} \ \mathscr{P}(d\omega) \Big|^2 < +\infty .$$

Hence we can select $N = N(H)$ so large that $\forall \ N'' > N_2$

$$\sum_{\substack{N \leqslant k, j \leqslant N' \\ k \neq j}} \Big| \int_H \frac{Z_k(\omega) \ \overline{Z_j(\omega)}}{\sigma_k \ \sigma_j} \ \mathscr{P}(d\omega) \Big|^2 \leqslant \frac{1}{4} A^2.$$

Replacing N' by N in (7) and (8) and taking account of

them and the above inequality we obtain (6).

Lemma 5[3]. Suppose that $\{Z_n(\omega)\}$ is an N-sequence of random variables defined on $(\Omega, \mathscr{A}, \mathscr{P})$. Then $\forall H \in \mathscr{A}$ and $\mathscr{P}(H) > 0$, $\exists N = N(H) \in \mathbb{N}$ and $e = e(H) > 0$ such that for any sequence $\{b_n\}$ of complex numbers and for any $N' > N$,

$$\int_H \Big| \sum_{n=N}^{N'} b_n Z_n(\omega) \Big|^2 \mathscr{P}(d\omega) \geqslant e \sum_{n=N}^{N'} |b_n|^2. \tag{9}$$

Proof. Since $Z_n(\omega)$ are equally distributed, by (2) we can choose $\delta > 0$ such that

$$\mathscr{P}(\{\omega|\ |Z_n(\omega)| < \delta) = \mu(|Z_n(\omega)| < \delta) < \frac{1}{2}\mathscr{P}(H).$$

Then

$$\int_H |Z_n(\omega)|^2 \mathscr{P}(d\omega) \geqslant \int_{H-\{\omega||Z_n(\omega)|<\delta\}} |Z_n(\omega)|^2 \mathscr{P}(d\omega)$$

$$\geqslant \delta^2 \mathscr{P}(H - \{\omega|\ |Z_n(\omega)| < \delta) \geqslant \frac{\delta^2}{2}\mathscr{P}(H).$$

Putting $e = \frac{1}{4}\delta^2\mathscr{P}(H)$ and taking account of the fact that $\sigma_1^2 = \sigma_2^2 = \ldots$ we deduce (9) from (6).

II. RANDOM ANALYTIC FUNCTION OF INFINITE ORDER (R) IN THE RIGHT-HALF PLANE

Consider the random Dirichlet series (1), $\{Z_n(\omega)\}$ being an N-sequence of random variables defined on $(\Omega, \mathscr{A}, \mathscr{P})$. Suppose that

---

3) This is an extension of Paley-Zygmund lemma [4]. Compare [13].

$$\overline{\lim_{n\to+\infty}} \frac{\log|b_n|}{\lambda_n} = 0 \tag{10}$$

and that $\exists\delta\in(0, 1)$ such that

$$\lim_{n\to+\infty} \frac{\log n}{\lambda_n^{\delta}} = 0 \text{ [4)]} . \tag{11}$$

Then in virtue of (5) the random series (1) represents a.s. an analytic function $f(s; \omega)$ in the right-half plane $\sigma > 0$ [3].

For $\sigma > 0$, $\omega\in\Omega$ and α and β satisfying $\alpha \leqslant \beta$, let

$$M(\sigma; \omega) = \sup_{-\infty<t<+\infty} \{|f(\sigma + it; \omega)|\},$$

$$m(\sigma; \omega) = \sup_{n\geqslant 1} \{|b_n Z_n(\omega)|e^{-\lambda_n\sigma}\},$$

$$M(\sigma, \alpha, \beta; \omega) = \sup_{\alpha\leqslant t\leqslant\beta} \{|f(\sigma + it; \omega)|\} .$$

Let

$$e_0 r = \log_0 r = r, \quad e_n^r = \exp(e_n^r - 1) \ (r > 0)$$

$$\log_n r = \log(\log_{n-1} r) \ (r \text{ sufficiently large})$$

$$(n \in \mathbb{N}).$$

We shall study the distribution of values (1) in terms of the (p, q) order (R) (p and $q \in \mathbb{N}$, $p \geqslant q$) introduced by O.P. Juneji and others [2].

---

4) In the case of infinite order (R) we can adopt this condition instead of $\overline{\lim_{n\to+\infty}} (n/\lambda_n) < +\infty$ or the corresponding condition in [9].

Let $\rho(r)$ be a positive differentiable function tending monotonically to $\rho \in (0, +\infty)$ $(r \to +\infty)$ and satisfying

$$\lim_{r \to +\infty} \rho'(r) \prod_{c=0}^{q} \log_c r = 0.$$

If $U(r) = e_p^{\rho(r) \log_q r}$ $(p, q \in \mathbb{N})$ satisfies

$$U\left(r + \frac{r}{\log^2 U(r)}\right) \sim U(r) \quad (r \to +\infty), \tag{12}$$

we say that $U(r)$ is a <u>type-function.</u> We obtained [6].

<u>Theorem A.</u> If for the series

$$g(r) = \sum_{n=1}^{+\infty} b_n e^{-\lambda_n r},$$

(2), (10) and (11) are varied and if $U(r) = e_p^{\rho(r) \log_q r}$ is a type-function ($p > q$ or $p = q > 1$ and $\rho > 1$), then $g(s)$ is analytic in $\sigma > 0$ and

$$\overline{\lim_{n \to +\infty}} \left(\log_q \frac{\lambda_n}{\overset{+}{\log} |b_n|} - \log_q W(\log_j \lambda_n)\right) \prod_{c=j+1}^{p+1} \log_c \lambda_n = 0 \tag{13}$$

$$\Longleftrightarrow \overline{\lim_{\sigma \to +0}} \frac{\overset{+}{\log}_{j+1} M_g(\sigma)}{\log_j U(1/\sigma)} = 1, \tag{14}$$

where $r = w(u)$ is the inverse function of $u =$

$e_{p-j}^{\rho(r) \log_q r}$, $1 \leqslant j \leqslant p$ and we take $\prod_{c=j+1}^{p-1} \log_c \lambda_n = \frac{1}{\log_p \lambda_n}$ where $j = p$ and

$$M_g(\sigma) = \sup_{-\infty<t<+\infty} \{|g(\sigma + i t)|\} \ (\sigma > 0).$$

From Theorem A we deduce

Theorem 1. If for the random series (1), $\{Z_n(\omega)\}$ is an N-sequence and (2), (10), (11) and (13) are verified, then (1) represents a.s. an analytic function f (s; w) in $\sigma > 0$ and we have

$$\overline{\lim_{\sigma\to+0}} \frac{\overset{+}{\log}_{j+1} M(\sigma \ ; \ \omega)}{\log_j \ U(1/\sigma)} = 1 \text{ a.s.},$$

where $1 \leqslant j \leqslant p$ and $U(1/\sigma)$ is defined in Theorem A.

Proof. Let δ be the number appearing in (11). Divide $\{b_n\}$ into two sets:

$$\{b_n | \log|b_n| \geqslant \lambda_n^{(1 + \delta)/2}\} \text{ denoted by } \{\overline{b}_n\}$$

and

$$\{b_n | \log|b_n| < \lambda_n^{(1 + \delta)/2}\} \text{ denoted by } \{b_n\}.$$

By (11), when n is sufficiently large, we have $\log n < \lambda_n^{\delta}$. Moreover, by Lemma 3, for $\omega\epsilon\Omega$, $\exists N(w) > 0$ a.s. such that $\forall n > N(w)$, (5) holds. Hence, for $\omega\epsilon\Omega$, $\exists N_1(\omega) > 0$ a.s, such that $\forall n > N_1(\omega)$,

$$\frac{\lambda_n}{\log|\bar{b}_n Z_n(\omega)|} = \frac{\lambda_n}{\log|\bar{b}_n|}$$

$$\times\left(1 - \frac{\log|Z_n(\omega)|}{\log|\bar{b}_n|(1 + \log|Z_n(\omega)|/\log|\bar{b}_n|)}\right)$$

$$\geq \frac{\lambda_n}{\log|\bar{b}_n|}\left(1 - \frac{k_o \log n}{\lambda_n^{(1+\delta)/2}(1 - k_o \log n/\lambda_n^{(1+\delta)/2})}\right)$$

$$\geq \frac{\lambda_n}{\log|\bar{b}_n|}\left(1 - 2\, k_o\, \lambda_n^{(\delta - 1)/2}\right). \tag{15}$$

Applying Lemma 3 in another way, we see that for $\omega\epsilon\Omega$, $\exists\, N_2(\omega) > 0$ a.s. such that $\forall\, n > N_2(\omega)$.

$$\frac{\lambda_n}{\log|\bar{b}_n|}\left(1 + 2\, k_o\, \lambda_n^{(\delta - 1)/2}\right)$$

$$\geq \frac{\lambda_n}{\log|\bar{b}_n Z_n(\omega)|}\, . \tag{16}$$

Similarly, for $\omega\epsilon\Omega$, $\exists N_3(\omega) > 0$ a.s. such that $\forall\, n > N_3(\omega)$,

$$\frac{\lambda_n}{\log|\underline{b}_n Z_n(\omega)|} \geq \frac{\lambda_n}{\lambda_n^{(1 + \delta)/2} + k_o \log n}$$

$$\geq \lambda_n^{(1 - \delta)/2}\, . \tag{17}$$

We remark that the inverse function $r = W(u)$ of $u = e_{p-j}^{\rho(r) \log_q r}$ has the form $r = e_q^{(1/\rho_1(r)) \log_{p-j} u}$, where $\rho_1(r) \to \rho\ (r \to +\infty)$ [7]. Let

$$\Theta\ (a_n) = (\log_q a_n - \log_q W(\log_j \lambda_n)) \prod_{c=j+1}^{p-1} \log_c \lambda_n \ .$$

It is easy to verify that $\forall \varepsilon > o$, $\Theta\ (\lambda_n^{\varepsilon}) \to +\infty\ (n \to +\infty)$. Hence by (17),

$$\lim_{n \to +\infty} \Theta\ (\frac{\lambda_n}{\log|\underline{b}_n\ Z_n(\omega)|}) = +\infty \quad \text{a.s.} \tag{18}$$

By (18) and (13) we see that $\{\overline{b}_n\}$ is an infinite sequence. Since

$$\varliminf_{n \to +\infty} \Theta\ (\frac{\lambda_n}{\log|\overline{b}_n\ Z_n(\omega)|})$$

$$\geq \varliminf_{n \to +\infty} (\frac{\lambda_n}{\log|\overline{b}_n|}\ (1 - 2\ k_o\ \lambda_n^{(\delta-1)/2}))\ \text{a.s.}$$

and for n sufficiently large,

$$\log\ (\frac{\lambda_n}{\log|\overline{b}_n|}\ (1 - 2\ k_o\ \lambda_n^{(\delta - 1)/2}))$$

$$= \log_{q-1}\ (\log \frac{\lambda_n}{\log|\overline{b}_n|} + \log(1 - 2\ k_o\ \lambda_n^{(\delta-1)/2}))$$

$$\geq \log_{q-1}\ (\log \frac{\lambda_n}{\log|\overline{b}_n|}\ (1 - 4\ k_o\ \lambda_n^{(\delta-1)/2} / \log \frac{\lambda_n}{\log|\overline{b}_n|})) \geq \ldots$$

$$\geq \log_q \frac{\lambda_n}{\log|\overline{b}_n|} - 2^{q+1}\ k_o\ \lambda_n^{(\delta-1)/2}\ .$$

Hence

$$\varliminf_{n\to+\infty} \Theta\left(\frac{\lambda_n}{\log|\overline{b_n}\, Z_n(\omega)|}\right) \geqslant \varliminf_{n\to+\infty} \left(\Theta\left(\frac{\lambda_n}{\log|\overline{b}_n|}\right) - 2^{q+1} k_o \frac{\prod_{c=j+1}^{p-1} \log_c \lambda_n}{\lambda_n^{(1-\delta)/2}}\right) = 0 \text{ a.s.}$$

Similarly we can prove

$$0 = \varlimsup_{n\to+\infty} \left(\Theta\left(\frac{\lambda_n}{\log|\overline{b}_n|}\right) + 2^{q+1} k_o \frac{\prod_{c=j+1}^{p-1} \log_c \lambda_n}{\lambda_n^{(1+\delta)/2}}\right)$$

$$\geqslant \varlimsup_{n\to+\infty} \Theta\left(\frac{\lambda_n}{\log|\overline{b}_n|}\left(1 + 2k_o \lambda_n^{(\delta-1)/2}\right)\right)$$

$$\geqslant \varlimsup_{n\to+\infty} \Theta\left(\frac{\lambda_n}{\log|\overline{b}_n\, Z_n(\omega)|}\right) \text{ a. s.}$$

Combining the above results with (18) we prove that

$$\lim_{n\to+\infty} \Theta\left(\frac{\lambda_n}{\log|b_n Z_n(\omega)|}\right) = 0 \text{ a. s.}$$

and Theorem 1 follows from Theorem A.

Theorem 2. If for the random series (1), $\{Z_n(\omega)\}$ is an N-sequence and (2), (10), (11) and (13) are verified, then for the function $f(s, \omega)$ represented by (1), $\forall\, t \in \mathbb{R}$,

$$\overline{\lim_{\sigma\to+0}} \frac{\log_{j+1} |f(\sigma + i\, t;\ \omega)|}{\log_j U(1/\sigma)}$$

$$= \overline{\lim_{\sigma\to+0}} \frac{\log_{j+1} M(\sigma;\ \omega)}{\log_j U(1/\sigma)} = 1 \text{ a.s.}, \tag{19}$$

where $1 \leq j \leq p$ and $U(1/\sigma)$ is defined in Theorem A.

<u>Proof.</u> Suppose that for some $t_o \in R$, $\exists E \in \mathscr{A}(\mathscr{P}(E) > 0)$ such that

$$\ell_j(\omega) = \overline{\lim_{\sigma\to+0}} \frac{\log_{j+1} |f(\sigma + i\, t_o;\ \omega)|}{\log_j U(1/\sigma)} < 1 .$$

Suppose that

$$E_n = \{\omega |\, \ell_j(\omega) < 1 - 1/n;\ \omega \in E\}\ (n \in \mathbb{N}).$$

Evidently $E = \bigcup_{n=1}^{+\infty} E_n$ and $\exists n_o \in \mathbb{N}$ such that $\mathscr{P}(E_{n_o}) > 0$ and $\forall \omega \in E_{n_o}$, $\ell_j(\omega) < 1 - \frac{1}{n_o}$. Set $H = E_{n_o}$.

Hence, $\forall \omega \in H$, $\exists \sigma_o(\omega) > 0$, $\forall \sigma \in (0, \sigma_o(\omega))$,

$$\left|\sum_{n=1}^{+\infty} b_n Z_n(\omega)\, e^{-\lambda_n(\sigma + it_o)}\right|$$

$$< e_{j+1}^{(1 - 1/n_o) \log_j (1/\sigma)} \tag{20}$$

Evidently, $H = \bigcup_{m=1}^{+\infty} \{\omega |\, \omega \in H,\ \sigma_o(\omega) > 1/m\}$. Therefore, $\exists m_o$ such that $\mathscr{P}(H') > 0$, where $H' = \{\omega |\, \omega \in H,\ \sigma_o(\omega) > 1/m_o\}$. Consequently, $\forall \sigma \in (0, 1/m_o)$, $\forall \omega \in H'$

$$\left| \sum_{n=N}^{+\infty} b_n Z_n(\omega) e^{-\lambda_n(\sigma + i t_o)} \right|$$

$$< e_{j+1}^{(1 - 1/m_o) \log_j U(1/\sigma)}$$

$$+ \left| \sum_{n=1}^{N-1} b_n Z_n(\omega) e^{-\lambda_n(\sigma + i t_o)} \right|$$

$$< B\, e_{j+1}^{(1 - 1/m_o) \log_j U(1/\sigma)}$$

$$(1 + \sum_{n=1}^{N-1} |Z_n(\omega)|) \quad (B = \text{const.} > 0),$$

where $N = N(H')$ is determined according to Lemma 5. Hence by Lemma 5,

$$\sum_{n=N}^{+\infty} |b_n|^2 e^{-2\lambda_n \sigma} \leqslant \frac{1}{e} \int_{H'} \left| \sum_{n=N}^{+\infty} b_n Z_n(\omega) \right.$$

$$\left. e^{-\lambda_n(\sigma + i t_o)} \right|^2 \mathscr{P}(d\omega)$$

$$\leqslant \frac{1}{e} \int_{H'} B^2 (e_{j+1}^{(1 - \varepsilon) \log_j U(1/\sigma)})^2$$

$$(1 + \sum_{n=1}^{N-1} |Z_n(\omega)|)^2 \mathscr{P}(d\omega)$$

$$\leqslant K\, (e_{j+1}^{(1 - 1/n_o) \log_j U(1/\sigma)})^2,$$

where K is a positive constant.

Hence, for $n \geqslant N$ and $\sigma \in (0, 1/m_o)$,

$$|b_n| e^{-\lambda_n \sigma} \leqslant \sqrt{K}\, e_{j+1}^{(1 - 1/n_0) \log_j U(1/\sigma)}$$

and consequently for $\sigma > 0$ sufficiently small,

$$m_g(\sigma) = \max_{n \geqslant 1} \{|b_n| e^{-\lambda_n \sigma}\}$$

$$\leqslant \sqrt{K}\, e_{j+1}^{(1 - 1/n_0) \log_j U(1/\sigma)}.$$

Therefore, we have

$$\overline{\lim_{\sigma \to +0}} \frac{\log_{j+1} m_g(\sigma)}{\log_j U(1/\sigma)} \leqslant 1 - 1/n_0$$

and by Lemma 5 in [6]

$$\overline{\lim_{\sigma \to +0}} \frac{\log_{j+1} M_g(\sigma)}{\log_j U(1/\sigma)} \leqslant 1 - 1/n_0 ,$$

where $M_g(\sigma)$ is defined in Theorem A. We thus obtain a contradiction to this theorem. By Theorem 1 our proof is finished.

Range all the rational numbers in a sequence $\{t_n\}$ $(n \in \mathbb{N})$. Since a countable number of almost sure events occur almost surely at the same time, we deduce from Theorem 2 the following corollaries.

Corollary 1. Under the hypotheses of Theorem 2, for $f(s; \omega)$ represented by (1), we have a.s.: $\forall t_n \in \{t_n\}$,

$$\overline{\lim_{\sigma \to +0}} \frac{\log_{j+1} |f(\sigma + i t_n ; \omega)|}{\log_j U(1/\sigma)} = 1 \quad (1 \leqslant j \leqslant p).$$

Corollary 2. Under the hypotheses of Theorem 2, for $f(s;\omega)$ represented by (1), we have a.s.: $\forall\alpha,\ \beta\in\mathbb{R}$ and $\alpha<\beta$,

$$\overline{\lim_{\sigma\to+0}}\ \frac{\log_{j+1} M(\sigma,\ \alpha,\ \beta;\ \omega)}{\log_j\ U(1/\sigma)} = 1 \quad (1\leqslant j\leqslant p)\ .$$

Theorem 3. If for the random series (1), $\{Z_n(\omega)\}$ is an N-sequence and (2), (10), (11) and (13) are verified, for $f(s:\omega)$ represented by (1), we have a.s.: $\forall\eta\in(0,\ \pi/2)$, $\forall t_o\in\mathbb{R}$,

$$\overline{\lim_{\sigma\to+0}}\ \frac{\log_j\ n(\sigma,\ \alpha,\ t_o,\ \eta;\ \omega)}{\log_j\ U(1/\sigma)} = 1 \quad \text{a.s.},$$

where α is any finite complex number, with one possible exception, $U(1/\sigma)$ is defined in Theorem A, $n(\sigma,\ \alpha,\ t_o,\ \eta;\ \omega)$ is the number of zeros of $f(s;\ \omega)-\alpha$ in the domain $\{s|R_e\ s>\sigma\}\cap\{s|\ |I_m(s-it_o)|<\eta\}$, $f(s;\ \omega)$ is represented by (1) and $1\leqslant j\leqslant p$.

Proof. By Corollary 2, $\forall j\in\mathbb{N}$ and verifying $1\leqslant j\leqslant p$ and $\forall m$,

$$\overline{\lim_{\sigma\to+0}}\ \frac{\log_{j+1} M(\sigma,\ t_o-\pi/4m,\ t_o+\pi/4m;\ \omega)}{\log_j\ U(1/\sigma)} = 1 \text{ a.s.}$$

Consequently, $\forall m$ and $\forall\delta>0$, a sequence of positive numbers $\{\sigma_n\}$ tending to zero such that for n sufficiently large,

$$\log_{j+1} M(\sigma_n,\ t_o-\pi/4m,\ t_o+\pi/4m;\ \omega)$$
$$>(1-\delta)\log_j U(1/\sigma_n) \text{ a.s.} \tag{21}$$

Suppose that $f'(\sigma_o + it_o) \neq 0$, $f(\sigma_o + it_o) \neq 0$ $(0 < \sigma_o < 1)$.

Set

$$\overline{\sigma_n} = \frac{\sigma_n \log^2 U(1/\sigma_n)}{1 + \log^2 U(1/\sigma_n)}. \tag{22}$$

It is easy to verify that

$$z = Z_n(s) = \frac{e^{-(\sigma_o + 2\overline{\sigma_n})m} - e^{-(s - it_o + 2\overline{\sigma_n})m}}{e^{-(\sigma_o + 2\overline{\sigma_n})m} + e^{-(s - it_o + 2\overline{\sigma_n})m}}$$

$$\frac{+ e^{-(2\sigma_o + s - it_o)m} - e^{-(\sigma_o + 2s - 2it_o)m}}{- e^{-(2\sigma_o + s - it_o)m} - e^{-(\sigma_o + 2s - 2it_o)m}}$$

is a mapping of the domain $\{s | \mathrm{Re}\, s > \overline{\sigma_n}\} \cap \{s \mid |I_m(s - it_o)| < \pi/2m\}$ into $|z| < 1$ and $z_n(\sigma_o + it_o) = 0$. Denote the inverse mapping by $s = s_n(z)$. Since

$$1 - |z_n(s)| = 1 - \sqrt{(M^2 + N^2)/(P^2 + Q^2)},$$

where

$$M = e^{-(\sigma_o + 2\overline{\sigma}_n)m} - e^{-(\sigma + 2\overline{\sigma}_n)m} \cos m(t - t_o)$$

$$+ e^{-(2\sigma_o + \sigma)m} \cos m(t - t_o) - e^{-(\sigma_o + 2\sigma)m}$$

$$\cos 2m(t - t_o),$$

$$N = -e^{-(\sigma + 2\overline{\sigma}_n)m} \sin m(t - t_o) + e^{-(2\sigma_o + \sigma)m}$$

$$\sin m(t - t_o) - e^{-(\sigma_o + 2\sigma)m} \sin 2\, m(t - t_o),$$

$$P = -e^{-(2\sigma_o + \sigma)m} \cos m(t - t_o) - e^{-(\sigma_o + 2\sigma)m}$$

$$\cos 2\, m(t - t_o) + e^{-(\sigma_o + 2\overline{\sigma}_n)m} + e^{-(\sigma + 2\overline{\sigma}_n)m}$$

$$\cos m(t - t_o)$$

$$Q = -e^{-(2\sigma_o + \sigma)m} \sin m(t - t_o) - e^{-(\sigma_o + 2\sigma)m}$$

$$\sin 2\, m(t - t_o) + e^{-(\sigma + 2\overline{\sigma}_n)m} \sin m(t - t_o),$$

we have

$$1 - |z_n(s)| = \frac{1 - (M^2 + N^2)/(P^2 + Q^2)}{1 + \sqrt{(M^2 + N^2)/(P^2 + Q^2)}}$$

$$\geqslant \frac{P^2 + Q^2 + M^2 + N^2}{2(P^2 + Q^2)} .$$

By simplification, we obtain

$$1 - |z_n(s)| \geqslant [(e^{-m\overline{\sigma}_n} - e^{-m\sigma_o})(e^{-2m\overline{\sigma}_n} - e^{-2m\sigma})$$

$$e^{-m\sigma_o - \sigma m} \cos m(t - t_o)]/(P^2 + Q^2)$$

Let

$$K_n = \max \{ |z_n(\sigma + i t)| \, | \sigma_n \leqslant \sigma \leqslant 1, \; |t - t_o| \leqslant \pi/4m \}$$

and suppose that at $\sigma_n' + i t_n'$ $(\sigma_n \leqslant \sigma_n' \leqslant 1, |t - t_o| \leqslant \pi/4m)$, $|z_n(\sigma_n' + i t_n')| = K_n$.

Hence

$$1 - K_n \geqslant c(e^{-2m\overline{\sigma_n}} - e^{-2m\sigma_n'}) \geqslant c(e^{-2m\overline{\sigma_n}} - e^{-2m\sigma_n})$$
$$\geqslant 2mc(\sigma_n - \overline{\sigma_n})\, e^{-2m\sigma_n},$$

where c is a positive constant. By (22),

$$1 - K_n \geqslant c\sigma_n / \log^2 U(1/\sigma_n).$$

Put $H_n = K_n + (1 - K_n)/2$. Then

$$1 - H_n = H_n - K_n = (1 - K_n)/2$$
$$> c\,\sigma_n / \log^2 U(1/\sigma_n). \tag{23}$$

From

$$T(H_n, f(s_n(z); \omega)) \geqslant \frac{H_n - K_n}{H_n + K_n} \log M(K_n, f(s_n(z); \omega)) \geqslant \frac{c\,\sigma_n}{\log^2 U(1/\sigma_n)} \log M(\sigma_n, t_o - \frac{\pi}{4m}, t_o + \frac{\pi}{4m}; \omega)$$

and (21) we deduce

$$\log_j T(H_n, f(s_n(z); \omega)) \geq \log_{j+1} M(\sigma_n, t_o - \pi/4m,$$

$$t_o + \pi/4m; \omega) + \log_j \frac{\sigma_n}{\log^2 U(1/\sigma_n)} - c$$

$$\geq (1 - 2\sigma) \log_j U(1/\sigma_n), \tag{24}$$

where $T(r, \phi)$ is Nevanlinna characteristic function and

$$M(r, f(s_n(z); \omega)) = \max_{|z|=r} \{|f(s_n(z); \omega)|\}.$$

Let $T_0(r, \phi)$ represent Ahlfors-Shimizu characteristic function of a meromarphic function $\phi(z)$ [8]:

$$T_o(r, \phi) = \int_o^r \frac{S(t, \phi)}{t} dt,$$

where

$$S(t, \phi) = \frac{1}{\pi} \iint_{|z| \leq t} \left(\frac{|\phi'(z)|}{1 + |\phi(z)|^2}\right)^2 \rho \, d\rho \, d\theta$$

$$(z = \rho e^{i\theta}).$$

The difference of $T(r, \phi)$ and $T_0(r, \phi)$ is a constant depending on $\phi(0)$.

$$\forall s \in \{s | \text{Re } s > \sigma_n\} \cap \{s || I_m(s - i t_o) | \leq \pi/4m\},$$

since

$$\text{Re } e^{-(\sigma_o + s - i t_o)m} > 0, \quad 0 < \overline{\sigma_n} \to 0 \quad \text{and}$$

$\operatorname{Re} e^{-(s - \sigma_0 - i t_0)n} > 0, \quad \forall n \in \mathbb{N},$ we see that for n sufficiently large,

$$|z_n(s)| = \left| \frac{(e^{-\sigma_0 m} - e^{-(s - i t_0)m})}{(e^{-\sigma_0 m} + e^{-(s - i t_0)m})} \right.$$

$$\left. \frac{(e^{-(\sigma_0 + s - i t_0)m} + e^{-2\overline{\sigma}_n m})}{(e^{-2\overline{\sigma}_n m} - e^{-(\sigma_0 + s - i t_0)m})} \right|$$

$$\geqslant |[e^{-\sigma_0 m} (1 - e^{-(s - \sigma_0 - i t_0)})$$

$$(1 + e^{-(s - \sigma_0 - i t_0)} + \ldots$$

$$+ e^{-(s - \sigma_0 - i t_0)(m - 1)}) e^{-2\overline{\sigma}_n n}]/4|$$

$$\geqslant \left| \frac{1 - e^{-(s - \sigma_0 - i t_0)}}{8 e^{\sigma_0 m}} \right|.$$

$\forall x \in (0, 1)$, we have

$$\frac{x \pi}{64m e^{\sigma_0 m}} \geqslant |z_n(s)|$$

and consequently

$$\frac{1}{2} > \frac{x\pi}{8m} \geqslant |1 - e^{-(s - \sigma_0 - i t_0)}|$$

and

$$\frac{x\pi}{4m} \geqslant |s - \sigma_o - i\, t_o|.$$

Therefore, by Theorem VI. 21 in [8],

$$T(H_n, f(s_n(z); \omega)) - c$$

$$\leqslant \int_{\frac{\pi}{64m\, e^{\sigma_o m}}}^{H_n} \frac{S(u, f(s_n(z); \omega))}{u}\, du$$

$$+ \int_0^{\frac{\pi}{64m\, e^{\sigma_o m}}} \frac{S(u, f(s_n(z); \omega))}{u}\, du$$

$$\leqslant S(H_n, f(s_n(z)); \omega)) \log \frac{64m\, e^{\sigma_o m}}{\pi}$$

$$+ \int_0^{\frac{\pi}{4m}} \frac{S(u, f(s_n(z); \omega))}{u}\, du$$

$$\leqslant S(H_n, f(s_n(z); \omega))\, A + B$$

$$\leqslant A \sum_{\ell=1}^{2} n\,(1, f(s_n(z); \omega) = a_\ell) + \frac{c_1}{1 - H_n}$$

$$\leqslant A \sum_{\ell=1}^{2} n\,(\overline{\sigma_n}, a_\ell, t_o, \pi/2m; \omega) + c_1/(1 - H_n),$$

where A, B, c and c_1 are positive constants, a_1 and a_2 are two different complex numbers and $n(r, \phi)$ is an usual function in the theory of meromorphic functions. Combining the above inequality with (23), (24) and (12), we have

$$\overline{\lim_{\sigma\to+0}} \frac{\log_j \sum_{\ell=1}^{2} n(\sigma, a_\ell, t_o, \pi/2m; \omega)}{\log_j U(1/\sigma)} \geqslant 1.$$

It is easy to prove that the above inequality must be an equality and the theorem follows.

As we deduce Corollary 1 from Theorem 2, under the hypotheses of Theorem 3, we arrange all the rational numbers in a sequence $\{t_n\}$ $(n \in \mathbb{N})$ and we have a.s.: $\forall t_n \in \{t_n\}$, $\forall \eta \in (0, \pi/2)$, $\forall \alpha \in \mathbb{C}$, with one possible exception,

$$\overline{\lim_{\sigma\to+0}} \frac{\log_j n(\sigma, \alpha, t_n, \eta; \omega)}{\log_j U(1/\sigma)} = 1.$$

Consequently, $\forall t \in \mathbb{R}$ the above equality still holds if we replace t_n by t. Hence we have

Corollary 3. Under the hypotheses of Theorem 3 we have a.s.: $\forall t_o \in \mathbb{R}$, $\forall \eta \in (0, \pi/2)$, $\forall \alpha \in \mathbb{C}$, with one possible exception,

$$\overline{\lim_{\sigma\to+0}} \frac{\log_j n(\sigma, \alpha, t_o, \eta, \omega)}{\log_j U(1/\sigma)} = 1$$

III. RANDOM ENTIRE FUNCTION OF INFINITE ORDER (R)

Consider the random Dirichlet series (1), $\{Z_n(\omega)\}$ being an N-sequence of random variables defined on $(\Omega, \mathscr{A}, \mathscr{P})$. Suppose that the conditions (10) and (11) are replaced by

$$\lim_{n\to+\infty} \frac{\log|b_n|}{\lambda_n} = -\infty \qquad (25)$$

and

$$\overline{\lim_{n\to+\infty}} \frac{\log n}{\lambda_n} < +\infty \qquad (26)$$

Then in virtue of (5) the series (1) represents a.s. an entire function $f(s;\omega)$ [13] [10].

For an entire function we define $\rho(-\sigma)$ and the type-function $U(-\sigma)$ as in II ($\sigma < 0$). Applying Theorem 4 in [6] we obtain the following results analogous to Theorem 1 and Corollary 2.

Theorem 4. If for the random series (1), $\{Z_n(\omega)\}$ is an N-sequence and (2), (25) and (26) are verified, if $U(-\sigma) = e_p^{\rho(-\sigma)\log_q(-\sigma)}$ is a type-function ($p > q+1$ or $p = q+1$ and $\rho > 1$) and if

$$\underline{\lim_{n\to+\infty}} \{\log_q \frac{\log|b_n|}{-\lambda_n} - \log_q W(\log \lambda_n)\}$$

$$\times \prod_{c=j+1}^{p-1} \log_c \lambda_n = 0, \qquad (27)$$

than for the random entire function represented by (1),

$$\overline{\lim_{\sigma\to-\infty}} \frac{\log_{j+1} M(\sigma;\omega)}{\log_j U(-\sigma)} = 1 \text{ a.s.},$$

where $1 \leqslant j \leqslant p$, $r = W(u)$ is the inverse function of $u = e_{p-j}^{\rho(r)\log_q r}$ and $M(\sigma;\omega)$ is defined as in II.

Theorem 5. If for the random series (1), $\{Z_n(\omega)\}$ is an N-sequence and if (2), (25), (26) and (27) are verified, then for $f(s; \omega)$ represented by (1), we have a.s.: $\forall\alpha, \beta \in \mathbb{R}$ and $\alpha < \beta$,

$$\overline{\lim_{\sigma\to-\infty}} \frac{\log_{j+1} M(\sigma, \alpha, \beta; \omega)}{\log_j U(-\sigma)} = 1 \quad \text{a.s}$$

where $1 \leq j \leq p$.

The proofs of the two above theorems are similar to those of Theorems 1 and 2 and Corollaries 1 and 2. We need only replace (13), $U(1/\sigma)$ and $\sigma \to +0$ by (27), $U(-\sigma)$ and $\sigma \to -\infty$.

Like Theorem 4.5 in [10] we have

Theorem 6. Under the hypotheses of Theorem 5, for $f(s; \omega)$ represented by (1), we have a.s.: $\forall\eta\in(0, \pi/2)$, $\forall t_o \in \mathbb{R}$,

$$\overline{\lim_{\sigma\to-\infty}} \frac{\log_j n(\sigma, \alpha, t_o, \eta; \omega)}{\log_j U(-\sigma)} = 1,$$

where α is any finite complex number, with one possible exception, $n(\sigma, \alpha, t_o, \eta: \omega)$ and $U(-\sigma)$ are defined as in Theorem 3 and $1 \leq j \leq p$.

In the proof of Theorem 3 we replace $\sigma_n \to +0$, $U(1/\sigma)$, (22) and (23) by $\sigma_n \to -\infty$, $U(-\sigma)$,

$$\overline{\sigma_n} = \sigma_n + \sigma_n/\log^2 U(-\sigma_n)$$

and

$$1 - H_n = H_n - K_n = (1 - K_n)/2$$
$$> c/\sigma_n^{3m} \log^2 U(-\sigma_n)$$

respectively and then we obtain the proof of Theorem 6.

BIBLIOGRAPHY

(1) J. P. Kahane, Some random series of functions, 2nd Ed., Cambridge, Cambridge University Press, 1985.

(2) O. P. Juneja and others, J. reine angew. Math., 282 (1976), 53-67.

(3) S. Mandelbrojt, Selecta, Paris, Gautheir-Villers, 1981.

(4) R. E. A. C. Paley and A. Zygmund, Proc. Camb. Phil. Soc., 28 (1932), 190-205.

(5) Sun Dao-chun, J. Math. Res. and Exp., 4 (1984), 35-40.

(6) __________, Hunan Ann. of Math., 5 (1985), 71-80.

(7) __________, Acta Math. Scientia, 2 (1986), 231-239.

(8) M. Tsuji, Potential theory in modern function theory, Tokyo, Maruzen, 1959.

(9) Yu Jeou-man, J. Math. Res. and Exp., 4(1984), 22.

(10) Yu Jia-rong (Yu Chia-yung), Ann. Ec. Norm. Sup., (3) 68 (1951), 65-104.

(11) __________, Acta Math. Sinica, 21 (1978), 98-118.

(12) __________, C. R. Aoad. Sc. Paris, Ser. I, 296 (1983), 187-190.

(13) __________, __________, 300 (1985), 521-522.

AN EXTREMAL PROBLEM FOR HARMONIC FUNCTIONS

Carl H. FitzGerald
Department of Mathematics
University of California, San Diego
La Jolla, CA 92093
U.S.A.

ABSTRACT

A modified form of the Schwarz reflection principle for harmonic functions is developed. In the new form, the condition that the harmonic function be zero on an interval is replaced by the condition that it has small magnitude there. The extremal problem that arises is solved in the sense that the extremal harmonic function is characterized, specifically, its boundary values are given. This characterization is used to describe explicitly the asymptotic behavior as the bound of the magnitude of u on the interval tends to zero.

1. INTRODUCTION

We intend to modify the Schwarz reflection principle by changing the condition that u equals to zero. Recall the standard form of the reflection principle:

> Let Ω be a domain in the upper half plane with an interval (a,b) on the boundary of Ω. If $u(x,y)$ is harmonic on Ω and continuous on $\Omega \cup (a,b)$ and if $u(x,o) = 0$ on (a,b), then u can be extended to be harmonic on $\Omega \cup (a,b) \cup \bar{\Omega}$ where $\bar{\Omega} = \{(x,y):(x,-y) \in \Omega\}$. Furthermore $u(x,-y) = -u(x,y)$ for $(x,y) = \Omega$, that is, u is odd in

the y variable.

If the condition "$u(x,0) = 0$" is replaced with "$|u(x,o)| \leq \varepsilon$," then the conclusion fails in two major ways. First, there may be no harmonic extension of u. For example, consider a boundary value problem on the upper half plane where the boundary values are uniformly bounded along the x-axis by a positive number ε, are continuous, but are nowhere differentiable.

Secondly, even when there is an extension, the function could be arbitrarily far from being odd in y. For example,

$$u(x,y) = \varepsilon/2[e^{-\lambda y} \cos \lambda x + e^{\lambda y} \cos \lambda x]$$

is even in y and is unbounded at any fixed point off the x-axis as $\lambda \to +\infty$.

We can overcome these difficulties by (1) assuming that $u(x,y)$ has an extension to the symmetric domain and (2) assuming $|u(x,y)|$ satisfies a uniform bound on the symmetric domain. Finally, we can focus on the even part of u since

$$u(x,y) = \frac{u(x,y) + u(x,-y)}{2} + \frac{u(x,y) - u(x,-y)}{2}.$$

<u>Extremal Problem:</u> Let $D = \{(x,y):|y| < 1\}$ and $M > \varepsilon > 0$. Let $\mathcal{U} = \{u(x,y):u$ is harmonic on D, satisfies $|u(x,y)| \leq M$ on D, satisfies $|u(x,0)| \leq \varepsilon$ on the x-axis, and is even in y and even in $x\}$. For fixed $0 < h < 1$, find

$$\underset{u \in \mathcal{U}}{\text{maximum}}\ u(o,h)$$

It is exactly this problem that this paper solves.

Throughout the paper the notation is fixed and the values of M, and h are fixed. The value of ε will be fixed until the last part of section 6 of the paper.

The condition that functions in $\mathcal{U}$ be even in x is a convenience for the proof, but will be shown later to be inessential.

Note that we are justified in writing "maximum" rather than "supremum" because $\mathcal{U}$ is a compact family. Further, we note that finding an extremal will solve our problem of modifying the Schwarz reflection principle since an extremal function will give an upper bound on how far from odd the harmonic function can be.

In the section that follows, we define a function which will be shown to be an extremal function. The verification that it is extremal is carried out in several sections by examining a modified problem and by a rather long proof. Finally we return to the original problem and study the behavior of our extremal function as ε tends to zero. We obtain explicit estimates that show how the modified form of the Schwarz reflection principle tends to the standard form as ε tends to zero.

2. PUTATIVE SOLUTION

We now construct a function which will be shown to be an extremal function. The construction shows that the extremal value can be expressed in terms of certain conformal maps, specifically, in terms of an elliptic function.

Consider the unit disk with two distinguished arcs, one above the other, and each of length $\pi\varepsilon/M$ (see Fig. 1). The bounded harmonic function, with boundary values M on the distinguished arcs and 0 on the rest of the boundary of the disk, has value $+\varepsilon$ at the center of the disk.

The Schwarz-Christoffel formula maps the unit disk onto a rectangle in such a way that the ends of distinguished arcs go to the corners of the rectangle. By appropriate choices of the constants, the rectangle can be assumed to have two horizontal sides corresponding to the distinguished arcs, the center of the rectangle at the origin, and vertical sides of length two units. Let f(w) be the conformal mapping as shown in the figure.

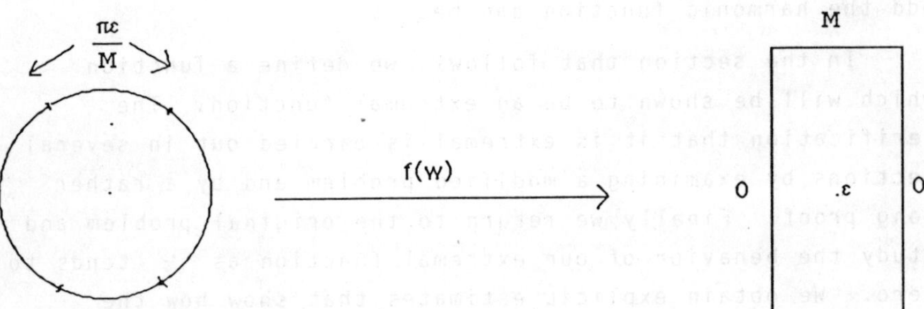

Figure 1

Let u be the bounded harmonic function in the rectangle of Fig. 1 that has boundary values M on the horizontal sides of the boundary and zero on the vertical sides. By the construction of the rectangle, the value of u at the origin is ε. By repeated Schwarz reflection across the vertical sides, we obtain a bounded harmonic function $\underline{u}$ on the strip D mentioned in the previous section. The boundary values of this harmonic function are alternately +M and -M. The intervals of constant boundary values are all the same length. The function is even across the real axis; the boundary values along the lower boundary line are the same as the values at the point directly above it (see Fig. 2). This harmonic function $\underline{u}(x,y)$ is the putative extremal.

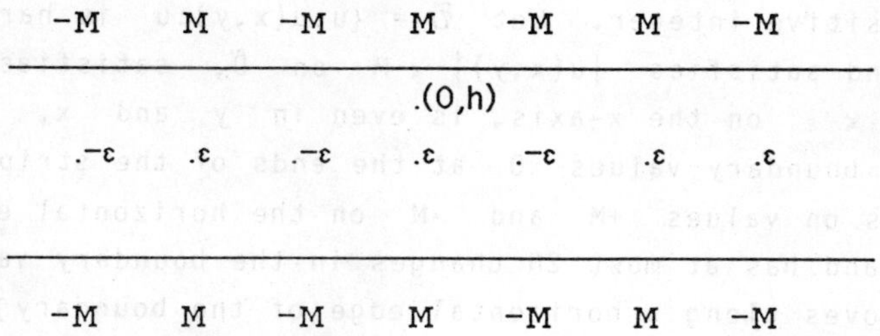

−M M −M M −M M −M

.(0,h)

.−ε .ε .−ε .ε .−ε .ε .ε

−M M −M M −M M −M

Figure 2

The value of $\underline{u}(0,h)$ follows from evaluation of the Poisson integral formula at $f^{-1}(0,h)$ with boundary values +M on the distinguished arcs and zero on the rest of the circle. Of course, f^{-1} is an elliptic function; hence the problem is of some difficulty computationally. But it is numerically solvable. We will later show that, for the asymptotic problem as $\varepsilon \to 0^+$, explicit results can be obtained and represented in elementary form.

Finally, notice the interesting fact that the extremal harmonic function does not depend on the location (0,h) at which the maximization takes place. Of course, the value of the maximum does depend on (0,h), but the extremal function depends only on M and ε.

3. MODIFIED PROBLEM

The function $\underline{u}$ defined in the previous section suggests a modified extremal problem which involves only finitely many parameters. This problem has an obvious candidate for its solution. And once this finite dimensional problem is solved, we will have solved the extremal problem of section 1; specifically, we will have shown the function $\underline{u}(x,y)$ of section 2 is extremal.

We consider a long horizontal strip of finite length: $\tilde{D} = \{(x,y): -L < x < L$ and $-1 < y < 1\}$. Let N be a large positive integer. Let $\tilde{\mathcal{U}} = \{u: u(x,y): u$ is harmonic on $\tilde{D}$ and satisfies $|u(x,y)| \leq M$ on $\tilde{D}$, satisfies $|u(x,0)| \leq \varepsilon$ on the x-axis, is even in y and x, and takes on boundary values 0 at the ends of the strip $\tilde{D}$, and takes on values $+M$ and $-M$ on the horizontal edges of $\tilde{D}$, and has at most $2N$ changes in the boundary values as one moves along a horizontal edge of the boundary$\}$.

We consider the extremal problem: $\underset{u \in \tilde{\mathcal{U}}}{\text{maximum}}\ u(o,h)$.

As before, the compactness of the family is clear so that there is an extremal function. Furthermore it is clear that the maximum over $\mathcal{U}$ will be approached if first L is chosen large and then N is chosen sufficiently large in terms of L.

In order to have an obvious candidate for the extremal, we make one final alteration to the problem. The value of L is chosen so that $2L$ is an odd multiple of the horizontal width of the rectangle in Fig. 1. With this restriction, the natural candidate for the extremal function over $\tilde{\mathcal{U}}$ is exactly the function $\underline{u}$ defined in section 2 restricted to $\tilde{D}$. The boundary values of candidate for the extremal is shown in Fig. 3.

The verification that we have the extremal function occupies the next four sections. Having solved this modified problem, we will have solved the original maximization problem.

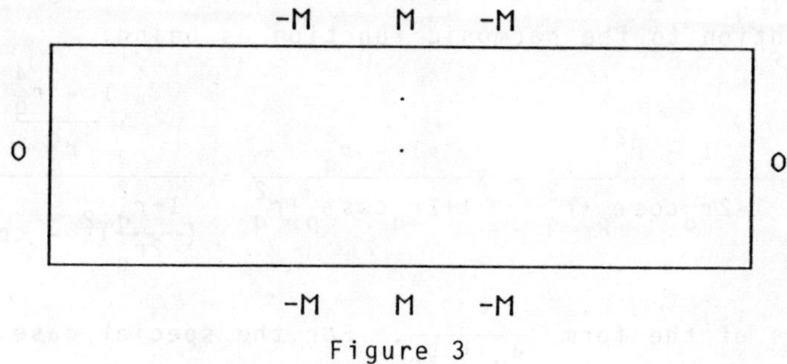

Figure 3

4. LEMMA

To study the effect of moving the location of the changes of boundary values, we introduce the following lemma. For convenience the problem has been pulled back to the unit disk.

<u>Lemma</u>: Consider K weighted delta functions on the first quarter of the unit circle. Let other weighted delta functions be placed in the other quarters of the circle so that the configuration is symmetrical both with respect to the x-axis and y-axis. Let points $P_1, \ldots, P_K$ be fixed inside the unit disk as follows: $P_1 = (0,H)$ for some $0 < H < 1$, $P_2 = (0,0)$, and $P_3, \ldots, P_K$ are distinct points along the open interval $(0,1)$ of the x-axis. Consider the harmonic function determined by the Poisson integral formula with boundary values being the weighted delta functions about the circle. This harmonic function is zero at $P_1, P_2, \ldots, P_K$ if and only if each of the weights on the delta functions is zero.

<u>Proof</u>: Consider the $p^{\underline{th}}$ delta function at $\exp i\theta_p$ for some $0 < \theta_p < \pi/2$. There are three other delta functions symmetrically about the circle (one of the θ_p can equal $\pi/2$ if the weight is counted half). Consider the effect of these delta functions on a point

$P_q = (0, r_q)$ for $2 < q \leqslant K$. The Poisson kernel gives the contribution to the harmonic function as being:

$$2\frac{1 - r_q^2}{1-2r_q\cos\theta_p + r_q^2} + 2\frac{1 - r_q^2}{1+2r_q\cos\theta_p + r_q^2} = \frac{\dfrac{1 - r_q^4}{r_q^2}}{\left(\dfrac{1+r_q^2}{2r_q}\right)^2 - \cos^2\theta_p},$$

which is of the form $\dfrac{c_q}{a_q + b_p}$. For the special case $q = 2$, the value of r_q is zero and the contribution is four regardless of the value of p. This value of the contribution is obtained as a limit as r_2 tends to zero: The values of a_2 and c_2 tend to plus infinity and $\dfrac{c_2}{a_2 + b_2}$ converges to four. In case $q = 1$, the calculation must be redone because the value of the angles is different. The contribution when $q = 1$ is

$$\frac{\dfrac{1 - H^4}{H^2}}{\left[-\left(\dfrac{1 + H^2}{2H}\right)^2 + 1\right] - \cos^2\theta_p}$$

which is in the form $\dfrac{c_1}{a_1 + b_p}$, that is, of the same form as before.

The Cauchy matrix $\left(\left(\dfrac{1}{a_q + b_p}\right)\right)$ has determinant [3p.54, 4p.98]:

$$\frac{\prod_{m>n} (a_m - a_n)(b_m - b_n)}{\prod_{p,q} (a_q + b_p)}.$$

The determinant is nonzero provided the values of a_q and b_p are distinct. It is finite provided $a_q + b_p$ is nonzero. These conditions are easily checked in this situation.

The determinant in our case is just the determinant of the indicated Cauchy matrix times $c_1 c_2 \ldots c_K$ considered in the limit as r_2 tends to zero and a_2 and c_2 tend to plus infinity. The determinant is nonzero. Thus the transformation from the weights on the delta functions on the first quarter of the circle to the values of the harmonic function at the specified points can be inverted and the lemma is proved.

5. LOCAL EXTREMA ON THE X-AXIS

Let $u(x,y)$ denote an extremal function for the modified maximization problem. We will examine the number of local extrema of $u(x,0)$ in the open interval $(-L,L)$ in comparison to the number of intervals with boundary values $+M$ or $-M$.

Suppose $u(x_o,0)$ is a local maximum of $u(x,0)$. For a small positive number η, the set $\{(x,y):u(x,y) > u(x_o,0) - \eta\}$ is open and contains $(x_o,0)$. Let S denote the connected component of the set which contains $(x_o,0)$. Then the maximum principle implies S has boundary points on the boundary of $\tilde{D}$. Since the function $u(x,y)$ is even in y, the set S is symmetrical about the x-axis.

For a local minimum $u(x_1,0)$, a set S is defined in a similar fashion except that the inequality is changed to $u(x,y) < u(x_1,0) + \eta$. Again the set is symmetrical about the x-axis.

For a local maximum point, the boundary values of u on $\tilde{D}$ which are assumed on the boundary of S must be either $+M$ or zero. Consider the $+M$ possibility first. If part of an interval of $+M$ values is on the boundary of S, then all of that interval must be on the boundary of S since u is close to $+M$ near any interior point of the interval.

Now the zero boundary values are considered. If u is positive at the local maximum, a small choice of η will eliminate the zero boundary values. If u is zero or negative at the local maximum point, there is an adjacent local minimum closer to the vertical interval on which the zero boundary values occur. Clearly u is negative at this local minimum; for a small choice of η, the corresponding set S' has $-M$ for its boundary values on the boundary of $\tilde{D}$, not zero. The set S' is symmetrical about the x-axis and separates S from the vertical line of zero boundary values. Hence, by picking η small, it can be assumed that no interior maximum has boundary values of S on the vertical lines of the boundary of $\tilde{D}$. Similarly no interior minimum has boundary values of the corresponding set S that are zero.

Finally we examine whether two local maxima points $(x_0,0)$ and $(x_2,0)$ of $u(x,0)$ can have a common interval of $+M$ boundary values on the boundary of their corresponding sets S. Note that there must be a local minimum point $(x_1,0)$ between the local maxima. Restrict η to be small enough that the corresponding set S for $(x_1,0)$ does not overlap those for $(x_o,0)$ and $(x_2,0)$. Then, if the sets S for $(x_o,0)$ and $(x_2,0)$ overlapped, their union would surround the set S for $(x_1,0)$ implying that that set had no points on the boundary of $\tilde{D}$. This contradiction shows each local maximum has (at least) one associated interval of $+M$ boundary values. Similarly, each local minimum has at least one associated interval of $-M$ boundary values as indicated in Fig. 4.

We conclude that the number of local extremal of $u(x,0)$ for $0 < x < L$ is less than or equal the number of intervals of $+M$ or $-M$ boundary values of u along the top edge of $\tilde{D}$ to the right of the y-axis. In particular, since the number of changes of sign of the boundary values has been required to be less than 2N along the upper edge, the number of local extrema of $u(x,0)$ is finite.

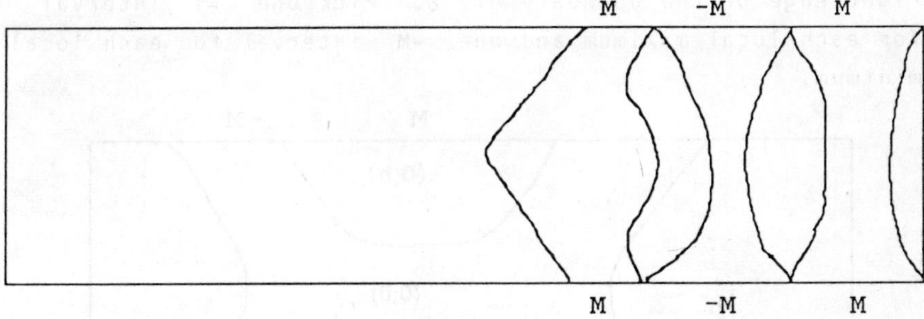

Figure 4

6. VARIATION OF THE EXTREMAL

We now deduce various properties of the extremal function $u(x,y)$ for the modified problem by variation of the intervals of $+M$ or $-M$ boundary values. The variations are specified by changes on the upper, right edge of the boundary of $\tilde{D}$; but it is to be understood that similar changes are made on the left half and on the lower edge in order to maintain the symmetries about the x and y axes. Also care must be used to maintain the constraint that $|u(x,0)| \leqslant \varepsilon$. We examine various cases. The first one is clearly not possible, but a careful proof is given so that the method of argument is clear for later cases.

Case I. Suppose that $(0,0)$ is a local minimum point of $u(x,0)$. Then the component of $\{(x,y): u(x,y) < u(0,0) + \eta\}$ that contains $(0,0)$ for some small $\eta > 0$ would appear as in Figure 5, except that the $+M$ interval might be of length zero, that is, not exist.

Let $P_2 = (0,0)$, and let $P_3, P_4, \ldots, P_K$ denote the local extreme points of $u(x,0)$ for $0 < x < L$. For $k > 2$, there corresponds to each P_k at least one new interval of $+M$ or $-M$ boundary values along the top right edge of the boundary of $\tilde{D}$. Pick one $+M$ interval for each local maximum and one $-M$ interval for each local minimum.

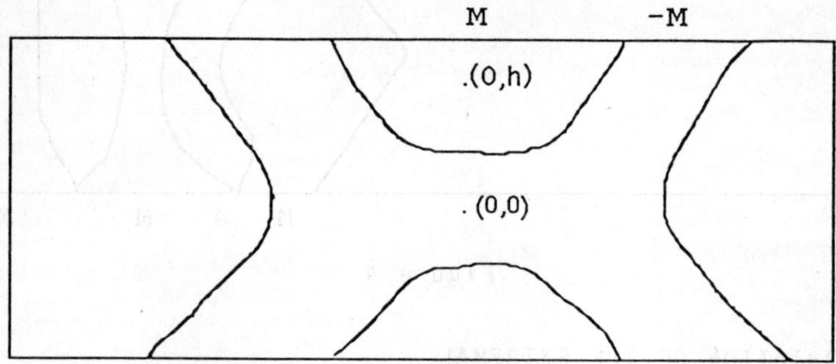

Figure 5

For $k = 2, 3, \ldots, K$, consider the rate of change of $u(P_k)$ as a function of the rates of change of the location of the left end points of the chosen intervals of boundary values. No other points of change of boundary values along the top right edge are allowed to move. We desire that the local minima are increased and the local maxima are decreased. The determinant involved is that of a submatrix of the one considered in the lemma. Clearly the determinant is nonzero. The only remaining question is whether the motions required can actually be carried out, specifically whether the motion of the left end of the first $-M$ interval is to the right (otherwise, if the $+M$ interval in Figure 5 has zero length, the motion would not be possible).

Let U be the harmonic function which is the rate of

change of u as the required changes in the boundary values take place. The function U has boundary values zero except for delta functions at the end points that move. Note the sign of the values $U(P_k)$: positive at local minimum points of $u(x,0)$ and negative at local maximum points of $u(x,0)$.

As before, sets S can be defined. By similar arguments, each set S from a positive value of $U(P_k)$ leads to a positive weight delta function and each negative value of $U(P_k)$ leads to a negative weight delta functions. For small $\eta > 0$, these sets S do not overlap. The positive value at $U(0,0)$ leads to the left end of the first chosen -M interval as shown in Figure 5. Thus that left endpoint is required to move to the right which is always possible.

Now we consider the harmonic function $\hat{u}(x,y)$ that results from moving the left endpoints as described for a short time. We are concerned whether the condition that $|\hat{u}(x,0)| \leqslant \varepsilon$ is satisfied. The local extreme points of $\hat{u}(x,0)$ are only slightly different from those of u. But U is positive in a fixed neighborhood of each local minimum point of u and negative in a fixed neighborhood of each local maximum point of u. Hence the extrema of maximum magnitude are decreased in magnitude and $|\hat{u}(x,0)| \leqslant \varepsilon$ as desired.

Finally we analyze $U(0,h)$. Let T denote the connected subset of $\{(x,y) : U(x,y) > U(0,0) - \eta > 0\}$ which contains $(0,0)$. Suppose that $U(0,h) \leqslant 0$. Then the point $(0,h)$ would not belong to T, and T would look like the set in Figure 5 separating $(0,h)$ from the delta functions. Consider the component of $\tilde{D}$ minus the closure of T which contains $(0,h)$. On this domain, the function U is harmonic with boundary values zero or positive. Since T contributes a nondegenerate part of the boundary, and U is positive there, the function U satisfies $U(0,h) > 0$. This contradicts the supposition that $U(0,h) \leq 0$. Hence $U(0,h) > 0$.

But this last inequality implies $\hat{u}(0,h) > u(0,h)$ which contradicts the choice of u as an extremal function. Hence $u(0,0)$ is not a local minimum. By the symmetry in x, $u(0,0)$ is a local maximum of $u(x,0)$.

The other cases use similar techniques of argument.

<u>Case II</u>. Suppose that two local extrema of $u(x,0)$ for $0 \leq x < L$ do not attain $u = +\varepsilon$ or $u = -\varepsilon$. The lemma can be applied to show that by moving the location at which the boundary values change the value at $(0,h)$ can be increased while the values are decreased in magnitude at those points $(x_k,0)$ at which $u(x_k,0) = \pm\varepsilon$. The result is a function $\hat{u}(x,y)$ which contradicts the choice of u as an extremal that maximizes $u(0,h)$. Hence there cannot be two such points.

<u>Case III</u>. Suppose that some extreme point $(x',0)$ of $u(x,0)$ for $0 \leq x < L$ is between two other such extreme points and suppose $|u(x',0)| < \varepsilon$.

As in Case I, let U be the rate of change as the boundary values change. Pick U to have large magnitude at $(x',0)$ and to be positive if $u(x',0)$ is a local maximum and negative if $u(x',0)$ is a local minimum. Ignore the adjacent extreme points. At the other local extreme points pick U to be positive if U has a local minimum and negative if u has a local maximum. Pick U to be positive at $(0,h)$. There are enough parameters to make these choices according to the lemma. If U is large enough magnitude at the extreme point $(x',0)$, then at the adjacent extreme points U will have the same sign as indicated in Figure 6.

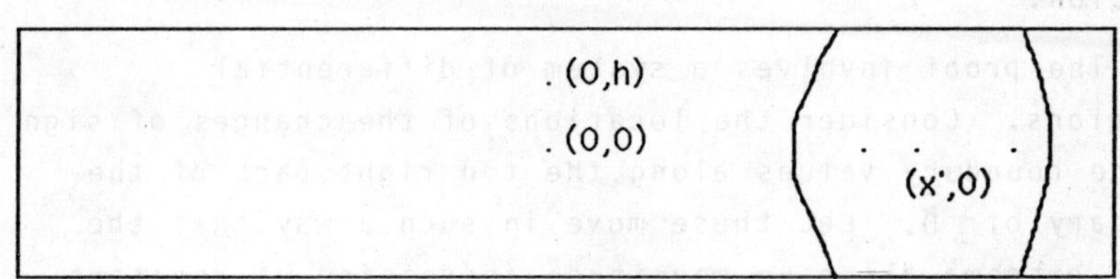

Figure 6

Let $\hat{u}(x,y)$ be the result of changing the boundary values at the required rates for a short time. If the time is short enough, the extremum near $(x',0)$ has not reached magnitude ε. The magnitudes of the other extrema are also less than ε in magnitude and $\hat{u}(0,h) > u(0,h)$ contradicting the choice of u.

Case IV. Suppose $u(x,0)$ is $+\varepsilon$ or $-\varepsilon$ at the extrema of $u(x,0)$ for $0 \leq x < L$ except perhaps for the right most. These values are indicated in Figure 7.

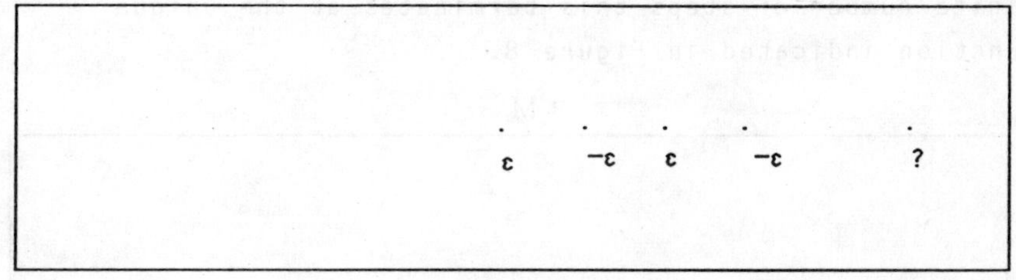

Figure 7

This remaining possibility characterizes a unique function; the function is the one indicated in Figure 3 with all intervals of $+M$ and $-M$ boundary values the same and the extreme points regularly spaced and the extreme values, including the one at the end, all of magnitude ε. In short, this is the conjectured extremal function.

The proof involves a system of differential equations. Consider the locations of the changes of sign of the boundary values along the top right part of the boundary of $\tilde{D}$. Let these move in such a way that the local extrema all have magnitude increasing at constant rate. The local extrema are those on the interval $[0,L)$ of the x-axis, except for the right most one. There are exactly the right number of parameters and the discussion of the determinant in the lemma shows that this construction is possible. We are considering the location of the changes of sign of boundary values as functions of ε. The system of differential equations has a solution until the right most change reaches the end of the boundary of $\tilde{D}$. At exactly this point, the last local extremum also disappears as it hits the boundary at $(L,0)$. The process can be started again with one less change of boundary values and one less local extremum of $u(x,0)$. After a finite number of steps this terminates at the unique function indicated in Figure 8.

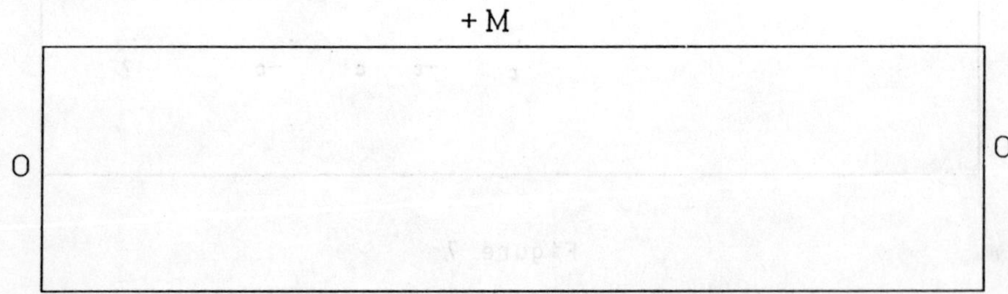

Figure 8

We start at the situation with only one boundary value $+M$ and run the differential equations in reverse.

Consideration of the Cauchy matrix as in the lemma shows that the Lipschitz condition is satisfied on each interval until a new change of sign of boundary values must be introduced. After a finite number of such intervals, the local extrema have again reached $\pm\varepsilon$ as indicated in Figure 7. But our construction shows there is only one configuration of the boundary values that can yield the values $\pm\varepsilon$ for the local extrema of $u(x,0)$ even ignoring the last one on the right. It is, of course, the conjectured extremal of Figure 3.

7. CONCLUSIONS OF THE PROOFS

In this section we show certain limitations on the functions considered in the modified problem were inessential. Then we show that the original problem is completely solved.

The hypothesis that $u(x,y)$ could have only a certain number of changes in its boundary values is inessential, except that the number cannot be made small in comparison to the length of the rectangle $\tilde{D}$. If it is made too small, there are no functions in the class $\tilde{\mathcal{U}}$.

In fact, $\tilde{\mathcal{U}}$ can be enlarged to allow any boundary values with magnitude less than or equal to M. We note that $\tilde{\mathcal{U}}$ is still compact. The first major step in the proof is the observation that extremal functions can have only a finite number of local extrema along the x-axis for $0 \leqslant x < L$. Otherwise they would accumulate. If they accumulated at a value of $x < L$, then $\frac{\partial u}{\partial x}$ would have to be identically zero on $(0,L)$. If they accumulated at $x = L$, then Schwarz reflection would make that an interior point; and again $\frac{\partial u}{\partial x}$ would be identically zero on $(0,L)$. Since $u(x,y)$ is even in y, $\frac{\partial u}{\partial y} = 0$ on $(0,L)$. Since u is harmonic on $\tilde{D}$, $u(x,y)$ is constant on $\tilde{D}$.

But u has boundary values 0 on the vertical edges of the boundary of $\tilde{D}$, hence $u(x,y) = 0$. This contradicts the extremal hypothesis on u.

The second major step is to note the constraint $|u(x,0)| \leq \varepsilon$ will be satisfied if we can control the finite number of local extrema of $u(x,0)$. If the number of these extrema is exceeded by the number of locations of jumps in the boundary values or places where the boundary values are not either $\pm M$, then the variational argument shows an extremal has not been reached. Hence even in this enlarged class $\tilde{\mathcal{U}}$, the extremal is the one found under the restrictive hypothesis.

Now we remove the hypothesis that the functions in $\tilde{\mathcal{U}}$ be even in x. Still the class is compact. If $\underset{\sim}{u}(x,y)$ is an extremal, then $\frac{1}{2}[\underset{\sim}{u}(x,y) + \underset{\sim}{u}(-x,y)]$ is in the class, even in x and attains the maximal value for the enlarged class at $(0,h)$ and consequently for the small class $\tilde{u}$. But then this function has boundary values $\pm M$ along the top and, in fact, $\tilde{u}$ was even in x to start with. Hence $\tilde{u}$ must be the extremal we already found.

We now turn to the original extremal problem for the class $\mathcal{U}$ of functions defined on D. Clearly the function $\underline{u}(x,y)$ indicated in Figure 2 is the limit of the extremal functions over $\tilde{\mathcal{U}}$ as L tends to plus infinity. Thus $\underline{u}(0,h)$ does attain the maximal value over $\mathcal{U}$ as desired; and $\underline{u}$ is an extremal function. It is not clear to the author whether $\underline{u}$ is the unique extremal function, although it appears to be so. If the class $\mathcal{U}$ is enlarged by dropping the restriction that the functions be even in x, the function $u(x,y)$ is still extremal.

The extremal function $\underline{u}(x,y)$ has the interesting property that it depends on M and ε, but not on h. Also it should be noted that the value $\underline{u}(0,h)$ could in principle be computed. In the next section such calculations are carried out as ε tends to zero.

8. ASYMPTOTIC BOUNDS

When ε is a small positive number, the functions in the class $\mathcal{U}$ are close to being odd in the y-variable. More explicitly, if u is in $\mathcal{U}$, then

$$\left|\frac{1}{2}\,[u(x,h) + u(x,-h)]\right| \leqslant \underline{u}(0,h) \tag{8.1}$$

where $\underline{u}$ is the extremal depicted Figure 2. As $\varepsilon \to 0^+$, the value $\underline{u}(0,h)$ tends to zero. In this section, the asymptotic behavior of $\underline{u}(0,h)$ as ε tends to zero will be calculated.

Figure 1 is helpful to interpret $\underline{u}(0,h)$. Consider the rectangle in Figure 1. Let u be the bounded harmonic function on the rectangle with the indicated boundary values. Then $u(0,h) = \underline{u}(0,h)$.

Unlike what is indicated in Figure 1, we will consider conformal maps of the upper half plane onto rectangles. We will postpone concern over the normalization.

The mappings can be obtained by the Schwarz-Christoffel formula [1]:

$$F(w) = \int_0^w \frac{dw}{\sqrt{(1-w^2)(1-k^2w^2)}}$$

for consistent choice of branch of the square root. For small $k > 0$, the interval $[-1,1]$ is carried to the narrow end of the rectangle image, the set $[1/k,+\infty] \cup [-\infty,-1/k]$ to the other narrow end. The positive imaginary axis is carried to a line parallel to the long sides of the rectangle and through the center of the rectangle.

The length of the image rectangle of the images of F is

$$|F(i\infty)| = \int_0^\infty \frac{d\tau}{\sqrt{(1+\tau^2)(1+k^2\tau^2)}} \tag{8.2}$$

$$= \int_0^{k^{-1/2}} \frac{d\tau}{\sqrt{1+\tau^2}}$$

$$+ \int_{k^{-1/2}}^\infty \frac{d\tau}{\tau(1+1/\tau^2)^{1/2}(1+k^2\tau^2)^{1/2}} + o(1) \text{ as } k\to 0^+$$

$$= \log(\sqrt{1+x^2}+x)\Big|_{x=0}^{x=k^{-1/2}} + \int_{k^{-1/2}}^\infty \frac{d\tau}{\tau\sqrt{1+k^2\tau^2}} + o(1)$$

$$= \log(2k^{-1/2}) - \log\frac{1/k+\sqrt{x^2+1/k^2}}{x}\Big|_{x=k^{-1/2}}^{x=\infty} + o(1)$$

$$= \log(2k^{-1/2}) + \log(2k^{-1/2}) + o(1)$$

$$= -\log k + 2\log 2 + o(1).$$

For fixed $0 < \lambda < 1$, a similar calculation traces the distance of the images of $i(1/k)^\lambda$ to the origin as $k \to 0^+$. This distance is

$$|F(i(1/k)^\lambda)| = -\lambda\log k + \log 2 + o(1). \tag{8.3}$$

In particular, for $\lambda = 1/2$, the points $i(1/k)^\lambda$ are mapped to approximately the center of the rectangle. More precisely, from (8.2) and (8.3),

$$-\lambda \log k + \log 2 + o(1)$$

$$= \frac{1}{2}[-\log k + 2\log 2 + o(1)]$$

$$-\lambda \log k = \frac{1}{2}[-\log k + o(1)]$$

$$\lambda = \frac{1}{2} + o(1/\log k).$$

More generally, to study the point h units above the real axis in D of the original problem, we need to consider the points $i(1/k)^{\lambda_h}$ which go $\frac{1-h}{2}$ units up the length of the rectangle (the length of the rectangle in Figure 1 is two units).

$$-\lambda_h \log k + \log 2 + o(1)$$

$$= (1-h)/2[-\log k + 2\log 2 + o(1)]$$

$$-\lambda_h \log k = -(1-h)/2\log k - h\log 2 + o(1)$$

$$\lambda_h = (1-h)/2 + h(\log 2)/\log k + o(1/\log k). \quad (8.4)$$

As the next step in solving the boundary value problem at $i(1/k)^{\lambda_h}$ in the upper half plane, we calculate the weight of the interval $[-1,1]$ with respect to $i(1/k)^{\lambda_h}$. The weight is

$$\frac{2}{\pi} \arctan 1/(\frac{1}{k})^{\lambda_h} = \frac{2}{\pi} k^{\lambda_h} + o(k^{\lambda_h}). \quad (8.5)$$

Since $(k)^{1/\log k} = e$, the weight of the interval $[-1,1]$, is

$$\frac{2}{\pi} k^{\frac{1-h}{2}} 2^h \{1 + o(1)\} = 2^{1+h} \pi^{-1} k^{\frac{1-h}{2}} \{1 + o(1)\}. \quad (8.6)$$

The weight of the set $[-\infty,-1/k] \cup [1/k,+\infty]$ can be similarly calculated; in fact, it can be calculated from formula (8.6) by replacing h by $-h$. As $k \to 0^+$, the weight on $[-\infty,-1/k] \cup [1/k,+\infty]$ decays more rapidly than that on $[-1,1]$. Of course, the intervals $[-1/k,-1]$ and $[1,1/k]$ do not contribute since they correspond to the sides of the rectangle with boundary values zero. Hence, it suffices to examine the contribution of $[-1,1]$.

We will now express k in terms of ε. Recall that the harmonic function u, defined on the rectangle, has value ε at the center of the rectangle. Formula (8.6) with $h = 0$ gives the weight on each end of the rectangle. The boundary values on these ends is M. Hence

$$\varepsilon = 2M\, 2\pi^{-1}\, k^{1/2}\{1+o(1)\}$$

$$k = 2^{-4}\, \pi^{2}\, M^{-2}\, \varepsilon^{2}\{1+o(1)\}. \qquad (8.7)$$

We can now substitute this expression for k into formula (8.6) for general h to obtain the weight on the closer narrow end of the rectangle. Simplifying and multiplying by the boundary value M, we obtain an asymptotic expression for the extremal function $\underline{u}(0,h)$:

$$\underline{u}(0,h) = 2^{-1+3h}\, \pi^{-h}\, M^{h}\, \varepsilon^{1-h}\{1+o(1)\} \qquad (8.8)$$

for fixed $0 < h < 1$ as $\varepsilon \to 0^+$.

9. CONCLUSION

A modified Schwarz reflection principle has been proved: Let $M > \varepsilon > 0$. Let $D = \{(x,y)$: x and y are real and $|y| < 1\}$. A function $\underline{u}(x,y)$ has been found such that if $u(x,y)$ is harmonic on D, $|u(x,y)| \leq M$ on D, and $|u(x,0)| \leq \varepsilon$, then

$$\left|\frac{1}{2}[u(x,y) + u(x,-y)]\right| \leq \underline{u}(0,|y|), \tag{9.1}$$

and this bound on the even part of u is sharp. Furthermore, the function $\underline{u}$ depends only on M and ε, not on $|y|$. Also the behavior of $\underline{u}(0,h)$ as $\varepsilon \to 0^+$ has been explicitly found and is given in equation (8.8).

Acknowledgments: The author gratefully acknowledges the opportunity for research afforded by Professor Lawrence Zalcman and Bar Ilan University. This work was partly motivated by ideas of Errett Bishop on constructive analysis [2]. It was rewarding to present this work at the Northwest University in Xian, The People's Republic of China and also at the Imperial College of Science and Technology in London, England.

REFERENCES

1. Ahlfors, Lars V., *Complex Analysis*, 3rd edition, McGraw-Hill Book Company (1979).

2. Bishop, Errett, *Foundations of Constructive Analysis*, McGraw-Hill Book Company (1967).

3. Gregory, Robert T. and David L. Karney, *A Collection of Matrices for Testing Computational Algorithms*, Wiley-Interscience (1969).

4. Polya, G. and G. Szego, *Aufgaben und Lehrsatze aus der Analysis*, Band II, Verlag von Julius Springer, Berlin (1925).

RECENT RESULTS IN THE THEORY OF EXTREME POINTS, SUPPORT POINTS AND SUBORDINATION

David J. Hallenbeck
Department of Mathematical Sciences
University of Delaware
Newark, DE 19716
U.S.A.

Let $\Delta = \{z : |z| < 1\}$ and let A denote the set of functions analytic in Δ. Let B_0 denote the subset of A consisting of all functions ϕ that satisfy the conditions $|\phi(z)| < 1$ and $\phi(0) = 0$. Let EB_0 denote the extreme points of B_0. Let $F \in A$ and let $s(F)$ denote the subset of A consisting of all functions f in A that are subordinate to F in Δ. This means that $f \in A$ and there exists $\phi \in B_0$ so that $f = F \circ \phi$.
Let $Es(F)$ denote the set of extreme points of $s(F)$.

In [2] Abu-Muhanna proved that if $F \in A$, F is univalent, F' is in the Nevanlinna class, and $F(\Delta)$ is a Jordan domain contained in a half plane then

$$Es(F) \subset \{F \circ \phi : \phi \in EB_0\}. \tag{1}$$

He conjectured in [2] that (1) holds whenever $F \in A$ and F is univalent. D.J. Hallenbeck proved in [10] that (1) holds for any univalent function F in A when $F(\Delta)$ is a Jordan domain. Recently Abu-Muhanna and Hallenbeck [4] proved that (1) holds whenever F is in A and F is univalent. So the conjecture made by Abu-Muhanna in [2] has been proved.

It is interesting to consider whether (1) could hold under other assumptions on F. In this direction, R. Younis [18] has proved that if F is a polynomial then

$$EHs(F) \subset \{F\circ\phi : \phi \in EB_0\} \tag{2}$$

where EHs(F) denotes the set extreme points of the closed convex hull of s(F). The argument depends in part on the fact that if $\phi \in B_0 \setminus EB_0$ then there exists $h \in B_0$ such that $\phi \pm e^{i\theta}h \in B_0$ for all $\theta \in [0,2\pi)$ [7, p.126]. This result suggests the conjecture that (2) holds whenever F is analytic in $\bar{\Delta}$.

If $F \in A$ and F is univalent in Δ then it is known [7, p.50] that F is in the Hardy class H^p for all $p < \frac{1}{2}$. Hence if $F \in A$, $f = F\circ\phi$ for $\phi \in B_0$ and F is univalent then by Littlewood's subordination inequality, $f \in H^p$ for all $p < \frac{1}{2}$. It follows that $\lim_{r\to 1} f(re^{i\theta}) = f(e^{i\theta})$ exists for almost all $\theta \in [0,2\pi)$. Let $D = F(\Delta)$ and let $\lambda(w,\partial D)$ denote the distance between w and ∂D. In [2] Y. Abu-Muhanna proved that if $F \in A$, F is univalent and F' is in the Nevanlinna class then

$$\int_0^{2\pi} \log\lambda(F(\phi(e^{it})),\partial D)dt = -\infty \tag{3}$$

if and only if $\phi \in EB_0$. It was conjectured in [2] that (3) holds for any univalent function F analytic in Δ if and only if $\phi \in EB_0$ (note that (3) holds trivially if $|\phi(e^{it})| = 1$ on a set of positive measure since F is univalent). D.J. Hallenbeck proved in [10] that if (3) holds and F is univalent and analytic in Δ then $\phi \in EB_0$. In [12] Hallenbeck proved that if F is analytic and univalent in Δ and $\phi \in EB_0$ then (3) holds. Hence if $F \in A$ and F is univalent then $\{F\circ\phi : \phi \in EB_0\} = \{F\circ\phi : \phi \in EB_0$ and (3) holds$\}$. This verifies the conjecture made in [2]. The equality of these two sets was known previously only when $F(\Delta)$ was a bounded convex domain with a smooth boundary and positive curvature [5]. This equality [10], [12] can be regarded as a generalization of the classical H^∞ result that $\phi \in EB_0$ if and only if $\int_0^{2\pi} \log(1 - |\phi(e^{it})|)dt = -\infty$.

In [16] Sheil-Small raised the problem of determining general conditions on F so that

$$EHs(F) = \{F(xz) : |x| = 1\}. \tag{4}$$

A classical example of this situation is provided by $F(z) = \frac{1+z}{1-z}$. In this case (4) follows for example from the Herglotz representation formula. It is known that $\{F(xz) : |x| = 1\} \subset EHs(F)$ for any non constant function F analytic in Δ [13]. When $F(z) = (\frac{1+cz}{1-z})^\alpha$ $|c| \leqslant 1$, $c \neq -1$ and $\alpha > 1$, Brannan, Clunie and Kirwan [6] proved that (4) holds.

Recently Y. Abu-Muhanna [3] has proved the striking result that (4) holds whenever F is univalent in Δ, $₵\backslash F(\Delta)$ is convex and $\partial F(\Delta)$ satisfies a certain smoothness condition at ∞. D. Wilken and S. Perera were able to remove this last assumption (private communication) and so (4) holds whenever F is univalent and $₵\backslash F(\Delta)$ is convex. Y. Abu-Muhanna conjectured in [3] that (4) holds whenever F is the universal covering map from Δ onto a domain D with $₵\backslash D$ bounded and convex. It was proved in [8] that (4) holds for $F(z) = \exp \frac{1+z}{1-z}$ and this can be regarded as evidence for Abu-Muhanna's conjecture.

We next consider the problem of determining $Es(F)$ when $F \in A$, F is univalent and $F(\Delta)$ is a convex domain. When $F(\Delta)$ is a wedge or a strip it is known [5] that

$$Es(F) = \{F \circ \phi : \phi \in B_0, \ \phi \text{ is inner}\}. \tag{6}$$

Abu-Muhanna and T.H. Mac Gregor also conjectured in [5] that (6) holds whenever $F(\Delta)$ is any (bounded or unbounded) convex polygon. This conjecture was proved false by J. Gevirtz in [9]. He proved that if $F(\Delta)$ is a convex domain other than a half plane, a strip or a wedge then there exists $f = F \circ \phi \in Es(F)$ such that ϕ is not inner. Also under the same assumption he constructed an extreme point $f = F \circ \phi$ such that $|\phi(e^{it})| < 1$ for almost all $t \in [0,2\pi)$. For example there are functions $\phi \in EB_0$ such that $|\phi(e^{i\theta})| < 1$ for almost all $\theta \in [0,2\pi)$. Whenever $F \in A$, $F(\Delta)$ is convex and not a half plane then it is known [15] that

$$\{F \circ \phi : \phi \in B_0, \ \phi \text{ is inner}\} \subset Es(F) \subset \{F \circ \phi : \phi \in EB_0\}. \tag{7}$$

The first inclusion follows in part from the fact that under these assumptions $F \in H^p$ for all $p > 1$ [13]. The second inclusion holds for any univalent function F which is analytic in Δ [4]. In [15] it was asked whether the set $Es(F)$ ever lies strictly between the minimal and maximal subsets exhibited in (7) above. K. Tkaczyńska has proved recently [17] that this always happens when $F(\Delta)$ is a convex domain other than a strip, a wedge or a half plane and $\partial F(\Delta)$ contains a line segment. It was also proved in [17] that if $f \in Es(F)$ then $\{f(e^{i\theta}) : \lim_{r \to 1} f(re^{i\theta}) \text{ exists}\} \cap E\partial D \neq \phi$ for any convex domain $D = F(\Delta)$ where $E\partial D$ denotes the set of extreme points of ∂D.

Finally in [17] it was proved that $Es(F)$ is maximal whenever $\partial F(\Delta)$ has curvature which is piecewise smooth and positive. This result was known previously only when the curvature was smooth and positive [5].

We now turn our attention from extreme points to support points. Recall that A is a locally convex linear topological space with respect to the topology given by uniform convergence on compact subsets of Δ. By a continuous, linear functional on A, we mean a complex-valued functional defined on A that is continuous and linear. A function f is called a support point of a compact subset F of A if $f \in F$ and if there is a continuous, linear functional J on A so that $\operatorname{Re} J(f) = \max\{\operatorname{Re} J(g) : g \in F\}$ and $\operatorname{Re} J$ is not constant on F. The set of support points of F is denoted by supp F. The set supp B_0 consists of all finite Blaschke products vanishing at the origin [14]. It is known that $\{F(xz) : |x| = 1\} \subset \operatorname{supp} s(F)$ whenever $F \in A$. We point out that one of the ingredients in the proof given by Abu-Muhanna that (4) holds whenever $\mathbb{C}\backslash f(\Delta)$ is convex [3] is the observation made in [15, p. 92] that $HF = H(\operatorname{supp} F \cap EHF)$ for any compact family F contained in A. D.J. Hallenbeck proved in [11] that if

$$Hs(F) = \{\int_{\partial\Delta} F(xz)\, d\mu(x) : \mu \text{ a probability measure on } \partial\Delta\}$$

and $F(z) = \dfrac{G(z)}{(z-x_0)^\alpha}$ where $|x_0| = 1$, G is analytic in Δ, continuous in $\bar{\Delta}$, $G(x_0) \neq 0$ and $\alpha > 1$ then

$$\operatorname{supp} s(F) = \{F(xz) : |x| = 1\}. \qquad (8)$$

It follows from the previous result [11], the results of Abu-Muhanna [3], and Perera and Wilken (private communication) that (8) holds whenever $\mathbb{C}\backslash F(\Delta)$ is convex, F is univalent and F has the form described above. Also included is example $F(z) = (\frac{1+cz}{1-z})^\alpha$ $|c| \leq 1$, $c \neq -1$, $\alpha > 1$ found previously in [14]. When $F \in A$ Abu-Muhanna [1] proved

$$\operatorname{supp} s(F) \subset \{\bar{F}\circ\phi : \phi \in \operatorname{supp} B_0\}. \qquad (9)$$

This inclusion was first proved by Hallenbeck and Mac Gregor in [14] under the additional assumption that $F'(z) \neq 0$ for $z \in \Delta$. D.J. Hallenbeck and T.H. Mac Gregor [14] proved that if $F \in A$ then $\operatorname{supp} s(F)$ is maximal whenever F is univalent and $F(\Delta)$ convex. K. Tkaczynska recently [17] proved that $\operatorname{supp} s(F)$ is maximal whenever F is in the closed convex hull of the normalized convex mappings. Finally, Y. Abu-Muhanna proved in [1] that $\operatorname{supp} s(F)$ is maximal whenever F is analytic in $\bar{\Delta}$. He conjectured [1] that $\operatorname{supp} s(F)$ is maximal whenever F is a univalent H^{∞} function. This conjecture remains open.

This completes our survey of results in the theory of extreme points, support points and subordination. In spite of intense recent activity interesting conjectures remain open.

REFERENCES

1. Y. Abu-Muhanna, Variability regions and support points of subordinate families, *J. London Math. Soc.* (2), 29 (1984), 477-484.

2. Y. Abu-Muhanna, On extreme points of subordination families, *Proc. Amer. Math. Soc.* 87 (1983), 439-443.

3. Y. Abu-Muhanna, Subordination and extreme points, to appear in Complex Variables.

4. Y. Abu-Muhanna and D.J. Hallenbeck, Subordination families and extreme points, to appear.

5. Y. Abu-Muhanna and T.H. Mac Gregor, Extreme points of families of analytic functions subordinate to convex mappings, *Math. Z.* 176 (1981), 511-519.

6. D. Brannan, J. Clunie and W. Kirwan, On the coefficient problem for functions of bounded boundary rotation, *Ann. Acad. Sci. Fenn. Ser. A.I.* 523 (1973).

7. P.L. Duren, *Theory of H^p Spaces*, Academic Press, New York (1970).

8. J. Feng, Thesis, State University of New York at Albany.

9. J. Gevirtz, On extreme points of families of analytic functions with values in a convex set, *Math. Z.* 193 (1986), 79-83.

10. D.J. Hallenbeck, Extreme points of subordination families with univalent majorants, *Proc. Amer. Math Soc.* 91 (1984), 54-58.

11. D.J. Hallenbeck, A note on support points of subordination families, to appear in *Proc. Amer. Math. Soc.*

12. D.J. Hallenbeck, A note on extreme points of subordination classes, to appear.

13. D.J. Hallenbeck and T.H Mac Gregor, Subordination and extreme point theory, *Pacific J. Math.* 50 (1974), 455-468.

14. D.J. Hallenbeck and T.H. Mac Gregor, Support points of families of analytic functions described by subordination, *Trans. Amer Math. Soc.* 278 (1983), 523-545.

15. D.J. Hallenbeck and T.H. Mac Gregor, *Linear Problems and Convexity Techniques in Geometric Function Theory*, Pitman Publishing Co., Boston (1984).

16. T. Shiel-Small, Extreme points of a class of subordinate functions, *Proc. Amer. Math. Soc.* 91, (1984), 73-74.

17. K. Tkaczyńska, Thesis, University of Delaware.

18. R. Younis, On extreme points of families described by subordination, to appear in *Proc. Amer. Math. Soc.*

SOME RESULTS ON UNIVALENT FUNCTIONS

Liu Shuqin
Department of Mathematics
Northwest University
Xian, China

In recent years, some of my colleagues and students in Northwest University have worked in the field of univalent functions and obtained some results of which the present article gives a brief account.

Contents

I. Coefficient problems
II. Meromorphic univalent functions in the open unit disk
III. Some special families
IV. Extreme points and support points
V. Integral means and coefficients of inverse functions

I. COEFFICIENT PROBLEMS

Let S be the class of functions $f(z)=z+a_2z^2+\ldots$ analytic and univalent in the open unit disk $D=\{z:|z|<1\}$. Denote by Σ the set of functions $g(z)=z+b_0+b_1z^{-1}+\ldots$ meromorphic and univalent in $E=\{z:|z|>1\}$. Let S_k be the subclass of S consisting of functions with k-fold symmetry.

If $f(z)=z+\sum_{n=1}^{\infty} a_{nk+1}^{(k)} z^{nk+1} \in S_k$, we know that

$|a_{k+1}^{(k)}| \leq 2/k$, $|a_{2k+1}^{(k)}| \leq \frac{2}{k}\exp(-2(k-1)/(k+1))+1/k$.

Milin [1] proved that $|a_{2n+1}^{(2)}| \leq 1.14$. For $n=3$, Rosenblat [2] gave $|a_7^{(2)}|<1.15$ and Ma [3] further improved it to $|a_7^{(2)}|<1.053$. Recently, Liu [4] got for $k \geq 2$ that

$$|a_{3k+1}^{(k)}| \leq 4(A^2+BC)^{3/2}/(9ACK^3 3^{\frac{1}{2}}),$$

where $A=3k+2$, $B=k^2+2k+2$ and $C=3k^2+3k+1$. In particular, we had

$$|a_7^{(2)}| < 1.05293, \quad |a_{10}^{(3)}| < 0.58994.$$

In 1983, Zalcman conjectured that for every $f(z)=z+a_2z^2+\ldots$ in S,

$$|a_n^2-a_{2n-1}| \leq (n-1)^2 \quad (n \geq 2), \tag{1}$$

which implies the famous Bieberbach conjecture. It is well known that (1) is valid for $n=2$. The conjecture is known to be true for the subclass S_R of functions in S with real coefficients. Brown and Tsao [5] proved that (1) holds for the family T of typically real functions and the class S^* of starlike functions. Ma [6] gave the following results last year.

THEOREM 1.1. Let $f(z)=z+a_2z^2+\ldots$ belong to the closed convex hull of C, which consists of close-to-convex functions, then

$$|a_n^2-a_{2n-1}| \leq (n-1)^2 \quad (n=4,5,\ldots).$$

In particular, this inequality is true for the class C.

It was conjectured by Kirwan and Schober [7] that for $g(z)=z+b_0+b_1z^{-1}+\ldots\in\Sigma$,

$$\mathrm{Re}(nb_1-b_n)\leqslant n \quad (n\geqslant 2). \tag{2}$$

In general, the problem is to find the smallest values for t such that

$$\mathrm{Re}(tb_1-b_n)\leqslant t. \tag{3}$$

For $n=2,3$, Garabedian and Schiffer [8] proved that $t=n$ is the least value for which (3) holds. In addition, Leung and Schober [9] verified that there is an explicit finite number t such that (3) is true.

Let $T\Sigma$ be the class of typically real meromorphic functions in E. Ma [10] got the following

THEOREM 1.2. Let $g(z)=z+b_0+b_1z^{-1}+\ldots\in T\Sigma$, $n\geqslant 2$ and $-\infty<t<+\infty$, then

$$t+4m(n,t)\leqslant tb_1-b_n\leqslant t+4M(n,t),$$

where

$$m(n,t)=\min_{0\leqslant x\leqslant\pi}(\sin x\,\sin nx-t\sin^2 x),$$

$$M(n,t)=\max_{0\leqslant x\leqslant\pi}(\sin x\,\sin nx-t\sin^2 x).$$

In particular, $tb_1-b_n\leqslant t$ if $t\geqslant n$.

COROLLARY 1.1. Let $g(z)=z+b_0+b_1z^{-1}+\ldots\in\Sigma$ with real coefficients, then $tb_1-b_n\leqslant t$ for $t\geqslant n$.

II. MEROMORPHIC UNIVALENT FUNCTIONS IN THE OPEN UNIT DISK

Consider the class $S(p)$, $0<p<1$, consisting of meromorphic univalent functions $f(z)$ in D with a simple pole at p and normalization $f(0)=f'(0)-1=0$. The "Koebe function" of $S(p)$ is

$$k_p(z) = z/(1-pz)(1-z/p).$$

Fenchel [11] got that $|f(z)|\geq -k_p(-|z|)$ for every $f\in S(p)$. The upper bound of $|f(z)|$ was given by Kirwan and Schober [12]. They proved that for $f\in S(p)$, $|f(z)|\leq k_p(|z|)$. The following theorems were given by Zhang [13].

THEOREM 2.1. Let $f(z)$ be analytic in $D\setminus\{p\}$, $z=p$ be a simple pole of $f(z)$ and $f(0)=f'(0)-1=0$. Then $f(z)\in S(p)$ if and only if $\frac{1}{\pi}\iint_{|\eta|<1} |f'(z)f'(\eta)/(f(z)-f(\eta))^2 -(z-\eta)^{-2}|^2 d\sigma_z \leq (1-|\eta|^2)^{-2} (|\eta|<1, \eta\neq p)$.

THEOREM 2.2. Let $f(z)\in S(p)$, $|z|=r<p$, then

$$(p-r)^2(1-pr)^2/p^2(1-r^2)^3 \leq |f'(z)| \leq k_p'(r),$$

$$|\arg f'(z)| \leq \log((1-r^2)k_p'(r)).$$

In [14], Ma gave a method of constructing variations for functions in $S(p)$ and obtained the following results.

THEOREM 2.3. Let $g\in S$ and $g_\lambda(z)=g(z)+\lambda(zg'(z)h(z) -h(0)g(z))+O(\lambda)$ be a variation of g within the class S. Then every $f\in S(p)$ has a variation

$$f_\lambda(z) = f(z)+\lambda(zf'(z)h(z)+ph(p)f^2(z)/ \alpha(f)-h(0)f(z))+O(\lambda)$$

within the class $S(p)$, where $\alpha(f)=\mathrm{Res}(f,p)$.

THEOREM 2.4. Let $f\in S(p)$ be an extremal function of a continuous functional ϕ on $S(p)$. If ϕ has a Fréchet differential $L(f,.)$, then $f(z)$ satisfies the differential equation

$$(\eta f'(\eta)/F(\eta))^2 P(f(\eta)) = q(\eta) \quad (\eta\in D\setminus\{p\}),$$

where $P(w)=L(f,f^2(z)/(w-f(z)))$, $q(\eta)\geq 0$ on $|\eta|=1$ and has at least one zero of even order on $|\eta|=1$. Moreover, if ϕ is a functional of degree n, then $P(w)$ is at most a rational of degree n.

THEOREM 2.5. Let $f\in S(p)$ be an extremal function of a continuous functional ϕ of finite degree on $S(p)$ and ϕ have a Fréchet differential $L(f,.)\neq 0$. Then the range $f(D)$ of f is dense in $\mathbb{C}$ and $\bar{\mathbb{C}}\setminus f(D)$ consists of finitely many analytic arcs which are bounded.

THEOREM 2.6. Let $f\in S(p)$ be an extremal function of a continuous functional ϕ on $S(p)$ and ϕ have a Fréchet differential $L(f,.)$ at f. Then we have

$$L(f,\ f'(z)+f^2(z)/\alpha(f)-1-f''(0)f(z))$$

$$= \bar{L}(f,\ z^2f'(z)+p^2f^2(z)/\alpha(f)),$$

where $\bar{L}(f,.) = \overline{L(f,.)}$.

THEOREM 2.7. Let $f\in S(p)$ and $|z|=r<1$, then

$$k_p'(-r) \leq |f'(z)| \leq 1/(1-r^2)(1-pr)^2(1-r/p)^2.$$

The lower bound is sharp.

III. SOME SPECIAL FAMILIES

It is known that if $p(t)$ is a continuous function in the interval $0 \leqslant t < \infty$ with possible exception of finite number of points of discontinuity of first kind and $|p(t)|=1$, then the solution $w=f(z,t)$ of the differential equation

$$\partial w/\partial t = -w(1+p(t)w^k)/(1-p(t)w^k) \quad (k=1,2,\ldots), \qquad (4)$$

satisfying the initial condition $w|_{t=0}=z$, has the following properties:

$$f(0,t)=0, \quad f'(0,t)=e^{-t},$$

$f(z,t)$ is analytic, univalent and k-fold symmetric in D for fixed t, and the limit function

$$f(z) = \lim_{t\to\infty} e^t f(z,t) \qquad (5)$$

exists with $f(0)=f'(0)-1=0$. The class of all such limit functions is dense in S_k.

Liu [15] studied some special cases of the function $p(t)$. The following results are given.

THEOREM 3.1. Let $p(t)=(e^{-t}+i(1-e^{-2t})^{\frac{1}{2}})^{k+2}$, then the function $f(z)$ determined from (4) and (5) is $zF(2/k, 1/k, 1+1/k; z^k)$, where F is the hypergeométric function. Furthermore, $f(z)$ is convex with k-fold symmetry, and the extremal function for the coefficient functionals on the class of convex functions with k-fold symmetry.

He also got that for $p(t)=e^{-i\alpha}$ $(\alpha\in\mathbb{R})$ and $p(t)=e^{-it\beta}$ $(\beta\in\mathbb{R})$, $f(z)$ is $z(1+e^{-i\alpha}z^k)^{-2/k}$,

$z(1+(k-i\beta)z^k/(k+i\beta))^{-2/(k-i\beta)}$ respectively.

Miller [16] introduced a subset $\Sigma^*(p,w_o)$ of $S(p)$. A function $f(z)=z+\ldots \in \Sigma^*(p,w_o)$ if and only if $f\in S(p)$ and satisfies

$$\mathrm{Re}(zf'(z)/(f(z)-w_o)+p/(z-p)-pz/(1-pz))<0 \quad (z\in D).$$

Miller proved that if $f\in\Sigma^*(p,w_o)$, then

$$-1+f(z)/w_o = p/(z-p)(1-pz)\ \exp(2\int_o^{2\pi} \log(1-e^{it}z)d\mu(t)),$$

where μ is a probability measure on $[0,2\pi]$. Zhang [17] showed that the inverse of the above result is also true. By using this integral representation, Zhang got the following fact.

THEOREM 3.2. If $f\in\Sigma^*(p,w_o)$, then we have sharp inequalities $p|w_o|(1-r)^2/(r+p)(1+pr)^2 \leqslant |f(z)-w_o| \leqslant p|w_o|(1+r)^2/|r-p|(1-pr)^2$.

Let $\alpha\geqslant 0$, $0\leqslant\beta<1$ and $J(\alpha,\beta)$ be the class of functions $f(z)=z+\ldots$ holomorphic and satisfying the conditions $f(z)f'(z)/z\neq 0$,

$$\mathrm{Re}\{(1-\alpha)zf'(z)/f(z) + \alpha(1+zf''(z)/f'(z))\} > \beta$$

in D. In $J(\alpha,\beta)$ the functions

$$k_\beta(0,z) = z(1-z)^{-2}(1-\beta),$$

$$k_\beta(\alpha,z) = \{\alpha^{-1}\int_o^z t^{1/\alpha-1}(1-t)^{-2(1-\beta)/\alpha}dt\}^\alpha \quad (\alpha>0)$$

play the role of extremal functions.

The class $J(\alpha,0)$ was studied by Miller and others [18,19,20]. Miller proved that for $f\in J(\alpha,0)$,

$$k_0'(\alpha,-r) \leqslant |f'(z)| \leqslant k_0'(\alpha,r)$$

if $\alpha\geqslant 1$ and $|z|=r<1$. He conjectured that the inequalities are true for $0<\alpha<1$. In 1983, Liu [21] obtained for the functions in $J(\alpha,\beta)$ the subordination

$$(f(z)/z)^{(1-\alpha)/2} f'(z)^{\alpha/2} \prec (1-z)^{-(1-\beta)},$$

and the inequalities

$$-k_\beta(\alpha,-r) \leqslant |f(z)| \leqslant k_\beta(\alpha,r),$$

$$k_\beta'(\alpha,-r) \leqslant |f'(z)| \leqslant k_\beta'(\alpha,r) \quad (\alpha\geqslant 1),$$

where $|z|=r<1$, and other results.

Ma [22] also studied the class $J(\alpha,\beta)$ and got the following results. In particular, he proved the Miller's conjecture.

THEOREM 3.3. Let $f\in J(\alpha,\beta)$, then we have

$$zf'(z)/f(z) \prec zk_\beta'(\alpha,z)/k_\beta(\alpha,z) \equiv \phi(z).$$

THEOREM 3.4. Let $f\in J(\alpha,\beta)$ and $|z|=r<1$. Then we have sharp inequalities

$$\phi(-r) \leqslant |zf'(z)/f(z)| \leqslant \phi(r),$$

$$|\arg zf'(z)/f(z)| \leqslant \max_{|z|=r} \arg\phi(z),$$

$k'_{\beta}(\alpha,-r) \leqslant |f'(z)| \leqslant k'_{\beta}(\alpha,r).$

THEOREM 3.5. Let $w=f(z)\in J(\alpha,\beta)$, then we get sharp result

$$f(D) \supset \{w: |w|<d(\alpha,\beta)\},$$

where

$$d(0,\beta) = 2^{-2(1-\beta)},$$

$$d(\alpha,\beta) = F(1/\alpha,\ 2(1-\beta)/\alpha,\ 1+1/\alpha;\ -1)^{\alpha} \quad (\alpha>0)$$

and F is the hypergeometric function.

THEOREM 3.6. Let $f\in J(\alpha,\beta)$, then $f\in J(0,s(\alpha,\beta))$, where

$$s(\alpha,\beta) = \min_{0<\theta\leqslant 2\pi} \operatorname{Re} \phi(e^{i\theta}).$$

He also gave some sharp coefficient estimates for $J(\alpha,\beta)$. In [23], he extended the above results to functions in $J(\alpha,\beta)$ with k-fold symmetry.

Let $f(z)=z+\ldots$ be analytic in D. Then f is said to be starlike of order α and type β if and only if

$$|zf'(z)/f(z)-1| < |(2\beta-1)zf'(z)/f(z)+1-2\beta\alpha|$$

in D. Denote the class of all such functions by $S^*(\alpha,\beta)$, which was first introduced by Juneja and Mogra [24]. They gave a representation formula, bound of $|f(z)|$ and some coefficient estimates as well as the radii of convexity of functions in $S^*(\alpha,\beta)$. The following results were given by Ma [25].

THEOREM 3.7. Let $f \in S^*(\alpha,\beta)$, $0 \leqslant \lambda \leqslant 1$, then

$$zf'(z)/f(z) \prec (1+(1-2\beta\alpha)z/(1+(1-2\beta)z),$$

$$f(z)/z \prec e^{(1-\alpha)z} \quad (\beta = \tfrac{1}{2}),$$

$$f(z)/z \prec (1+(1-2\beta)z)^{2\beta(1-\alpha)/(1-2\beta)} \quad (\beta \neq \tfrac{1}{2}),$$

$$\phi(f) \equiv (f(z)/z)^{\lambda(2\beta-1)/2\beta(1-\alpha)} ((2\beta-1)zf'(z)/$$

$$2\beta(1-\alpha)f(z) + (1-2\beta\alpha)/2\beta(1-\alpha)^{1-\lambda} \prec (1+(1-2\beta)z)^{-1}.$$

From this theorem, the author got sharp bounds of $|f(z)|$, $|zf'(z)/f(z)|$, $|f'(z)|$, $|\arg zf'(z)/f(z)|$, $|\arg f(z)/z|$, $|1-1/\phi(f)|$, $|\arg\phi(f)|$ and $|\arg f'(z)|$ $(\beta = \frac{1}{2}/\alpha)$. He also obtained a covering theorem and the best estimates of $|a_3 - a_2^2|$.

For $\alpha>0$, define $S^*(\alpha)$ to be the set of functions $f(z)=z+\ldots$ analytic in D and satisfying $|\arg zf'(z)/f(z)| < \frac{1}{2}\alpha\pi$. Feng [26] determined the extreme points of closed convex hull of $S^*(2)$ and gave sharp coefficient inequality of functions in $S^*(2)$. The following results on $S^*(\alpha)$ were given by Ma [27].

THEOREM 3.8. Let $f \in S^*(\alpha)$ maximizes some continuous functional L on $S^*(\alpha)$, and L have a Fréchet differential $I(f,.)$ at f. If $I(f,h) \neq ah(0)+bh'(0)$ $(a,b \in \mathbb{C})$, then

$$f(z) = z\exp\left(\int_0^z (p(t)^{\alpha}-1)/t\,dt\right),$$

where

$$p(z) = \sum_{k=1}^{n} \lambda_k (1+x_k z)/(1-x_k z), \quad \lambda_k > 0,$$

$$\sum_{k=1}^{n} \lambda_k = 1, \ |x_k| = 1 \text{ and } n = 1,2,\ldots$$

The author proved that the converse of Theorem 3.8 is also true.

THEOREM 3.9. Let $f(z)=z+a_2z^2+\ldots\in S^*(\alpha)$ and

$$F(z) = z\exp\left(\int_0^z ((1+t)^{\alpha}/(1-t)^{\alpha}-1)/t\,dt\right) = z+A_2z^2+\ldots,$$

then we have

$$|a_n| \leq A_n \quad (\alpha\geq 1,\ n=2,3,\ldots), \quad |a_2|\leq 2\alpha \quad (0<\alpha<1),$$

$$|a_3| \leq 3\alpha^2 \quad (1/3\leq\alpha<1), \quad |a_3|\leq\alpha \quad (0<\alpha<1/3).$$

Let T_k be the class of typically real functions $f(z)=z+\ldots$ in D with k-fold symmetry. Robertson [48] and Goluzin [43] proved the following results independently. If $f(z)\in T_1$, then

$$f(z) = \int_0^{\pi} z/(1-2z\cos\theta+z^2)\,d\mu(\theta),$$

where $\mu(\theta)$ is a probability measure on $[0,\pi]$. In [44], Hu got a similar result for T_2. If $f(z)\in T_2$, then

$$f(z) = \int_0^{\pi} z(1+z^2)/(1-2z^2\cos 2\theta+z^4)\,d\mu(\theta).$$

Liu [45] went a step further for the class T_k $(k\geq 2)$.

THEOREM 3.10. [45] Let $f\in T_k$, $k\geq 2$, then there is a probability measure μ on $[0,\pi]$ such that

$$f(z) = \int_0^{\pi} z(1+z^k \sin(k-1)\theta/\sin\theta)/$$

$$(1-2z^k \cos k\theta + z^{2k}) d\mu(\theta).$$

IV. EXTREME POINTS AND SUPPORT POINTS

Let A be the family of functions analytic in D. Then A is a locally convex linear topological space under the topology of uniform convergence on compact subsets of D. If $B \subset A$, we denote by HB, EB and $\mathrm{Supp}B$ the closed convex hull, the set of extreme points and the set of support points of B respectively.

Duren, Leung and Schiffer [28] proved the following theorem.

THEOREM 4.1. Let $f(z)$ be a support point of S which maximizes $\mathrm{Re}\{L\}$, where L is a linear functional of rational type whose corresponding rational function $\phi(w) = L(f^2/(f-w))$ has no zeros of order two. If the arc Γ omitted by f has a radial angle of $\pm\pi/4$ at its tip, then Γ is a half-line.

They said that the hypothesis that ϕ has no double zeros is probably not essential. Ma [29] eliminated this hypothesis and gave the following results.

THEOREM 4.2. Let $f \in \mathrm{Supp}S$ maximizes $\mathrm{Re}\{L\}$ and $\phi(w)=L(f^2/(f-w))$ be analytic in $|w| > \frac{1}{4}$. Suppose for every $\theta \in \mathbb{R}$ that $\mathrm{Im}\,\phi(te^{i\theta})$ has only one zero or $\mathrm{Im}\,\phi(te^{i\theta}) \equiv 0$ in $\frac{1}{4} < t < +\infty$. Then $\mathbb{C} \backslash f(D)$ has monotonic argument.

THEOREM 4.3. Let $f \in \mathrm{Supp}S$ maximizes $\mathrm{Re}\{L\}$. If $L(f^3)/L(f^2)^2 \bar{\in} \mathbb{R}$, then $\mathbb{C} \backslash f(D)$ has monotonic radial angle near ∞.

Duren [30] posed the two-functional conjecture: If a function $f \in S$ maximizes $\mathrm{Re}\{L\}$ and $\mathrm{Re}\{M\}$ for two essentially different linear functionals L and M (one not a positive multiple of the other), then f is a rotation of the Koebe function. He verified the conjecture for some special cases. Ma [29] further proved this conjecture for the following cases.

1. $L(f)=f(z_1)$, $M(f)=f(z_2)$, $(z_1 \neq z_2,\ z_1, z_2 \in D\setminus\{0\})$,
2. $L(f)=f'(z_1)$, $M(f)=f'(z_2)$, $(z_1 \neq z_2,\ z_1, z_2 \in D\setminus\{0\})$,
3. $L(f)=f(z_1)$, $M(f)=f'(z_2)$, $(z_1,\ z_2 \in D\setminus\{0\})$.

In [31], Ma discussed the extreme points of $S(p)$ and got the following.

THEOREM 4.4. Let $f \in ES(p)$, then the area of $\mathbb{C}\setminus f(D)$ is zero.

Let $T\Sigma'$ denote the class of nonvanishing functions in $T\Sigma$. Next theorem is due to Ma [10].

THEOREM 4.5. Let Λ be the set of probability measures on $[0,\pi]$, then we have

$$T\Sigma' = \{f(z)=z(1-1/z^2)^2 \int_0^\pi (1-2\cos x/z+1/z^2)^{-1} d\mu(x) : \mu \in \Lambda\},$$

$$ET\Sigma' = \{z(1-1/z^2)^2(1-2y/z+1/z^2)^{-1} : -1 \leq y \leq 1\}.$$

By using the Schiffer's boundary variation, Chen [32] showed

THEOREM 4.6. Let $f \in \mathrm{Supp}S(p)$ associated with the functional L, then $\mathbb{C}\setminus f(D)$ consists of finitely many analytic arcs which satisfy the differential equation

$$L(f^2/(f-w))dw^2/w^2 > 0.$$

Ma [33] also studied $\mathrm{Supp}J(\alpha,\beta)$. His result is as follows.

THEOREM 4.7. Let $k_n(z) = z\prod_{k=1}^{n}(1-x_k z)^{-t_k}$ and

$f_n(z) = (\alpha^{-1}\int_0^z k_n(\zeta)^{1/\alpha}\zeta^{-1}d\zeta)^{\alpha}$. Then $\text{Supp}J(\alpha,\beta) \subset \{f_n(z): |x_k|=1, t_k>0, t_1+\ldots+t_n=2(1-\beta), n=1,2,\ldots\}$.

Let $F(z)\in A$ and $s(F)$ be the set of functions which are subordinate to $F(z)$. If $\mathscr{F}\subset A$, define $s(\mathscr{F}) = \{g: g\prec f \text{ for some } f\in\mathscr{F}\}$. For some special F, $EHs(F)$ and $\text{Supps}(F)$ have been investigated. For example, Hallenbeck [34] proved that for $B_0=s(z)$,

$$\text{Supp}B_0 = \{xz\prod_{k=1}^{n}(z+\alpha_k)/(1+\bar{\alpha}_k z): |x|=1, |\alpha_k|=1,$$

$$n=1,2,\ldots\},$$

and Hoffman [35] gave that

$$EB_0 = \{f\in B_0: \int_0^{2\pi}\log(1-|f(e^{i\theta})|)d\theta = -\infty\}.$$

Luo [36] showed the following theorems.

THEOREM 4.8. Let $F(z)$ be analytic and univalent in D, and $C=\partial F(D)$ be a Dini-smooth Jordan arc. Then

$$\text{Supps}(F) = \{F\circ\phi: \phi\in\text{Supp}B_0\}.$$

THEOREM 4.9. Let $\mathscr{F}$ be a compact subset of A. If for every $f\in\mathscr{F}$, $f'(z)\neq 0$ and L is a continuous linear functional on A with $L(f)\neq af(0)$ $(a\in\mathbb{C})$. Then

$$\text{Supp}\{s(\mathscr{F}), L\} \subset \{F\circ\phi: \phi\in\text{Supp}B_0, F\in\mathscr{F}\},$$

where $\text{Supp}\{s(\mathscr{F}),L\}$ is the set of support points of $s(\mathscr{F})$ associated with L.

THEOREM 4.10. Let $B(H^p)=\{f\in H^p: \|f\|_p<1\}$, $\mathscr{F}$ be a compact subset of $B(H^p)$ and $z\in\mathscr{F}$. Then we have

$\{F\circ\phi\colon \phi\in \mathrm{Supp}B_0,\ F\in\mathscr{F}\}\subset \mathrm{Supps}(\mathscr{F})$.

Next, we give some results on multivalent functions.

Let p, q be integers and $1\leq q\leq p$. Denote by $S(p,q)$ the class of analytic functions $f(z)=z^q+a_{q+1}z^{q+1}+\ldots$ which satisfy

$$\mathrm{Re}(zf'(z)/f(z)) > 0 \quad (z\in D) \quad \text{and}$$

$$\int_0^{2\pi} \mathrm{Re}(zf'(z)/f(z))d\theta = 2p\pi \quad (z=re^{i\theta},\ p<r<1)$$

for some $p = p(f)$.

Defined $S(p,q)_{\mathbb{R}} = \{f\in S(p,q)\colon f$ has real coefficients$\}$.

We say $f(z)=z^q+\ldots\in K(p,q)$ if and only if $zf'(z)/q\in S(p,q)$.

If $f(z)=z^q+\ldots$ is analytic in D, and there exists a $g\in S(p,q)$ such that $\mathrm{Re}(e^{i\alpha}zf'(z)/g(z))>0$ $(p<|z|<1)$ for some $\alpha(0<\alpha\leq 2\pi)$ and ρ $(0<\rho<1)$, we write $f(z)\in C(p,q)$.

Hallenbeck and Livingstone [37] determined the closed convex hull and extreme points of the closed convex hull of above classes for $p=q$. Luo [36] also got the following

THEOREM 4.11. $\mathrm{Supp}S(p,p)=EHS(p,p)$, $\mathrm{Supp}K(p,p)=EHK(p,p)$, $\mathrm{Supp}S(p,p)_{\mathbb{R}}=EHS(p,p)_{\mathbb{R}}$, $\mathrm{Supp}C(p,p)\subset EHC(p,p)$.

Ma [38] investigated the extreme points of the class of analytic functions with positive real part and restricted second coefficient, which we omit here.

V. INTEGRAL MEANS AND COEFFICIENTS OF INVERSE FUNCTIONS

Let $f\in S$, $\phi(x)$ be nondecreasing and convex in

$(-\infty,+\infty)$, Baernstein [39] proved that

$$\int_{-\pi}^{\pi} \phi(\pm\log|f(re^{i\theta})|)d\theta \leq \int_{-\pi}^{\pi} \phi(\pm\log|k(re^{i\theta})|)d\theta \quad (0<r<1),$$

where $k(z)$ is the Koebe function. Brown [40] obtained for $f \in C$,

$$\int_{-\pi}^{\pi} \phi(\pm\log|f'(re^{i\theta})|)d\theta \leq \int_{-\pi}^{\pi} \phi(\pm\log|k'(re^{i\theta})|)d\theta \quad (0<r<1).$$

Ma [41] has extended these results to the classes of functions with k-fold symmetry.

Let $\Sigma(p)$, $0<p<1$, be the class of nonvanishing meromorphic univalent functions $f(z)$ in D with a simple pole at p and $f(0)=1$. Define $\Sigma(p,q)$, $0<p<1$ and $0<q<1$, to be the family of meromorphic univalent functions $f(z)$ with a simple pole at p and satisfying $f(0)=1$, $f(z_0)=0$ at some z_0 $(|z_0|=q)$.

In [41], the author got the following theorems.

THEOREM 5.1. Let $f(z)\in\Sigma(p)$, $F(w)$ be the inverse function of $f(z)$, and $K_p(w)$ be the inverse of $k_p(z)$. If $F(w) = A_1(w-1)+A_2(w-1)^2+\ldots$ and $K_p(w)=L_1(w-1)+L_2(w-1)^2+\ldots$ in the neighbourhood of $w=1$. Then we have sharp inequalities $|A_n|\leq|L_n|$ $(n=1,2,\ldots)$. If in the neighbourhood of $w=\infty$, $F(w)=p+B_1w^{-1}+\ldots$ and $K_p(w)=p+M_1w^{-1}+\ldots$, then $|B_n|\leq|M_n|$ $(n=1,2)$.

Similar results were given for the class $\Sigma(p,q)$.

THEOREM 5.2. Let $f(z)\in S_k$, $F(w)$ be the inverse function of $f(z)$ and $G_k(w)$ be the inverse of $g_k(z)=$

$z(1+z^k)^{-2/k}$. Suppose that $F(w)=w+C_{k+1}w^{k+1}+\ldots$ and $G_k(w)$ $=w+D_{k+1}w^{k+1}+\ldots$, then

$$|C_{nk+1}| \leqslant |D_{nk+1}| \quad (n=1,2,\ldots).$$

Let $S_o(b)=\{f(z)=1+\ldots$: $f(z)$ is analytic and univalent in D, $f(z)\neq 0$, $|f'(0)|=b\}$, where $0<b\leqslant 4$.

THEOREM 5.3. [42] Let $f(z)\in S_o(b)$, $F(w)=b_1(w-1)+\ldots$ be the inverse function of $f(z)$. If $G_b(w)=B_1(w-1)+\ldots$ is the inverse function of $g_b(z)=(1+(b-2)z+z^2)/(1-z)^2\in S_o(b)$. Then we get $|b_n|\leqslant|B_n|$ $(n=1,2,\ldots)$. This result is best possible.

Let I be the set of functions $f(z)=z+\ldots$ analytic and satisfying the condition $\mathrm{Re}f'(z)>0$ in D. Libera and Zlotkiewicz [46] gave the sharp estimates of coefficients of inverse functions for I. Zhang [47] got sharp coefficient bounds of Schwarz derivatives of inverse functions for I.

THEOREM 5.4. [47] Let $f(z)\in I$, $F(w)$ be the inverse function of f and the Schwarz derivative $\{F,w\}$ have the following form near 0.

$$\{F,w\} = a_o+a_1w+\ldots.$$

If $G(w)$ is the inverse of $g(z)=\int_o^z (1-t)/(1+t)dt$ and in the neighbourhood of $w=0$,

$$\{G,w\} = A_o+A_1w+\ldots.$$

Then

$$|a_n| \leqslant A_n \quad (n=0,1,\ldots).$$

The equality holds if and only if $F(w)$ is the rotations of $G(w)$.

REFERENCES

1. V.I. Milin, in Metric Questions in the Theory of Functions (edited by G.D. Suvorov; "Naukova Dumka": Kiev, 1980 (in Russian).

2. Rosanblat, Rev. Ci. Lima. 40 (1938), 177-179.

3. Ma Wancang, Coefficients of odd univalent functions (to appear).

4. Liu Shuqin, J. Northwest University (1980), No. 4, 15-22 (in Chinese).

5. J.E. Brown and A. Tsao, Math. Z. 191 (1986), 467-474.

6. Ma Wancang, On the Zalcman conjecture for close-to-convex functions (to appear).

7. W.E. Kirwan and G. Schober, Math. Z. 180 (1982), 19-40.

8. P.R. Garabedian and M. Schiffer, Ann. of Math. 61 (1955), 116-136.

9. Y.J. Leung and G. Schober, Proc. Amer. Math. Soc. 94 (1985), 659-664.

10. Ma Wancang, On the Kirwan conjecture for typically real meromorphic functions (to appear).

11. W. Fenchel, Preuss. Acad. Wiss. Phys. - Math. 22/23 (1931), 431-436.

12. W.E. Kirwan and G. Schober, J. Analyse Math. 30 (1977), 330-348.

13. Zhang Yulin, J. Northwest University (1983), No. 2, 52-55 (in Chinese).

14. Ma Wancang, Schiffer differential equation for meromorphic univalent functions and its applications (to appear).

15. Liu Shuqin, J. Northwest University (1980), No. 4, 14-22 (in Chinese).

16. J. Miller, Proc. Amer. Math. Soc. 31 (1972), 446-452.

17. Zhang Yulin, On the integral representation of the functions in the class $\Sigma^*(p,w_o)$ (to appear).

18. S.S. Miller, P.T. Mocanu and M.O. Reade, Proc. Amer. Math. Soc. 37 (1973), 553-554.

19. S.S. Miller, Proc. Amer. Math. Soc. 38 (1973), 311-313.

20. S.S. Miller, P.T. Mocanu and M.O. Reade, Mathematica, 20 (43) (1978), 25-30.

21. Liu Liquan, Acta Math. Sinica, 26 (1983), 179-186 (in Chinese).

22. Ma Wancang, Acta Math. Sinica, 29 (1986), 207-212 (in Chinese).

23. Ma Wancang, On α-convex functions of order β with k-fold symmetry (to appear).

24. O.P. Juneja and M.L. Mogra, Rev. Roum. Math. Pures et Appl. 13 (1978), 751-765.

25. Ma Wancang, Kexue Tongbao, 29 (1984), 1404-1405; Pure and Applied Math. 2 (1986), No. 1, 35-43 (in Chinese).

26. D.A. Brannan and J.G. Clunie, eds., Aspects of contemporary complex analysis, Academic Press, New York, 1980, pp. 137-146.

27. Ma Jinxi, On extremal problems of the class of analytic functions (to appear).

28. P.L. Duren, Y.J. Leung and M.M. Schiffer, Complex Variables: Theory and Application, 1 (1983), 267-277.

29. Ma Jinxi, Properties of support points of the class S (to appear).

30. K.F. Darth, D.A. Brannan and W.K. Hayman, Research Problems in Complex Analysis, Bull. London Math. Soc. 16 (1984), 490-517.

31. Ma Jinxi, Extreme points and support points of a class of meromorphic univalent functions (to appear).

32. Chen Hongbing, Boundary variations of univalent functions (to appear).

33. Ma Jinxi, On extreme problems of the class of α-convex functions of order B, Pure and Applied Math. (to appear)(in Chinese).

34. D.J. Hallenbeck and T.H. MacGregor, Trans. Amer. Math. Soc. 278 (1983), 523-546.

35. K. Hoffman, Banach Spaces of analytic functions, Prentice-Hall, Englewood Cliffs, N.J., 1962.

36. Luo Donghan, On the support points of the families of analytic functions (to appear).

37. D.J. Hallenbeck and A.E. Livingston, Trans. Amer. Math. Soc. 221 (1976), 339-359.

38. Ma Wancang, On extreme points of the class of analytic functions with positive real part and restricted second coefficient (to appear).

39. A. Baernstein, Acta Math., 133 (1974), 139-169.

40. J.E. Brown, Math. Z. 178 (1981), 353-358.

41. Ma Wancang, Pure and Applied Math. 1 (1985), No. 1, 84-91 (in Chinese).

42. Ma Wancang, On the coefficients of inverse of nonvanishing univalent functions (to appear).

43. G.M. Goluzin, Mat. Sb. 27 (69) (1950), 201-218 (in Russian).

44. Hu Jiagan, Acta Math. Sinica, 6 (1956), 651-664 (in Chinese).

45. Liu Shuqin, J. Northwest University (1980), No. 3, 8-10 (in Chinese).

46. R.J. Libera and E.J. Zlotkiewicz, Proc. Amer. Math. Soc. 92 (1984), 58-60.

47. Zhang Yulin, J. Northwest University (1987), No. 2 (to appear)(in Chinese).

48. M.S. Robertson, Bull. Amer. Math. Soc. 41 (1935), 565-572.

THE SUCCESSIVE COEFFICIENTS OF UNIVALENT FUNCTIONS

Hu Ke
Department of Mathematics
Jiangxi Normal University
Nanchang, China

Let S be the class of functions $f(z)$ which are analytic, univalent in the unit disk $|z| < 1$ and have Taylor expansion of the form $f(z) = z + \sum_{n=2}^{\infty} a_n z^n$. For $f(z) \in S$ and $p > 0$, let

$$\{\frac{f(z)}{z}\}^p = 1 + \sum_{n=1}^{\infty} D_n(p) z^n .$$

On the other hand, define the classes of functions:

$$S^* = \{f(z) | f(z) \in S \text{ and } \mathrm{Re}\{\frac{zf'(z)}{f(z)} > 0\}\},$$

$$S_c = \{f(z) | f(z) = z + \sum_{n=2}^{\infty} a_n z^n \text{ analytic in } |z| < 1, \text{ and there is a function } g(z) \in S^* \text{ and a real number } \alpha \text{ such that } \mathrm{Re}\{e^{i\alpha} z\frac{f'(z)}{g(z)}\} > 0\},$$

$$S(\alpha) = \{f(z) | f(z) \in S,\ \lim_{\rho \to 1}(1-\rho)^2 \max_{0 \le \theta \le 2\pi} |f(\rho e^{i\theta})| = \alpha > 0\}.$$

It is well known that $S^* \subset S_c \subset S$.

The purpose of the present paper is to give a brief

survey of the results obtained in the study of some problems concerning the coefficients $D_n(p)$ $(n=1,2,\ldots)$ for different values of p.

Problem 1. Find the best possible real number α such that the inequality

$$||D_n(p)| - |D_{n+1}(p)|| \leqslant An^{\frac{\alpha-1}{2}}, \quad (n=1,2,\ldots) \tag{1}$$

holds for $f \in S$, where A is an absolute constant.

This problem was first studied by Goluzin [1] and he showed

$$||D_n(1)| - |D_{n+1}(1)|| \leqslant An^{\frac{1}{4}}, \quad (n=1,2,\ldots) \tag{2}$$

$$||D_n(\tfrac{1}{2})| - |D_{n+1}(\tfrac{1}{2})|| \leqslant An^{-\frac{1}{4}}, \quad (n=1,2,\ldots) \tag{3}$$

In particular for $f \in S^*$, he obtained the result

$$||D_n(1)| - |D_{n+1}(1)|| \leqslant A, \quad (n=1,2,\ldots). \tag{4}$$

In 1963, Hayman [1] proved (4) for $f \in S$. In 1956, the author [1] proved for $f \in S^*$ and $p > 0$,

$$||D_n(p)| - |D_{n+1}(p)|| \leqslant An^{p-1}, \quad (n=1,2,\ldots). \tag{5}$$

In particular for $f(z) = z/(1-z^2) \in S$, both sides of (5) have the same order with respect to n. In 1985, the author [2] and Dong Xinhan proved the following theorem by different methods at the same time:

If $\quad f_k(z) = z + \sum_{n=1}^{\infty} D_n(\frac{1}{k}) z^{nk+1} \in S_c$, then $\left||D_n(\frac{1}{k})| - |D_{n+1}(\frac{1}{k})|\right| \leqslant An^{\frac{1}{k}-1}$, $(n=1,2,\ldots)$.

The case $k = 2$ was proved by Song Ruiya.

For $f \in S$, Lucas [1] obtained in 1969,

$$\left||D_n(p)|-|D_{n+1}(p)|\right| \leqslant An^{(t(p)-1)/2}, \ (n=1,2,\ldots), \qquad (6)$$

where $t(p) = (2\sqrt{p}-1)^2$ and $1/4 < p < 1$.

In 1984, the author [3] showed that we can use $T(p) = (4p-1)/(2p+t(p))$ instead of $t(p)$ in (6). It is clear $t(p) > T(p)$ for $1/4 < p < 1$, so our result is better than that of Lucas.

In the case $p = 1$, Hayman's result is precise. But Problem 1 remains open for general value of p. For $p = 1/2$, Duren and Hayman conjectured that $\alpha = 0$, in other words,

$$\left||D_n(\tfrac{1}{2})| - |D_{n+1}(\tfrac{1}{2})|\right| \leqslant An^{-\frac{1}{2}}, \ (n=1,2,\ldots). \qquad (7)$$

Concerning this problem, the best result known is $(\alpha-1)/2 = (T(1/2)-1)/2 = -0.42677$. But we can prove that if $f \in S$ (even if f is mean univalent), then

$$\left||D_n(\tfrac{1}{2})|^2 - |D_{n+1}(\tfrac{1}{2})|^2\right| \leqslant 2en^{-\frac{1}{2}}, \ (n=1,2,\ldots). \qquad (8)$$

This result was conjectured by P.L. Duren.

After the solution of Problem 1 for the case $p = 1$, it is natural to go still further to study the following.

Problem 2. Find the best possible numbers A, B such that

$$-B \leq d_n = |a_{n+1}| - |a_n| \leq A, \ (n=1,2,\ldots). \tag{9}$$

This problem is also interesting and is closely related to a problem of Littlewood. Let $f_2(z) = z + \sum_{n=1}^{\infty} b_{2n+1} z^{2n+1}$. Littlewood showed $|b_n| \leq A$ and conjectured $|b_n| \leq 1$. This conjecture is false, because someone has found a function for which $|b_5| = \frac{1}{2} + e^{-\frac{2}{3}} > 1$. Littlewood's problem is to find $\sup|b_n|$. The author [5] proved $|b_n| \leq 1.1305$, $(n=3,5,\ldots)$. In the particular case $a_{2m} = 0$, Problem 2 reduces to Littlewood's problem. So we can claim $A > 1$. Pommerenke, Milin and Ilina have studied this problem.

Grimspan showed $d_n \in (-2.97, 3.614)$. Ye Zhongqin improved this to $d_n \in (-2.943, 3.394)$. The author [4] obtained

$$-2.793 < |a_{n+1}| - |a_n| < 3.26. \tag{10}$$

Pommerenke made the conjecture that for $f \in S^*$,

$$\big||a_{n+1}| - |a_n|\big| \leq 1, \ (n=1,2,\ldots). \tag{11}$$

In 1978, Leung [1] proved that this conjecture is true.

Although we know $A > 1$, $B > 1$ in inequalities (9), we may ask the question: Is it true that for a fixed function f,

$$\overline{\lim_{n\to\infty}}\big||a_{n+1}| - |a_n|\big| \leq 1? \tag{12}$$

This question is completely settled by Hamilton [1] in 1982. This is a very important and beautiful result.

The author and Dong Xinhan [1] gave an alternative proof of Hamilton's theorem.

Problem 3. Study the asymptotic behavior of $||D_n(p)| - |D_{n+1}(p)||$ for functions $f \in S(\alpha)$.

Hayman [2] and Milin [1] showed that if $f \in S(\alpha)$, then

$$\lim_{n\to\infty} \frac{||D_n(p)|-|D_{n+1}(p)||}{d_n(2p-1)} = \alpha^p, \quad (\tfrac{3}{4} \leqslant p < 1) \tag{13}$$

where $d_n(h) = \dfrac{h(h+1)\ \dots\ (h+n-1)}{n!}$.

Whether the relation (13) is true or not for $\frac{1}{2} < p < \frac{3}{4}$ is still unknown. But we obtained [6].

Theorem 1. If $f \in S_c \cap S(\alpha)$, then the relation (13) holds for $p > 1/2$. The equality occurs if the function $f(z) = (z - z^2\cos\theta)/(1-e^{i\theta}z)^2$.

The author [7] improved the result of Milin [1] and proved the following inequality for $f \in S(\alpha)$:

$$||D_n(\tfrac{1}{2})| - |D_{n+1}(\tfrac{1}{2})|| \leqslant A_n^{-\frac{1}{2}-\frac{1}{300}} \tag{14}$$

where A depends only on α.

In [7] we also proved the following result:

Let $f(z) \in S(\alpha)$, then there exists a θ_o such that

$$f(z) = \frac{z}{(1-e^{i\theta_o}z)^2}\left\{\frac{1}{2\pi}\int_o^{2\pi}\frac{1-|z|^2}{1-2|z|\cos(\theta-t)+|z|^2}g(t)dt\right\}^{\frac{1}{p}},$$

$$(p > 0) \tag{15}$$

for some $g \in L[0,2\pi]$.

It is remarkable that Milan created Milan-Tauberian theorem, when he proved inequality (13). By this theorem we can prove not only (13) but also Hayman's coefficients regularity theorem [2]. We proved the following slightly more general theorem [7].

Theorem 2. Let $w(z) = \sum_{k=1}^{\infty} A_k z^k$, $\varphi(z) = e^{w(z)} = 1 + \sum_{k=1}^{\infty} D_k^{(1)} z^k$.

Let

$$h^{-p} \sum_{k=1}^{\infty} k^{p-1} |A_k|^p x^k = \log \frac{1}{1-x} + c + O(1), \quad (x \to 1-0)$$

where c is an absolute constant, and

$$\left|\frac{D_m^{(1)}}{d_m(h)}\right|^p - \left|\frac{D_n^{(1)}}{d_n(h)}\right|^p \to 0, \quad \text{as} \quad n \to \infty, \quad \frac{m}{n} \to 1,$$

then $\lim_{n\to\infty} \left|\frac{D_n^{(1)}}{d_n(h)}\right| \leqslant e^{c/p}$.

By Theorem 2, we can prove not only Hayman's regularity theorem and (12), but also Theorem 1. To prove either Theorem 1 or inequality (14), we need the following theorem of Bazilevich:

Let $f(z) \in S(\alpha)$ and $\log\frac{f(z)}{z} = 2 \sum_{n=1}^{\infty} \nu_n z^n$. Then

$$\sum_{k=1}^{\infty} k\left|\nu_k - \frac{1}{k}\right|^2 \leqslant \frac{1}{2} \log\frac{1}{\alpha}, \tag{16}$$

where $\theta_0 = 0$ is a direction of maximum growth for $f \in S$.

An interesting problem is to find generalizations of (15) and (16) and their applications. B.G. Ebe defined the class $S(\theta_1, \theta_2, \ldots, \theta_k)$ of $f(z) \in S$ as follows: There are $c>0$, $\delta>0$ and a sequence $r_n \to 1-0$, satisfying

$$\text{(i)} \quad (1-r_n)^{2/k} \left| f(r_n e^{i\theta_n^{(\nu)}}) \right| \geq c,$$

$$\text{(ii)} \quad \delta < \theta_n^{(\nu)} - \theta_n^{(\mu)} < 2\pi - \delta, \quad (1 < \nu < \mu \leq k).$$

The author and Dong Xinhan [1] improved (15) and (16) as follows:

Theorem 3. Let $f \in S(\theta_1, \theta_2, \ldots, \theta_k)$. Then there are k numbers $\theta_1, \theta_2, \ldots, \theta_k$ such that

$$\sum_{n=1}^{\infty} n \left| \gamma_n - \frac{1}{k} \sum_{\nu=1}^{k} e^{-in\theta_\nu} \right|^2 \leq \frac{1}{2} \log \frac{1}{c}$$

$$+ \frac{1}{2k^2} \sum_{\mu \neq \nu} \log \frac{2}{|e^{i\theta_\mu} - e^{i\theta_\nu}|^2}, \tag{17}$$

$$f(z) = z \prod_{\nu=1}^{k} (1-ze^{-i\theta_\nu})^{-\frac{2}{k}} \left\{ \frac{1}{2\pi} \int_0^{2\pi} \frac{1-r^2}{1-2r\cos(\theta-t)+r^2} g(t)\,dt \right\}^{\frac{1}{p}} \tag{18}$$

where $p > 0$, $z = re^{i\theta}$, $g \in L[0,2\pi]$.

From (16) we can immediately obtain the striking Hamilton's theorem: If $f(z) \in S(\theta_1, \theta_2)$, then

$$\varlimsup_{n \to \infty} \left| |a_{n+1}| - |a_n| \right| \leq 1.$$

Problem 4. Study the mean values of $|D_k(p)| - |D_{k+1}(p)|$ $(k=1,2,\ldots,n)$.

Problem 1 has not been solved completely for $1/4<p<1$ and $f \in S$. However the author got the following results:

(1) If $f \in S$, then

$$\frac{1}{n} \sum_{k=1}^{n} \frac{\left||D_k(p)|-|D_{k+1}(p)|\right|}{d_k(p)} \leqslant A, \ (0 < p < 1). \tag{19}$$

(2) If $f \in S(\alpha)$, then

$$\frac{1}{n} \sum_{k=1}^{n} \frac{|D_k(p)|-|D_{k+1}(p)|}{d_k(2p-1)} \to \alpha^p, \quad (p > \tfrac{1}{2}). \tag{20}$$

In the case $p = 1/2$, inequality (19) is due to Milin.

Remark. Recently we have proved that the inequalities (13), (15) and (20) also hold for mean univalent functions.

REFERENCES

G.M. Goluzin

1. Geometric theory of functions of complex variable, Moscow, 1952.

D.H. Hamilton

1. The successive coefficients of univalent functions, J. London Math. Soc., 25 (1982), 147-212.

Hu Ke

1. On the coefficients of starlike functions, J. Fudan University, (1956), 77-81.
2. On the coefficients of close-to-convex univalent functions, J. of Jiangxi Normal University, 4 (1986), 1-6.
3. On successive coefficients of univalent functions, Proc. AMS (1985), Vol. 95, No. 1, 37-41.
4. Adjacent coefficients of univalent functions (to

appear).

5. Coefficients of odd univalent functions, Proc. AMS (1986), Vol. 96, No. 1, 183-186.

6. Some theorems of analytic functions, P. of Math. Res. and Expos., 4 (1984), 1924.

7. Some properties of univalent functions, J. of Jiangxi Normal University, 3 (1986), 1-7.

8. Asymptotic behavior of analytic functions, J. of Jiangxi Normal University, 2 (1986), 1-4.

9. On a theorem of Milin, J. Math. Res. and Exp., (1986), No. 4, 77-80.

Hu Ke and Dong Xinhan

1. The asymptotic behavior of univalent functions, Proc. AMS (1987), Vol. 100 (1), 75-82.

W.K. Hayman

1. On successive coefficients of univalent functions, J. London Math. Soc., 38 (1963), 228-243.

2. The asymptotic behavior of p-valent functions, Proc. London Math. Soc., 5 (1955), 257-284.

K.W. Lucas

1. On successive coefficients of a really mean p-valent functions, J. London Math. Soc., 44 (1969), 631-642.

Y.J. Leung

1. Successive coefficients of starlike functions, Bull. London Math. Soc., 10 (1978), 193-196.

L.M. Milin

1. Univalent functions and Orthonormal Systems, "Nanka" Moscow, 1971, English Transl. in Transl. Math. Monographs, Vol. 49, AMS Providence R. 1 (1977), V. 1, Milin, Sib., Mz, Vol. 22, No. 2, 139-147.

ON THE EXISTENCE THEOREM OF QUASICONFORMAL HOMEOMORPHISMS

He Cheng-qi
Fudan University
Shanghai, China

ABSTRACT

Let $\mu(z)$ be a measurable function on the complex plane $\mathbb{C}$, and set

$$K(z) = \lim_{r\to 0} \operatorname{ess\,sup}_{|\zeta - z|<r} (1+|\mu(\zeta)|)/(1-|\mu(\zeta)|).$$

If $K(z)$ satisfies: (1) $K(z)$ is locally integrable in $\mathbb{C}$; (2) When $\delta \to 0$ or $\beta \to +\infty$, $\int_\delta^\beta \frac{dr}{r\int_0^{2\pi} K(z_0+re^{i\theta})d\theta}$ tends to infinity for every complex number z_0. Then there exists a homeomorphism $f(z)$ which is a generalized regular solution of the Beltrami equation

$$w_{\bar{z}}(z) - \mu(z)w_z(z) = 0.$$

If a homeomorphism $f_0(z)$ is a generalized regular solution of the equation, and $K\circ f_0^{-1}(w)$ is locally integrable in $\mathbb{C}$ then every generalized regular solution $f(z)$ of the equation can be represented as

$$f(z) = F\circ f_0(z),$$

where F is an entire function.

Let $f(z)$ be an orientation preserving homeomorphism of any domain D, and denote by $K(D)$ and $K(z)$ their maximal dilatation and locally maximal dilatation respectively. It is well known that [2] $K(z)$ is upper semi-continuous, i.e., $K(z) \geq \lim_{r\to 0} \sup_{|\zeta - z|<r} K(\zeta)$ and $K(D) = \sup_{z\in D} K(z)$.

If $K(D)$ is a finite number, $f(z)$ is a quasiconformal mapping. This restriction is essential. In this paper, we shall give up this restriction under certain conditions. We shall discuss the regularity of the homeomorphisms in §1, the compactness of the family of the homeomorphisms in §2, the corresponding generalized Beltrami equations in §3. We shall obtain the existence of homeomorphic solutions of the equations and the uniqueness of the solutions under certain conditions.

§1. REGULARITY

Let $f(z)$ be an orientation preserving homeomorphism of any domain D with $K(z)$ as its locally maximal dilatation. If $K(z)$ is locally integrable in D, i.e.,

$$\iint_G K(z)\,d\sigma_z < +\infty$$

holds for every Borel set G such that $\bar{G} \subset D$, then $f(z)$ will be called a quasiconformal homeomorphism as in [1].

Lemma 1. Suppose $f(z)$ is a quasiconformal homeomorphism, then $f(z)$ has ACL property. In other words, $f(z)$ has locally integrable generalized derivatives.

Proof. Set $R_d = R[a,b;\,c,d] = \{z = x+iy;\ a\leq x\leq b,\ c\leq y\leq d\}$, such that $\bar{R}_d \subset D$. Since $K(z)$ is integrable, $K(x+iy)$ is integrable with respect to x in $[a,b]$ for almost every $y \in [c,d]$. Set $A(t) = \text{mes } f\circ R_t$, $(c<t<d)$.

$A'(t)$ exists almost everywhere in $[c,d]$ because of the monotonicity of $A(t)$. Take $t_o \in [c,d]$ such that $A'(t_o)$ exists and $K(x+it_o)$ is integrable in $[a,b]$. It is obvious that the exceptional set e is of linear measure zero. We take $\delta > 0$, such that $t_o+\delta<d$ and consider a group of small intervals $[x_k', x_k'']$ $(k=1,2,...,n)$ in $[a,b]$ which don't intersect each other. Then redivide these intervals:

$$x_k' = x_{k_o} < x_{k_1} < \dots < x_{k_{n_k}} = x_k'' \text{ , } k=1,...,n.$$

We demand that $n(\delta)\to 0$ when $\delta\to 0$, where $n(\delta) = \underset{1\leq k\leq n}{\text{Max}} \underset{1\leq i\leq n_k}{\text{Max}} (x_{k_i}-x_{k_{i-1}})$. Set $R_{k_i} = R(x_{k_{i-1}}, x_{k_i}; t_o, t_o+\delta)$, and denote by $S_{k_i}(\delta)$ the distance between the two image curves of the vertical sides of R_{k_i}. It follows from the Rengel Inequality that

$$\frac{S_{k_i}^2(\delta)}{\text{mes } f\circ R_{k_i}} \leq \text{mod } f\circ R_{k_i} \tag{1.1}$$

As $K(R_{k_i}) = \sup\limits_{z\in R_{k_i}} K(z)$, we have

$$\begin{aligned} \text{mod } f\circ R_{k_i} &\leq K(R_{k_i}) \cdot \text{mod } R_{k_i} \\ &= K(R_{k_i}) \frac{x_{k_i} - x_{k_{i-1}}}{\delta} . \end{aligned} \tag{1.2}$$

Combine (1.1) with (1.2), and sum up on both sides,

$$\sum_{k=1}^{n}\sum_{i=1}^{n_k}\frac{S_{k_i}^2(\delta)}{\text{mes } f\circ R_{k_i}}$$

$$\leqslant \frac{1}{\delta}\sum_{k=1}^{n}\sum_{i=1}^{n_k} K(R_{k_i}).(x_{k_i}-x_{k_{i-1}}). \qquad (1.3)$$

By Schwarz Inequality,

$$(\sum_k\sum_i S_{k_i}(\delta))^2 \leqslant \sum_k\sum_i \frac{S_{k_i}^2(\delta)}{\text{mes } f\circ R_{k_i}} \cdot \sum_k\sum_i \text{mes } f\circ R_{k_i}.$$

Substituting (1.3) and $\sum_k^n\sum_i^n \text{mes } f\circ R_{k_i} \leqslant A(t_o+\delta)-A(t_o)$ into the above inequality, we obtain

$$(\sum_{k=1}^{n}\sum_{i=1}^{n} S_{k_i}(\delta))^2$$

$$\leqslant \frac{A(t_o+\delta)-A(t_o)}{\delta}\sum_{k=1}^{n}\sum_{i=1}^{n_k} K(R_{k_i})(x_{k_i}-x_{k_{i-1}}). \qquad (1.4)$$

As $K(z)$ is upper semi-continuous, for every $\varepsilon>0$, there exists $\delta_\varepsilon>0$ such that when $\delta<\delta_\varepsilon$,

$$K(R_{k_i}) = \sup_{z\in R_{k_i}} K(z) < K(x_{k_i}+it_o) + \varepsilon.$$

Substitute it into (1.4),

$$(\sum_{k=1}^{n}\sum_{i=1}^{n_k} S_{k_i}(\delta))^2$$

$$\leqslant \frac{A(t_o+\delta)-A(t_o)}{\delta}\sum_{k=1}^{n}\sum_{i=1}^{n_k} K(x_{k_i}+it_o)(x_{k_i}-x_{k_{i-1}})+\varepsilon(b-a).$$

As $\delta \to 0$, $\sum_{i=1}^{n_k} S_{k_i}(\delta)$ tends to the length of a curve, which is not less than the length of the corresponding chord, and

$$\frac{A(t_0+\delta) - A(t_0)}{\delta} \to A'(t_0),$$

$$\sum_{k=1}^{n} \sum_{i=1}^{n_k} K(x_{k_i}+it_0)(x_{k_i}-x_{K_{i-1}}) \to \sum_{k=1}^{n} \int_{x_K'}^{x_K''} K(x+it_0)dx.$$

Hence

$$\left(\sum_{k=1}^{n} |f(x_k'+it_0) - f(x_k''+it_0)|\right)^2$$

$$\leqslant A'(t_0). \sum_{k=1}^{n} \int_{(x_k',x_k'')} K(x+it_0)dx + \varepsilon(b-a).$$

Since ε is arbitrary, we have

$$\left(\sum_{k=1}^{n} |f(x_k'+it_0) - f(x_k''+it_0)|\right)^2$$

$$\leqslant A'(t_0) \cdot \int_{\sum_{k=1}^{n} (x_k',x_k'')} K(x+it_0)dx.$$

From the absolute continuity of the integral, we obtain the absolute continuity of $f(x+it_0)$ with respect to x in $[a,b]$, where $t_0 \in [c,d]\setminus e$, e is a set of linear measure zero. With the same reason, we can obtain the absolute continuity of $f(t_0+iy)$ with respect to y in $[c,d]$, which completes the proof of the lemma.

Let $f(z)$ be a quasiconformal homeomorphism in D, whose locally maximal dilatation is $K(z)$. Set

$$E = \{z;\ K(z) = +\infty\}. \tag{1.5}$$

Because $K(z)$ is integrable, E is a set of measure zero. We represent E as $E = \bigcap_{n=1}^{\infty} \{z;\ K(z) \geqslant n\}$. As $\{z;\ K(z) \geqslant n\}$ are closed sets, from the upper semi-continuity of $K(z)$, we know that E is a closed set of measure zero. Lemma 1 tells us that $f(z)$ has ACL property not only in $D\setminus E$, but also in D. It follows from the local quasiconformality of $f(z)$ in $D\setminus E$ that $f_z(z)$ and $f_{\bar{z}}(z)$ exists almost everywhere in D, and $f_z(z)$ does not vanish. Set

$$D(z) = \lim_{r\to o} \operatorname*{ess\,sup}_{|\zeta - z| < r} \frac{1+|\mu(\zeta)|}{1-|\mu(\zeta)|}, \tag{1.6}$$

where $\mu(z) = f_{\bar{z}}(z)/f_z(z)$ is the complex dilatation of $f(z)$. Referring to the proof in [2, P5], it is not difficult to obtain

$$K(z) = D(z). \tag{1.7}$$

As to the Jacobian of $f(z)$, $J_f(z) = |f_z(z)|^2 - |f_{\bar{z}}(z)|^2$, by a routine discussion, we have

$$\iint_e J_f(z)d\sigma_z \leqslant \text{mes } f\circ e \tag{1.8}$$

where e is an arbitrary Borel set. Generally speaking, the equality in (1.8) does not hold. But we have the following result.

<u>Lemma 2</u>. Let $f(z)$ be a quasiconformal homeomorphism, whose locally maximal dilatation is $K(z)$. If $K\circ f^{-1}(w)$ is locally integrable, then $f(z)$ has locally square integrable generalized derivatives, and

$$\iint_e J_f(z)d\sigma_z = \text{mes } f\circ e. \tag{1.9}$$

Proof. From (1.7), we know that

$$|f_z(z)|^2 + |f_{\bar{z}}(z)|^2 \leq (K(z) + 1/K(z)).J_f(z)$$

holds almost everywhere. To obtain the existence of locally square integrable generalized derivatives, we need only to prove that

$$\iint_R K(z)J_f(z)d\sigma_z < +\infty \tag{1.10}$$

holds for every rectangle R, $\bar{R}\subset D$.

Successively dividing R into two rectangles vertically and horizontally, we obtain 4^n small rectangles R_j, $j=1,\ldots,4^n$. Construct a function

$$K_n(z) = \inf_{z\in R_j} K(z), \ z\in R_j, \ j=1,2,\ldots,4^n.$$

It is obvious that $\{K_n(z)\}$ is an increasing sequence and

$$K(z) = \lim_{n\to\infty} K_n(z)$$

holds almost everywhere. For any fixed n, it follows from (1.8) that

$$\iint_R K_n(z)J_f(z)d\sigma_z = \sum_{j=1}^{4^n} K_n(z) \iint_{R_j} J_f(z)d\sigma_z$$

$$\leq \sum_{j=1}^{4^n} K_n(z) \iint_{f\circ R_j} d\sigma_w = \iint_{f\circ R} K_n\circ f^{-1}(w)d\sigma_w.$$

As $\{K_n\circ f^{-1}(w)\}$ is monotone increasing and

$\iint_{f\circ R} K_n\circ f^{-1}(w)d\sigma_w \leq \iint_{f\circ R} K\circ f^{-1}(w)d\sigma_w < +\infty$, we use Levi lemma and obtain $\iint_R K(z)J_f(z)d\sigma_z = \lim_{n\to\infty} \iint_R K_n(z)J_f(z)d\sigma_z \leq \iint_{f\circ R} K\circ f^{-1}(w)d\sigma_w$, which shows that (1.10) is proved. Then by a routine method, we can prove (1.9). The proof of the lemma is complete.

As a corollary, we have: If both the locally maximal dilatation $K(z)$ of a quasiconformal homeomorphism $f(z)$ and $K\circ f^{-1}(w)$ are locally integrable, then $f(z)$ has absolute continuity with respect to the two-dimensional measure, i.e., $w = f(z)$ maps sets of measure zero onto sets of measure zero.

§2. COMPACTNESS

Lemma 3. Let $f(z)$ be a quasiconformal homeomorphism in D, whose locally maximal dilatation is $K(z)$. Then the inequalities

$$\int_\delta^\beta \frac{dr}{r\int_0^{2\pi} K(z_0+re^{i\theta})d\theta} \leq \text{mod } f\circ R$$

$$\leq \left[\int_0^{2\pi} \frac{d\theta}{\int_\delta^\beta \frac{K(z_0+re^{i\theta})}{r}dr}\right]^{-1} \tag{2.1}$$

holds for every annulus $R = \{Z;\ \delta<|z-z_0|<\beta\}$, $\bar{R}\subset D$.

This result is known to us when $\sup_{z\in D} K(z) < +\infty$ [4]. Now we adopt a simpler method to prove it under weaker conditions.

Proof. Denote by $\Gamma = \{\nu\}$ the family of closed curves which separate the two boundary components of $f\circ R$.

Obviously the images of the circles $\{|z-z_0| = r;\ \delta<r<\beta\}$ under the mapping $f(z)$ form a sub-family of Γ, and we denote it by $\{\nu_r\}$. Suppose $\rho(z)$ is proper and note the ACL property of $f(z)$, we have

$$L_{\nu_r}(\rho) = \int_{\nu_r} \rho(w)|dw|$$

$$= \int_{|z-z_0|=r} \rho\circ f(z)\cdot|f_z(z)dz+f_{\bar z}(z)d\bar z|,$$

where $z = z_0+re^{i\theta}$, $\delta<r<\beta$, $0\leqslant\theta\leqslant 2\pi$. Thus

$$L_{\nu_r}(\rho) \leqslant \int_0^{2\pi} \rho\circ f(z)(|f_z(z)| + |f_{\bar z}(z)|)rd\theta.$$

Since $|f_z(z)|+|f_{\bar z}(z)| = K^{\frac{1}{2}}(z)J_f^{\frac{1}{2}}(z)$ by (1.7), using Schwarz inequality, we obtain

$$L^2_{\nu_r}(\rho) \leqslant \int_0^{2\pi} \rho^2\circ f(z)J_f(z)d\sigma_z\cdot\int_0^{2\pi} K(z)rd\theta.$$

It follows that $\iint_R \rho^2\circ f(z)J_f(z)d\sigma_z \leqslant \iint_{f\circ R} \rho^2(w)d\sigma_w = A(\rho)$. Thus we have

$$\inf_\nu L^2_\nu(\rho)\cdot\int_\delta^\beta \frac{dr}{r\int_0^{2\pi} K(z_0+re^{i\theta})d\theta}$$

$$\leqslant \iint_R \rho^2\circ f(z)J_f(z)d\sigma_z \leqslant A(\rho).$$

Hence $\lambda_\Gamma = \sup_\rho \dfrac{\inf_\nu L^2_\nu(\rho)}{A(\rho)} \leqslant \left[\int_\delta^\beta \dfrac{dr}{r\int_0^{2\pi} K(z_0+re^{i\theta})d\theta}\right]^{-1}.$

As $1/\lambda_\Gamma$ = mod $f \circ R$, we obtain the left side inequality of (2.1)

$$\int_\delta^\beta \frac{dr}{r\int_0^{2\pi} K(z_0+re^{i\theta})d\theta} \leqslant \text{mod } f\circ R.$$

Consider the image curves of the radii of R under the mapping $f(z)$ with a similar method, we can obtain the right side inequality of (2.1)

$$\text{mod } f\circ R \leqslant \left[\int_0^{2\pi} \frac{d\theta}{\int_\delta^\beta \frac{K(z_0+re^{i\theta})}{r}dr}\right]^{-1}.$$

The proof is complete.

It is not difficult to know, by a careful analysis of the proof, that there exists another group of estimations under the conditions of the lemma. Set $R = \{x+iy, 0\leqslant x\leqslant a, 0\leqslant y\leqslant b\}$, $R\subset D$, then we have

$$\int_0^a \frac{dx}{\int_0^b K(x+iy)dy} \leqslant \text{mod } f\circ R \leqslant \left[\int_0^b \frac{dy}{\int_0^a K(x+iy)dx}\right]^{-1}.$$

<u>Lemma 4</u>. Let $\{f_n\}$ be a sequence of orientation preserving homeomorphisms of the complex plane $\mathbb{C}$ onto itself, which satisfy the normal conditions $f_n(0)=0$, $f_n(1)=1$, and $f_n(\infty)=\infty$. Suppose $K_n(z)$ are the locally maximal dilatations of $f_n(z)$. If there exists a function $K(z)$, which fulfils the following two conditions:

(1) $K_n(z) \leqslant K(z)$ for all n and z, and $K(z)$ is locally integrable,

(2) When $\delta\to 0$ or $\beta\to+\infty$, $\int_\delta^\beta \frac{dr}{r\int_0^{2\pi} K(z_0+re^{i\theta})d\theta}$ tends to ∞ for all z_0.

Then every subsequence of $\{f_n(z)\}$ has a locally uniformly convergent subsequence, whose limit is also an orientation preserving homeomorphism of $\mathbb{C}$ onto itself.

Proof. For any complex number z_0 and $0<\beta<1$, there must be one (denote it by e) of the two points 0 and 1, which lies outside $\{|z-z_0|<\beta\}$. Suppose $R_\delta = \{z;\ \delta<|z-z_0|<\beta\}$, where $0<\delta<\beta$. For any z_1 which lies in $\{|z-z_0|<\delta\}$, $f_n(z_0)$, $f_n(z_1)$ and e, ∞ belong to the two components of the complement of $f_n\circ R_\delta$ respectively. Let $S(a,b)$ be the spherical distance between a and b, and set $\eta_n = \operatorname{Min}\{S(e,\infty),\ S(f_n(z_0)),\ f_n(z_1))\}$. By a theorem in [3, P36], we have the following estimation

$$\operatorname{mod} f_n\circ R_\delta \leqslant \pi^2/2\eta_n^2. \tag{2.2}$$

As $K_n(z)\leqslant K(z)$, and $K(z)$ is locally integrable, we can apply lemma 3 to have

$$\int_\delta^\beta \frac{dr}{r\int_0^{2\pi} K(z_0+re^{i\theta})d\theta} \leqslant \operatorname{mod} f_n\circ R_\delta. \tag{2.3}$$

Combine (2.2) with (2.3), then

$$\eta_n \leqslant \pi/\sqrt{2}\left[\int_\delta^\beta \frac{dr}{r\int_0^{2\pi} K(z_0+re^{i\theta})d\theta}\right]^{-1} \tag{2.4}$$

for every n. We know that, according to condition (2), the right side of (2.4) tends to zero when $\delta\to 0$. Hence

$\eta_n = S(f_n(z_0), f_n(z_1))$ at that time. Noting that the right side of (2.4) is an infinitely small quantity which does not depend on n, we obtain the equicontinuity of $\{f_n(z)\}$ near z_0.

If $z_0 = \infty$, we change the symbols in the above proof properly, and apply the condition (2) that

$\int_\delta^\beta \frac{dr}{r\int_0^{2\pi} K(z_0+re^{i\theta})d\theta}$ tends to infinity when $\beta \to +\infty$. Thus we can also prove that $\{f_n(z)\}$ have equicontinuity near $z_0=\infty$. It follows from the Arzera-Ascoli theorem that there is a subsequence in $\{f_n(z)\}$ which converges locally uniformly. Let the limit mapping be $f(z)$.

Obviously, $f(z)$ is a continuous mapping of $\mathbb{C}$. It is also not difficult to prove that $f(z)$ is injective. In fact, for arbitrary z_0 and r_1, r_2, $0<r_1<r_2<+\infty$, we set $R = \{z;\ r_1<|z-z_0|<r_2\}$. Applying Lemma 3, we can derive the following estimation

$$\int_{r_1}^{r_2} \frac{dr}{r\int_0^{2\pi} K_n(z_0+re^{i\theta})d\theta} \leqslant \operatorname{mod} f_n\circ R.$$

By condition (1), there exists a positive number

$$\varepsilon = \int_{r_2}^{r_1} \frac{dr}{r\int_0^{2\pi} K(z_0+re^{i\theta})d\theta},$$

such that $0<\varepsilon\leqslant\operatorname{mod} f_n\circ R$ for every n. According to Lemma 4 of [1], we know that the limit mapping $f(z)$ is injective. Now we only need to prove that the image of $f(z)$ is $\mathbb{C}$ itself.

For any annulus $R_\beta = \{z;\ 1<|z|<\beta\}$, we have the estimation

$$\int_1^\beta \frac{dr}{r\int_0^{2\pi} K_n(re^{i\theta})d\theta} \leqslant \text{mod } f_n \circ R_\beta .$$

By the stability of the module, mod $f_n \circ R_\beta$ tends to mod $f \circ R_\beta$ when $n \to \infty$. Thinking of condition (1), we have

$$\int_1^\beta \frac{dr}{r\int_0^{2\pi} K(re^{i\theta})d\theta} \leqslant \text{mod } f \circ R_\beta .$$

The left side of the above inequality tends to infinity as $\beta \to +\infty$. Thus we obtain

$$\lim_{\beta\to\infty} \text{mod } f \circ R_\beta = +\infty,$$

which shows that $f(\infty)=\infty$. Hence the image of $\mathbb{C}$ under the mapping $f(z)$ is $\mathbb{C}$, and the proof is completed.

§3. THE EXISTENCE AND THE UNIQUENESS

Suppose $f(z)$ is a quasiconformal homeomorphism. By Lemma 1 we know that $f(z)$ has locally integrable generalized derivatives and $f(z)$ is locally quasiconformal everywhere except for a closed set of measure zero. Hence a quasiconformal homeomorphism is a generalized regular solution of a generalized Beltrami equation

$$w_{\bar z}(z) - \mu(z)w_z(z) = 0$$

where $\mu(z) = f_{\bar z}(z)/f_z(z)$ is the complex dilatation of f(z). We know that $K(z) = \lim_{r\to 0} \operatorname{ess\,sup}_{|\zeta - z|<r} \frac{1+|\mu(\zeta)|}{1-|\mu(\zeta)|}$ is the local maximal dilatation of $f(z)$ and it is only locally integrable.

If $K\circ f^{-1}(w)$ is locally integrable too, if follows from Lemma 2 that the quasiconformal homeomorphism is a generalized regular solution of the Beltrami equation, which has locally square integrable generalized derivatives. Hence this solution is absolutely continuous with respect to the two-dimensional measure.

What we are concerned about now is the converse problem, i.e., whether there exist generalized regular solutions for these generalized Beltrami equations, and under what conditions the solutions are unique. The following theorem is the main result of this paper.

<u>Theorem</u>. Let $\mu(z)$ be a measurable function on the complex plane $\mathbb{C}$, and set

$$K(z) = \lim_{r\to 0} \operatorname{ess\,sup}_{|\zeta - z|<r} \frac{1+|\mu(\zeta)|}{1-|\mu(\zeta)|}. \tag{3.1}$$

If $K(z)$ is upper semi-continuous, it fulfils the following conditions

(1) $K(z)$ is locally integrable in $\mathbb{C}$;

(2) When $\delta\to 0$ or $\beta\to+\infty$, $\int_\delta^\beta \frac{dr}{r\int_0^{2\pi} K(z_0+re^{i\theta})d\theta}$ tends to infinity for every complex number z_0.

Then there exists a homeomorphism $f(z)$ which is the generalized regular solution of the generalized Beltrami equation

$$w_{\bar{z}}(z) - \mu(z)w_z(z) = 0. \tag{3.2}$$

In other words, the homeomorphism $f(z)$ has locally integrable generalized derivatives $f_{\bar{z}}$ and f_z which satisfy equation (3.2).

If homeomorphism f_o is a generalized regular solution of (3.2), and $K \circ f_o^{-1}(w)$ is locally integrable on $\mathbb{C}$, then every generalized regular solution $f(z)$ of the equation has the representation

$$f(z) = F \circ f_o(z) \tag{3.3}$$

where F is an entire function.

For the existence problem of the theorem, Tang Tong-gao made some researchers [5], the systematic approach was due to O. Lehto [1]. But they did not touch the regularity as well as the uniqueness of the solutions. Certainly, the uniqueness problem remains to be solved thoroughly.

Proof. With the notation (1.5), $E = \{z; K(z) = +\infty\}$ is a closed set of measure zero. Its complement E^c can be approximated from inside by a series of open sets $\{G_n\}$. We define $\mu_n(z)$ which equals $\mu(z)$ on G_n and equals zero outside G_n. Then the Beltrami equation with $\mu_n(z)$ as its coefficient has a univalent generalized regular solution, which we denote by $f_n(z)$. It is known that $f_n(z)$ is a quasiconformal map and the image of $\mathbb{C}$ under the mapping $w = f_n(z)$ is $\mathbb{C}$ itself. We may assume that $0, 1, \infty$ are its fixed points. Because $\mu_n(z) \leqslant \mu(z)$, the dilatation $K_n(z)$ defined as (3.1) must satisfy the inequality $K_n(z) \leqslant K(z)$. Since $K(z)$ has properties (1) and (2), it follows from Lemma 4 that there must be in $\{f_n(z)\}$ a subsequence which converges locally uniformly, and its limit is a homeomorphism of $\mathbb{C}$ onto itself, which we denoted by $f(z)$.

For any $z \in E^c$, there exists a neighborhood $U(z)$

such that $U(z) \subset G_n$ when n is sufficiently large. Hence $K_n(z) = K(z)$ in $U(z)$ and the locally maximum dilatation of $f(z)$ in $U(z)$ is $K(z)$. As $K(z)$ is locally integrable in $\mathbb{C}$, by Lemma 1, $f(z)$ has generalized derivatives f_z and $f_{\bar{z}}$ which are locally integrable in $\mathbb{C}$. It is obvious that they satisfy the Beltrami equation (3.2) almost everywhere in $\mathbb{C}$, which completes the existence proof.

One fact should be noticed before proving the uniqueness: If homeomorphism $f(z)$ is a generalized regular solution of equation (3.2) then the image of the complex plane $\mathbb{C}$ under $w = f(z)$ is $\mathbb{C}$ itself. In fact, $f(z)$ is a quasiconformal homeomorphism then. Hence for every annulus $R_\beta = \{z;\ 1<|z|<\beta\}$, we have the following estimation

$$\int_1^\beta \frac{dr}{r\int_0^{2\pi} K(re^{i\theta})d\theta} \leqslant \operatorname{mod} f\circ R_\beta .$$

Using the condition (2) of $K(z)$, we obtain $\lim\limits_{\beta\to+\infty} \operatorname{mod} f\circ R_\beta = +\infty$, which shows that $f(z)$ maps $\mathbb{C}$ onto itself.

Now we turn to the discussion of the uniqueness. It follows from Lemma 2 that $w = f_0(z)$ is absolutely continuous with respect to the 2-dimensional measure, and the image $\tilde{E}$ of E under $f_0(z)$ is also a closed set of measure zero. We now discuss the composite mapping

$$f\circ f_0^{-1}(w). \tag{3.4}$$

Because $f(z)$ is a quasiconformal homeomorphism of $\mathbb{C}$ onto itself, and $f(z)$, $f_0(z)$ have the same complex dilatation $\mu(z)$ in the neighborhood of every point in

E^c, $f\circ f_0^{-1}(z)$ is locally conformal everywhere in $\tilde{E}^c$. Hence the corresponding locally maximal dilatation is almost everywhere equal to 1. Since the locally maximal dilatation of $f\circ f_0^{-1}(w)$ is locally integrable, by Lemma 1, $f\circ f_0^{-1}(w)$ must have locally integrable generalized derivatives, which satisfy the Cauchy-Riemann equations almost everywhere in the complex plane. Thus we prove that $f\circ f_0^{-1}(w)$ is an analytic function which we denote by $F(w)$. Hence

$$f(z) = F\circ f_0(z).$$

If we demand that $f(z)$ is homeomorphic and has three fixed points, then $F(w)$ degenerates into the identical mapping. Hence $f(z) = f_0(z)$, which completes the proof of uniqueness. The theorem is then proved.

REFERENCES

(1) O. Lehto, Homeomorphism with a given dilatation, Proceedings of the 15th Scandinavian Congress, Oslo, 1969, 58-73.

(2) O. Lehto, On the existence of q.c. mappings, Ann. Acad. Sci. Fenn. 274, 1960, 1-23.

(3) O. Lehto and K.I. Virtanen, Quasikonforme Abbildungen, Springer-Verlag, Berlin-Eidelberg-New York, 1965.

(4) He Cheng-qi, Distortion theorem of the module for quasiconformal mappings, Acta. Math. Sinica. Vol. 15, No. 4 (1965).

(5) Tang Tong-gao, The existence theorem of the homeomorphic regular solutions for the degenerate Beltrami equation systems, Fudan Journ. Vol. 10, No. 2-3 (1965).

MODULI FOR RIEMANN SURFACES*

Irwin Kra
State University of New York
Stony Brook, New York 11794
USA

Let (p,n) be two non-negative integers with

$$2p - 2 + n > 0.$$

Let $T(p,n)$ be the Teichmüller space of compact Riemann surfaces of genus p with n distinguished points (usually viewed as punctures). Starting with the fundamental work of Maskit [M1], [M2], we constructed in [K] intrinsic coordinates for $T(p,n)$. In this note we discuss some properties of these coordinates and outline how to use these intrinsic coordinates to construct the deformation spaces used by Bers [B1], [B2] in his work on compactified moduli space. Our constructions are in the same spirit as those of Bers [B1], [B2] and Earle-Marden [EM], but differ significantly in technical aspects. Details will appear elsewhere.

Let S be a (topological) surface of finite type (p,n). Let $\Sigma = \{a_1,\ldots,a_{3p-3+n}\}$ be a maximal partition of S.

* Research partially supported by NSF grant DMS8701774.

THEOREM 1. <u>The partition</u> Σ <u>canonically determines global coordinates</u> $\tau_1, \ldots, \tau_{3p-3+n}$ <u>on the Teichmüller space</u> $T(p,n)$ <u>with the following properties</u>:

(a) <u>For each</u> $j = 1,2,\ldots,3p-3-n$, <u>we have</u>

$$\operatorname{Im} \tau_j > \frac{1}{2} .$$

(b) <u>The image</u> $\tau(T(p,n))$ <u>contains the product of half planes</u>

$$\{\tau \in \mathbb{C}^{3p-3+n};\ \operatorname{Im} \tau_j > 2\} .$$

(c) <u>The Dehn twist about the curve</u> a_k <u>is represented in these coordinates by</u>

$$\tau_j \to \tau_j \qquad \text{for } j \neq k,$$

$$\tau_k \to \tau_k + 2 .$$

(d) <u>Let</u> $c : [0,1] \longrightarrow T(p,n)$ <u>be a continuous path.</u> <u>Let</u> $\ell_k(s)$ <u>be the hyperbolic length on</u> $c(s)$ <u>of the unique geodesic freely homotopic to the curve</u> a_k. <u>Then</u>

$$\lim_{s\to 1} \operatorname{Im} \tau_k(s) = \infty$$

<u>whenever</u>

$$\lim_{s\to 1} \ell_k(s) = 0 .$$

(Here $\tau_k(s)$ is the k-th component of $c(s)$.)

Before describing a converse to part (d) of the above theorem, we discuss the nature of our coordinates and introduce some <u>forgetful</u> maps.

Every point $\tau \in T(p,n)$ is represented by a regular torsion free terminal b-group $\Gamma(\tau)$ of type (p,n) in the following sense. The group $\Gamma(\tau)$ has a simply connected invariant component $\Delta(\tau)$. The Riemann surface corresponding to the point $\tau \in T(p,n)$ is $S(\tau) = \Delta(\tau)/\Gamma(\tau)$. The natural projection $\pi(\tau)$: $\Delta(\tau) \longrightarrow S(\tau)$ has the property that for $j = 1,2,\ldots,3p-3+n$, every component of $\pi(\tau)^{-1}(a_j)$ is an open Jordan curve in $\Delta(\tau)$ precisely invariant under a cyclic accidental parabolic subgroup of $\Gamma(\tau)$. To each j, we can assign a modular subgroup $G_j(\tau)$ of $\Gamma(\tau)$. The modular subgroup is again a regular terminal b-group. It is of type (1,1) or (0,4). The coordinate τ_j is (essentially) the trace of an element of $G_j(\tau)$ in the (1,1) case, and the cross ratio of four fixed points of parabolic elements of $G_j(\tau)$ in the (0,4) case.

The domain $\Delta(\tau)$ and the groups $\Gamma(\tau)$ can, of course, be chosen to vary holomorphically with $\tau \in T(p,n)$. One can hence construct a fiber space

$$\pi : V(p,n) \longrightarrow T(p,n),$$

whose fiber of the point $\tau \in T(p,n)$ is the surface $S(\tau)$. This is the punctured Teichmüller curve (see, for example, Earle-Kra [EK]).

Let J be a subset of $\{1,2,\ldots,3p-3+n) = \mathbb{Z}_{3p-3+n}$. Consider the surface with nodes obtained by shrinking each curve a_j with $j \in J$ to a point. Call the resulting surface S_J. The curves a_k, $k \in \mathbb{Z}_{3p-3+n} - J$ form a maximal partition on each of the parts of S_J. Note that each part of S_J is of hyperbolic type. Let us denote by $T_J(p,n)$ the subspace of $T(p,n)$ obtained by forgetting the coordinates τ_j with $j \in J$. We thus have a well defined holomorphic surjection.

$$\rho_J : T(p,n) \longrightarrow T_J(p,n) .$$

As a consequence of an isomorphism theorem due to Maskit [M3], we can show that

$$T_J(p,n) \approx \{\text{ the product of the Teichmüller spaces of the parts of } S_J\} .$$

The coordinates on each of the factors of $T_J(p,n)$ are, of course, the canonical coordinates described by Theorem 1 for the corresponding parts of S_J.

It is easy to see that

$$\dim T_J(p,n) = 3p-3+n - |J| ,$$

where $|J|$ is the cardinality of the set J.

PROBLEM 1. Determine all holomorphic sections s of the <u>forgetful</u> map ρ_J; that is, all holomorphic maps $s : T_J(p,n) \longrightarrow T(p,n)$ with $\rho_J . s$ = the identity on $T_J(p,n)$.

PROBLEM 2. Describe the fibers of ρ_J; that is, the preimages of points. For $|J| = 1$, are the fibers always connected?

We are now ready to state the converse to Theorem 1.d.

THEOREM 2. <u>With the hypothesis of</u> Theorem 1, <u>also assume that</u>

$$\lim_{s\to 1} \operatorname{Im} \tau_k(s) = \infty, \ k \in J \ ,$$

<u>and that</u>

$$\lim_{s\to 1} \rho_J(c(s))$$

<u>exists</u> (<u>in</u> $T_J(p,n)$). <u>Then</u>

$$\lim_{s\to 1} \ell_k(s) = 0, \ \underline{\text{all}} \ \ k \in J \ .$$

We proceed to describe $D(p,n;\ \Sigma)$, the <u>strong deformation space</u> (as defined by Bers [B2]) corresponding to an <u>arbitrary</u> partition Σ on S. Assume that the partition Σ consists of K curves, $0 \leq K \leq 3p-3+n$:

$$\Sigma = \{a_1,\ldots,a_K\} \ .$$

Complete Σ to a maximal partition Σ_o on S :

$$\Sigma_o = \{a_1,\ldots,a_{3p-3+n}\} \ .$$

We work with the coordinates on $T(p,n)$ determined by Σ_o. We factor the Teichmüller space $T(p,n)$ by $\mathbb{Z}_K$ (viewed as the (abelian) group of automorphisms generated by the K Dehn twists about the curves in Σ (see Theorem 1.c). The factor space

$$T(p,n)/\mathbb{Z}_K = \tilde{D}(p,n;\ \Sigma)$$

embedds in $\Delta_*^K \times U^{3p-3+n-K}$ by

$$\tau_j \to t_j = e^{\pi i \tau_j}, \ \ j = 1,\ldots,K,$$

$$\tau_j \to \tau_j, \ \ j = K+1,\ldots,3p-3+n$$

(here

$$\Delta = \text{unit disk} = \{t \in \mathbb{C} : |t| < 1\},$$
$$\Delta^* = \Delta - \{0\},$$
$$U = \text{the upper half plane } \{\tau \in \mathbb{C} ; \operatorname{Im} \tau > \tfrac{1}{2}\}).$$

We need to add certain boundary points to $\tilde{D}(p,n;\Sigma)$ to obtain $D(p,n;\Sigma)$. For each subset J of $\mathbb{Z}_K$, we let the Dehn twists about the curves a_k, $k \in \mathbb{Z}_K - J$, act on $T_J(p,n)$ and view the factor space $D_J(p,n)$ as a subset of $\Delta^K \times U^{3p-3+n-K}$, where the coordinates t_j, $j \in J$, are zero. We define

$$D(p,n;\Sigma) = \bigcup_{J \subset \mathbb{Z}_K} D_J(p,n).$$

Note that (here ϕ is the empty set)

$$D_\phi(p,n) = \tilde{D}(p,n;\Sigma)$$
$$D_{\mathbb{Z}_K}(p,n) \cong T_{\mathbb{Z}_K}(p,n).$$

In particular, if $K = 3p-3-n$, then the space $D_{\mathbb{Z}_K}(p,n)$ consists of a single point (the origin in Δ^{3p-3+n}). Note also that

$$D(p,n,\phi) = T(p,n).$$

THEOREM 3. <u>The strong deformation space</u> $D(p,n;\Sigma)$ <u>is a domain of holomorphy.</u>

It is possible to construct a fiber space $V(p,n;\Sigma)$ over $D(p,n;\Sigma)$ whose fiber over a point $t \in D(p,n;\Sigma)$ is the Riemann surface (possibly with nodes) represented by t. The properties of this fibration as well as proofs of the theorems in this paper will appear elsewhere.

REFERENCES

[B1] L. Bers, On spaces of Riemann surfaces with nodes, Bull. Amer. Math. Soc., 80 (1974), 1219-1222.

[B2] ________, Finite dimensional Teichmüller spaces and generalizations, Bull. Amer. Math. Soc., 5 (1981), 131-172.

[EK] C. J. Earle and I. Kra, On sections of some holomorphic families of closed Riemann surfaces, Acta Math., 137 (1976), 49-79.

[EM] C. J. Earle and A. Marden, Geometric complex coordinates for Teichmüller space (to appear).

[K] I. Kra, Non-variational global coordinates for Teichmüller spaces, MSRI Proceedings of Workshop on Geometric Function Theory (to appear).

[M1] B. Maskit, Moduli of marked Riemann surfaces, Bull. Amer. Math. Soc., 80 (1974), 773-777.

[M2] ________, On the classification of Kleinian groups: I - Koebe groups, Acta Math. 135 (1975), 249-270.

[M3] ________, Isomorphisms of function groups, J. d'Analyse Math., 32 (1977), 63-82.

SOME RESULTS ON ANALYTIC MAPPINGS BETWEEN TWO ULTRAHYPERELLIPTIC SURFACES

Kiyoshi Niino
Faculty of Technology
Kanazawa University
2-40-20, Kodatsuno
Kanazawa 920
Japan

In 1925 R. Nevanlinna [11] established the value distribution theory of meromorphic functions $f:\mathbb{C} \to P_1\mathbb{C}$. He proved so-called the first fundamental theorem, the second fundamental theorem and the defect relation. His theory is a generalization of Picard-Borel theorem.

After Nevanlinna there are many extensions of his theory. The first is extension to algebroid functions $f:A \to P_1\mathbb{C}$, where A is the proper existence domain of the algebroid function, by H. L. Selberg [19](cf. [20]), E. Ullrich [21] and G. Valiron [22].

The second is one to the system of entire functions $f:\mathbb{C} \to f = (f_o(z), f_1(z),..., f_n(z))$ by H. Cartan [4]. By the results of Valiron [22] the case of algebroid functions is similar to the case of systems of entire functions.

The third is one to holomorphic curves $x:\mathbb{C} \to P_n\mathbb{C}$. This study was initiated by H. and J. Weyl [24] and its second fundamental theorem like Nevanlinna's was proved by L. V. Ahlfors [1] (cf. [23], [25]). This is essentially same as systems of entire functions. However at the same time they considered the associated holomorphic curves of rank k $(1 \le k \le n-2)$ ${}_kx:\mathbb{C} \to G(n,k) \subseteq P_{\ell(k)-1}\mathbb{C}$ $\left(\ell(k) = \binom{n+1}{k+1}\right)$ of holomorphic curve $x:\mathbb{C} \to P_n\mathbb{C}$.

The fourth is one to analytic mappings between two Riemann surfaces $\phi: R \to S$, where R is an open Riemann surface, by L. Sario (cf. [18]). These are main cases of one complex variable. It was also extended to the case of several complex variables.

We now pay attention to analytic mapping $\phi: R \to S$ between two Riemann surfaces. Sario established the second fundamental theorem like Nevanlinna's. When S is a closed Riemann surface, he obtained a defect relation like Nevanlinna's from the second fundamental theorem. However if S is open and of infinite genus, then a defect relation could not be obtained from the second fundamental theorem. This was shown by Rodin-Sario [17]. It seems to me from the result of Rodin-Sario and our Theorem 4 that when S is of infinite genus, we shall not be able to have a defect relation like Nevanlinna's, that is, we shall not be able to establish Nevanlinna theory when S is of infinite genus. So we have to try another treatment.

Now I shall talk here the study on analytic mappings created by M. Ozawa. First of all we have a problem: <u>when do there exist any non-constant analytic mappings of a given R into a given S?</u>

If $R \notin O_{AB}$, then there are many non-trivial analytic mappings. So we may assume $R \in O_{AB}$, that is, there is no bounded analytic function on R.

We have a criterion of non-existence of analytic mappings between two Riemann surfaces. In order to state the criterion, we now introduce Picard constant $P(R)$ of R. Let R be an open Riemann surface. Let $\mathfrak{M}(R)$ be the family of non-constant meromorphic functions on R. Let $P(f)$ be the number of values which are not taken by $f \in \mathfrak{M}(R)$. We define

$$P(R) = \sup\{P(f);\ f \in \mathfrak{M}(R)\}.$$

Then Ozawa gave the following criterion:

THEOREM 1 (Ozawa [14]). *If* $P(R) < P(S)$, *then there is no non-constant analytic mapping of* R *into* S.

Proof. Contrarily suppose that there is an analytic mapping φ of R into S. For any $f \in \mathfrak{M}(S)$, $f(\varphi) \in \mathfrak{M}(R)$ and we have $P(f) \leq P(f(\varphi)) \leq P(R)$. It follows that $P(S) \leq P(R)$, which is a contradiction. q.e.d.

In general $P(R) \geq 2$. In fact, there is a non-constant analytic function f on every open Rieman surface by the existence theorem due to Behnke-Stein and then the composition of the exponential function and f does not take at least two values. It is very difficult in general to calculate $P(R)$ of a given open Riemann surface R. Because it depends on the theory of value distributions on R.

From the value distribution theory of algebroid functions every n-valued algebroid function has at most $2n$ Picard exceptional values. Hence we have $P(R_n) \leq 2n$ for an n-sheeted algebroid surface R_n, which is the proper existence domain of an n-valued algebroid function.

Now an n-sheeted covering surface R_n is called regularly branched if all its branch points are of order n-1. We can characterize a regularly branched algebroid surface by maximal Picard constant. We have

THEOREM 2 (Aogai [2], Ozawa [15] in the case when n=2, Hiromi-Niino [6] in the case when n=3). *Let* R *be an n-sheeted regularly branched. If* $P(R) > \frac{3}{2}n$, *then* $P(R) = 2n$ *and* R *can be defined by an algebroid function* y *such that*

$y^n = (e^{H(z)} - \alpha)(e^{H(z)} - \beta)^{n-1}$, $H(0) = 0$, $\alpha\beta(\alpha-\beta) \neq 0$, *where* $H(z)$ *is a non-constant entire function and* α *and* β *are constants.*

Now I turn my talk to analytic mappings between two algebroid surfaces. For simplicity we now confine ourselves to ultrahyperelliptic surfaces. An ultrahyperelliptic surface is the simplest example of a Riemann surface of infinite genus.

Let R and S be ultrahyperelliptic surfaces defined by $y^2 = G(z)$ and $u^2 = g(w)$, respectively, where G and g are entire functions having an infinite number of simple zeros and no other zeros. Put $p=(z,y)$ and $q=(z,-y) \in R$. Let f be an arbitrary function on R. We put $f_1(z) \equiv \frac{1}{2}(f(p) + f(q))$ and $f_2(z) \equiv \frac{1}{2y}(f(p)-f(q))$. Then f_1 and f_2 are single-valued functions of z. Hence we have the representation of function f on R such that

$$f(p) = f_1(z) + f_2(z)y, \quad p = (z,y) \in R. \qquad (*)$$

Let φ be an analytic mapping of R into S and $\mathfrak{P}_R$ (resp. $\mathfrak{P}_S$) be the projection map of R (resp. S) into z-plane (resp. w-plane) such that $(z,y) \to z$ (resp. $(w,u) \to w$). Put $\Phi = \mathfrak{P}_S \circ \varphi = h \circ \mathfrak{P}_R$. Then Φ is a holomorphic function. Hence from (*) we have $h(z) = h_1(z) + h(z)\sqrt{G(z)}$, where h_1 and h_2 are single-valued entire functions of z. Let Ψ be an analytic mapping of S into the complex plane

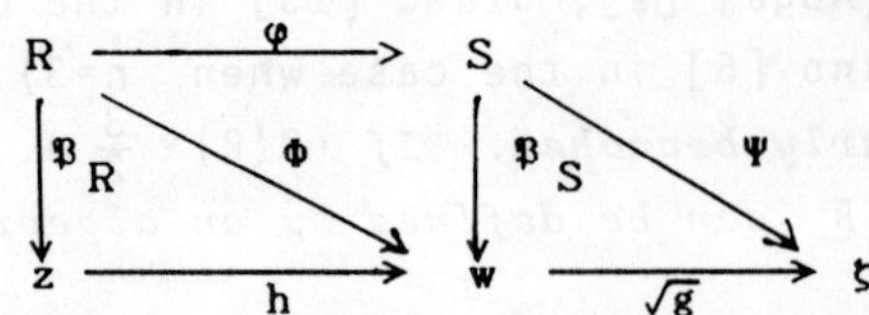

satisfying $\Psi = \sqrt{g}\circ\mathfrak{P}_S$. Then $\Psi\circ\varphi = \sqrt{g}\circ\mathfrak{P}_S\circ\varphi = \sqrt{g}\circ h\circ\mathfrak{P}_R$ is a holomorphic function on R. Hence we have

$$\sqrt{g}\circ h(z) = f_1(z) + f_2(z)\sqrt{G(z)},$$

where f_1 and f_2 are single-valued entire functions of z. Therefore we have

$$\sqrt{g}\circ(h_1(z) + h_2(z)\sqrt{G(z)}) = f_1(z) + f_2(z)\sqrt{G(z)},$$

that is,

$$g\circ(h_1(z) + h_2(z)\sqrt{G(z)}) = (f_1(z) + f_2(z)\sqrt{G(z)})^2$$

Assume that $h_2(z) \not\equiv 0$. Let w_μ be a zero of g. Then w_μ is a simple zero of g by the assumption. Therefore w_μ is either a Picard's value or a multiple value of two-valued algebroid function $h_1 + h_2\sqrt{G}$. From value distribution theory of algebroid functions, a two-valued algebroid function has at most four Picard values and has at most eight totally ramified values. On the other hand there are an infinite number of such points w_μ. This is a contradiction. Hence $h_2 \equiv 0$, that is, $h(z) \equiv h_1(z)$ is a simple-valued entire function of z. Then we have

$$g(h(z)) = f_1(z)^2 + f_2(z)^2G(z) + 2f_1(z)f_2(z)\sqrt{G(z)}.$$

Since $g(h(z))$ is single-valued, we have $f_1f_2 \equiv 0$. If $f_2 \equiv 0$, then $g(h(z)) = f_1(z)^2$. This again contradicts the Nevanlinna's ramification relation for h. Therefore $f_1 \equiv 0$, that is, $g(h(z)) = f_2(z)^2G(z)$. Thus we have a desired functional equation

$$f(z)^2G(z) = g(h(z)) \qquad (**)$$

with a suitable entire function f.

Conversely if G and g satisfy the equation (**) with suitable entire functions h and f. Let $\varphi = \mathfrak{P}_S^{-1} \circ h \circ \mathfrak{P}_R$. We take z_o and w_o such that $g(w_o) \neq 0$, $h(z_o) = w_o$. Take one branch of $\mathfrak{P}_S^{-1}(w_o)$ and we do analytic continuation of φ from z_o. Then φ is well-defined mapping of R into S and evidently φ is analytic for every ordinary point. If p_1 is a branch point of R, then $z_1 = \mathfrak{P}_R p_1$ is a zero point of $G(z)$. Thus $g \circ h(z_1) = 0$. This shows that $h(z_1)$ is a zero point of g with odd multiplicity. Thus $\mathfrak{P}_S^{-1} h(z_1)$ is a branch point of S. Let t and T be the local parameters around p_1 and $\mathfrak{P}_S^{-1} h(z_1)$, respectively. Then we may put $z - z_1 = t^2$, $h(z) - h(z_1) = T^2 = a_k(z-z_1)^{2k+1}(1+\ldots)$. Hence we have $T^2 = a_k t^{4k+2}(1+t^2+\ldots)$ and so $T = a_k^{1/2} t^{2k+1}(1+t^2+\ldots)$. This implies the analyticity of $T(t)$. Hence φ is an analytic mapping of R into S.

Thus we have

THEOREM 3 (Ozawa [16]). *There is a non-trivial analytic mapping* φ *of* R *into* S, *if and only if there is a non-constant entire function* $h(z)$ *satisfying*

$$f(z)^2 G(z) = g(h(z)), \qquad (**)$$

where f *is a suitable entire function.*

We call $h(z)$ the <u>projection</u> of the analytic mapping φ. From this when we investigate the property of an analytic mapping of R into S, it is sufficient to investigate the property of its projection.

Let $N(r,R)$ be the quantity $N(r,\mathfrak{X})$ defined in Selberg's paper [20]. In this case $2N(r,R) = N(r,0,G)$. This is essentially the same as the integrated Euler characteristic in Sario (cf. [18]).

We have

THEOREM 4 (Ozawa [16]). *If there is an analytic mapping* φ *of* R *into* S, *then*

$$\varlimsup_{r\to\infty} \frac{N(r,R)}{T(r,h)} = \infty.$$

If $N(r,0,G)$ is of finite order ρ_G, then we may assume that G is the canonical product of ρ_G over these zeros; a similar remark applies to g. Then we have

THEOREM 5 (Hiromi-Mutō [15]). *Assume that* $\rho_G < +\infty$ *and* $0 < \rho_g < \infty$. *If there is an analytic mapping* φ *of* R *into* S, *then* $\rho_G = p\rho_g$, *where* p *is an integer and the projection* h *of* ψ *is a polynomial of degree* p.

Now we consider the family of analytic mappings of R into S. Let $\mathfrak{H}(R,S)$ be the family of projections of analytic mappings of R into S and $\mathfrak{H}_p(R,S)$ (resp. $\mathfrak{H}_T(R,S)$) be the subfamily of $\mathfrak{H}(R,S)$ consisting of polynomials (resp. transcendental entire functions). With respect to the structure of $\mathfrak{H}(R,S)$, the following theorems are obtained:

THEOREM 6 (Mutō [8]). $\mathfrak{H}(R,S)$ *is at most a countable set.*

THEOREM 7 (Niino [12],[13]). *If* $\mathfrak{H}_p(R,S)$ *is not empty, then it consists of polynomials of the same degree.*

THEOREM 8 (Mutō [9]). *If* $\mathfrak{H}(R,S) \neq \phi$, *then* $\mathfrak{H}(R,S) = \mathfrak{H}_p(R,S)$ *or* $\mathfrak{H}(R,S) = \mathfrak{H}_T(R,S)$.

THEOREM 9 (Niino [12],[13]). *Assume that* $h(z) = a_p z^p + \dots + a_o$ *and* $k(z) = b_p z^p + \dots + b_o$, $a_p b_p \neq 0$, *belong to* $\mathfrak{H}(R,S)$. *If* $|a_p| < |b_p|$, *then we have*

(i) $\rho_g = \rho_G = 0$ *and* p *is even.*

(ii) $k(z) = \frac{b_p}{a_p} h(z) + A$, *where* A *is constant.*

(iii) $\mathfrak{H}(R,S) = \mathfrak{H}_p(R,S)$ *and* $\mathfrak{H}(R,S)$ *consists of just two elements* h *and* k, *that is,* $\mathfrak{H}(R,S) = (h,k)$.

(iv) g *satisfies the functional equation*

$$g(\lambda w + A) = B(\lambda w + A - \alpha_1)g(w),$$

where $\lambda = \frac{b_p}{a_p}$ *and* α_1 *and* B *constants such that*

$\alpha_1 \neq -\frac{A}{\lambda - 1}$, $g\left(\frac{\alpha_1 - A}{\lambda}\right) \neq 0$.

(v) g *has just an infinite number of simple zeros,* $\{\alpha_j\}_{j=1}^{\alpha}$ *such that*

$$\alpha_{j+1} = \lambda^j \alpha_1 + A(\lambda^j - 1)/(\lambda - 1) \qquad (j \geq 1)$$

and $k(z) = \alpha_1 + P(z)^2$, *where* P *is a polynomial of degree* p/2.

(vi) *Examples of this situation indeed occur.*

THEOREM 10 (Baker [3]). *Assume that* $\rho(g) < +\infty$, *there exist two polynomials* h *and* k *of degree* p *belonging to* $\mathfrak{H}(R,S)$ *and the leading coefficients of* h *and* k *are the same in modulus. Then one of the following three cases occurs:*

(i) $k(z) = Lh(z) + M$, *where* L *is a root of unity and* M *is a constant.*

(ii) p *is even and there is a polynomial* r *such that*

$$h(z) = r(z)^2 + A_0 \quad \text{and} \quad k(z) = \{r(z)+\beta\}^2 + D_0,$$

where A_0, D_0 *and* β *are constants.*

(iii) *the ratio of the leading coefficients of* h *and* k *is a primitive s-th root of unity, the* (ps)*-th iterate* ψ_{ps} *of the expansion* ψ *of* $k^{-1} \circ h$ *about* ∞ *satisfies* $\psi_{ps}(z) \equiv z$. *Case* (iii) *can occur only if* $\rho_G \geq 2$.

Further examples of each of the cases exist.

Now, we assume that $P(R) = P(S) = 4$. Then we have

$$F(z)^2 G(z) = (e^{H(z)} - \alpha)(e^{H(z)} - \beta), \quad H(0) = 0, \quad \alpha\beta(\alpha-\beta) \neq 0,$$

$$f(w)^2 g(w) = (e^{L(w)} - \delta)(e^{L(w)} - \delta), \quad L(0) = 0, \quad \gamma\delta(\gamma-\delta) \neq 0,$$

where F and f are suitable entire functions, H and L are non-constant entire functions, and α, β, γ and δ are constants. With the above notations we have

THEOREM 11 (Hiromi-Ozawa [7]). *Suppose that* $P(R) = P(S) = 4$. *Then there exists an analytic mapping* φ *of* R *into* S *if and only if there exists an entire function* $h(z)$ *such that either*

$$H(z) = L(h(z)) - L(h(0)), \ \alpha e^{L(h(0))} = \gamma, \ \beta e^{L(h(0))} = \delta$$

or

$$H(z) = -L(h(z)) + L(h(0)); \ \alpha\gamma = e^{L(h(0))}, \ \beta\delta = e^{L(h(0))}.$$

THEOREM 12 (Mutō-Niino [10]). *Suppose that* $P(R) = P(S) = 4$ *and* $\mathfrak{H}(R,S) \neq \phi$. *Then the followings hold:*

(I) *If* $\mathfrak{H}_T(R,S) \neq \phi$, *then* $\mathfrak{H}_T(R,S)$ *consists of transcendental entire functions of the same order, the same type and the same class.*

(II) *If* $\mathfrak{H}_P(R,S) \neq \phi$ *and* h *and* k *are two elements of* $\mathfrak{H}_P(R,S)$, *then either*

(i) $k(z) = Lh(z) + M$, *where* L *is a root of unity and* M *is a constant*, *or*

(ii) *there is a polynomial* r *such that*
$$h(z) = r(z)^2 + A_0 \quad \text{and} \quad k(z) = (r(z)+\beta)^2 + D_0,$$

where A_0, D_0 *and* β *are constants.*

Open problems:

1. Is Baker's Theorem 10 also true without the condition $\rho_G < +\infty$?

2. In general does $\mathfrak{H}_T(R,S)$ consist of entire functions of the same degree, the same type and the same class without the maximal condition $P(R) = P(S) = 4$?

REFERENCES

[1] Ahlfors, L.V., *The theory of meromorphic curves*, Acta Soc. Sci. Fenn. Ser. A. Tom. 3, No. 4(1941), 31 pp.

[2] Aogai, H., *Picard constant of a finitely sheeted covering surface*, Kōdai Math. Sem. Rep. 25(1973), 219-224.

[3] Baker, I.N., *Analytic mappings between two ultrahyperelliptic surfaces*, Aequationes Math. 14(1976), 461-472.

[4] Cartan, H., *Sur les zéros des combinaison linéaires de p fonctions holomorphes données*, Mathematica 7(1933), 5-31.

[5] Hiromi, G. and H. Mutō, *On the existence of analytic mappings, I,* Kōdai Math. Sem. Rep. 19(1967), 236-244.

[6] Hiromi, G. and K. Niino, *On a characterization of regularly branched three-sheeted covering Riemann surfaces,* Kōdai Math. Sem. Rep. 17(1965), 250-260.

[7] Hiromi, G. and M. Ozawa, *On the existence of analytic mappings between two ultrahyperelliptic surfaces,* Kōdai Math. Sem. Rep. 17(1965), 281-306.

[8] Mutō, H., *Analytic mappings between two ultrahyperelliptic surfaces,* Kōdai Math. Sem. Rep. 22(1970), 53-60.

[9] Mutō, H., *On the family of analytic mappings among ultrahyperelliptic surfaces,* Kōdai Math. Sem. Rep. 26(1975), 454-458.

[10] Mutō, H. and K. Niino, *A remark on analytic mappings between two ultrahyperelliptic surfaces,* Kōdai Math. Sem. Rep. 26(1974/75), 103-107.

[11] Nevanlinna, R., *Zur Theorie der meromorphen Funktionen,* Acta Math. 46(1925), 1-99.

[12] Niino, K., *On the family of analytic mappings between two ultrahyperelliptic surfaces,* Kōdai Math. Sem. Rep. 21(1969), 182-190.

[13] Niino, K., *On the family of analytic mappings between two ultrahyperelliptic surfaces, II,* Kōdai Math. Sem. Rep. 21(1969), 491-495.

[14] Ozawa, M., *On complex analytic mappings,* Kōdai Math. Sem. Rep. 17(1965), 93-102.

[15] Ozawa, M., *On ultrahyperelliptic surfaces,* Kōdai Math. Sem. Rep. 17(1965), 103-108.

[16] Ozawa, M., *On the existence of analytic mappings,* Kōdai Math. Sem. Rep. 17(1965), 191-197.

[17] Rodin, B. and L. Sario, *Existence of mappings into noncompact Riemann surfaces,* J. Analyse Math. 17(1966), 219-224.

[18] Sario, L. and K. Noshiro, *Value distribution theory,* D. Van Nostrand Company, Inc., Princeton, 1966.

[19] Selberg, H.L., *Uber die Wertverteilung der algebroiden Funktionen,* Math. Z. 31(1930), 709-728.

[20] Selberg, H.L., *Algebroide Funktionen und Umkehrfunktionen Abelsher Integral*, Avh. Norske Vid.-Akad. Oslo No. 8, 72 pp.

[21] Ullrich, E., *Uber den Einfulss der Verzweigtheit einer Algebroide auf ihre Wertverteilung*, J. Reine Angew. Math. 167 (1931), 198-220.

[22] Valiron, G., *Sur la dériveé des fonctions algébroides*, Bull. Soc. Math. France 59(1931), 17-39.

[23] Weyl, H., *Meromorphic functions and analytic curves*, Ann. of Math. Studies, No. 12, Princeton Univ. Press, Princeton, 1943.

[24] Weyl, H. and J., *Meromorphic curves*, Ann. of Math. 39(1938), 516-538.

[25] Wu, H.-H., *The equidistribution theory of holomorphic curves*, Ann. of Math. Studies, No. 64, Princeton Univ. Press, Princeton; Univ. of Tokyo Press, Tokyo, 1970.

YOSIDA FUNCTIONS

David Minda[1,2]

Department of Mathematical Sciences
University of Cincinnati
Cincinnati, Ohio 45221-0025

ABSTRACT

A meromorphic function f defined on the complex plane $\mathbb{C}$ is called a Yosida function if the spherical derivative $f^{\#} = |f'|/(1 + |f|^2)$ is uniformly bounded on $\mathbb{C}$. Let $\mathscr{Y}$ denote the set of all Yosida functions. A meromorphic function f defined on the open unit disk $\mathbb{D}$ is called a normal function if $\sup\{(1 - |z|^2)f^{\#}(z) : z \in \mathbb{D}\}$ is finite. $\mathscr{N}$ is the class of all normal functions. We establish a number of results for the family $\mathscr{Y}$; many of these results stem from corresponding results for $\mathscr{N}$.

It is known that a Yosida function has order at most 2 and that this bound is sharp. For a holomorphic Yosida function we show that the order is at most 1. The holomorphic Yosida function $\exp(z)$ demonstrates that this bound is best possible. A consequence of the proof is that the only holomorphic Yosida functions that omit the value 0 are $A\exp(cz)$, where $A \neq 0$ and c are constants. Also, we give a characterization of

---

[1]Research supported in part by NSF Grant No. DMS-8521158

[2]The presentation of this paper at the Symposium on Complex Analysis at Northwest University, Xian, Peoples Republic of China, May 21-27, 1987 was funded by a travel grant from the Taft Faculty Committee of the University of Cincinnati.

Yosida functions: $f \notin \mathcal{Y}$ if and only if the Riemann image surface of f asymptotically contains the Reimann image surface of a nonconstant Yosida function. This implies that the spherical Bloch constant for $\mathcal{Y}$ actually equals the spherical Bloch constant for the larger family of all meromorphic functions on $\mathbb{C}$. Finally, we study the Marden constants for $\mathcal{Y}$ and present a plausible extremal function for both the Bloch and Marden constants for $\mathcal{Y}$. This function is a natural analogue of the Ahlfors-Grunsky function which is conjectured to be extremal for the classical Bloch constant.

1. INTRODUCTION

A meromorphic function f defined on the complex plane $\mathbb{C}$ is called a Yosida function if the spherical derivative $f^{\#}(z) = |f'(z)|/(1+|f(z)|^2)$ is uniformly bounded on $\mathbb{C}$. Let $\mathcal{Y}$ denote the set of all Yosida functions and $\|f\| = \sup \{f^{\#}(z) : z \in \mathbb{C}\}$. This class of functions was introduced by Yosida [11] who established some basic properties of the class $\mathcal{Y}$. For example, the order of any Yosida function is at most two and this is best possible. Also, $f \in \mathcal{Y}$ if and only if the family $\{f(z+a) : a \in \mathbb{C}\}$ is normal on $\mathbb{C}$. We establish a number of new results for $\mathcal{Y}$; many of these stem from related results for the family $\mathcal{N}$ of normal functions on the open unit disk $\mathbb{D}$. Recall that a meromorphic function g on $\mathbb{D}$ is normal if $(1-|z|^2)g^{\#}(z)$ is uniformly bounded on $\mathbb{D}$.

We begin by deriving a new characterization of Yosida functions. Roughly speaking, we show that a meromorphic function f is not Yosida if and only if the Riemann image surface of f asymptotically contains the Riemann image surface of a nonconstant Yosida function. This is analogous to a characterization of normal functions by Lohwater and Pommerenke [3]. An immediate consequence of this characterization is the fact that the spherical Bloch constant for the family of nonconstant Yosida functions is actually equal to the spherical Bloch constant $\mathscr{B}_1(\mathbb{C})$ for the full family of nonconstant meromorphic functions on $\mathbb{C}$. This leads to a sufficient condition for a normal function. A meromorphic function g on $\mathbb{D}$ is normal if there exists an $\varepsilon > 0$ such that the Riemann image surface of g (viewed as spread over the Riemann sphere $\mathbb{P}$) contains no unramified disk with angular radius greater than $\mathscr{B}_1(\mathbb{C}) - \varepsilon$.

Next, we discuss the Marden constants for nonconstant Yosida functions. We also consider locally schlicht Yosida functions. Generally speaking, Marden constants deal with the largest disk of univalence in the domain of the function. Marden constants for Bloch and normal functions were studied in [6]. Recently, Yamashita [10] studied related constants involving the largest disk of starlikeness for Bloch, normal and Yosida functions.

Our third topic is consideration of a specific function that might be extremal for both the spherical Bloch constant and the Marden constant for $\mathscr{Y}$. This function provides an upper bound for both constants. This plausible extremal function is the natural analog of the Ahlfors-Grunsky function [1] which is conjectured to be extremal for the classical Bloch constant. We give a geometric and an analytic discussion of this function.

Finally, we obtain several results for holomorphic Yosida functions. We obtain explicit upper bounds for $|f(z)|$ and $|f'(z)|$ for any holomorphic Yosida function. In particular, we show that the order of a holomorphic Yosida function is at most one. All of the results are sharp for the functions $A\exp(cz)$, where $A \neq 0$ and c are constants. In fact, we can even show that these are the only holomorphic, nonvanishing Yosida functions.

2. A CHARACTERIZATION OF YOSIDA FUNCTIONS

We give a curious characterization of non-Yosida functions in terms of Yosida functions.

Theorem 1. Let f be a meromorphic function on $\mathbb{C}$. Then $f \notin \mathcal{Y}$ if and only if there exist sequences $\{z_n\}$ and $\{\rho_n\}$ with $z_n \in \mathbb{C}$, $\rho_n > 0$ and $\rho_n \to 0$ such that $\lim f(z_n+\rho_n\zeta) = g(\zeta)$, locally uniformly on $\mathbb{C}$, where g is a nonconstant Yosida function.

Proof. We begin by establishing the sufficiency. Suppose $f(z_n+\rho_n\zeta) \to g(\zeta)$ locally uniformly on $\mathbb{C}$, where g is a nonconstant Yosida function and $\{z_n\}$ and $\{\rho_n\}$ are as in the statement of the theorem. We want to show $f \notin \mathcal{Y}$. Assume that $f \in \mathcal{Y}$, say $f^{\#}(z) \leqslant M$ for $z \in \mathbb{C}$. Then

$$g^{\#}(z) = \lim \rho_n f^{\#}(z_n+\rho_n\zeta) \leqslant \lim \rho_n M = 0 .$$

Thus, $g^{\#} = 0$, or g is constant. From this contradiction we conclude $f \notin \mathcal{Y}$.

The proof of the necessity is a bit more involved. Suppose $f \notin \mathcal{Y}$. Define

$$M(R) = \max_{|z|\leqslant R} \frac{f^{\#}(z)(R^2 - |z|^2)}{R} .$$

First, we note that $M(R)/R \to \infty$ as $R \to \infty$. If not, then there would exist a sequence $\{R_n\}$ with $R_n \to \infty$ and $M(R_n)/R_n \to M < \infty$. Fix $z \in \mathbb{C}$. For all n such that $|z| < R_n$ we have

$$f^{\#}(z) \leqslant \frac{R_n M(R_n)}{R_n^2 - |z|^2} \leqslant \frac{M(R_n)/R_n}{1 - (|z|/R_n)^2} .$$

By letting $n \to \infty$ we obtain $f^{\#}(z) \leq M$. This would hold for all $z \in \mathbb{C}$, contradicting $f \notin \mathscr{Y}$. Next, take any sequence $\{R_n\}$ with $R_n \to \infty$. Select z_n with $|z_n| < R_n$ and

$$M(R_n) = \frac{f^{\#}(z_n)(R_n^2 - |z_n|^2)}{R_n} .$$

Set

$$\rho_n = \frac{1}{f^{\#}(z_n)} = \frac{R_n^2 - |z_n|^2}{R_n M(R_n)} .$$

Note that

$$\rho_n = \frac{1 - (|z_n|/R_n)^2}{M(R_n)/R_n} \leq \frac{1}{M(R_n)/R_n} \to 0. \tag{1}$$

Also,

$$\frac{\rho_n}{R_n - |z_n|} = \frac{R_n + |z_n|}{R_n M(R_n)} \leq \frac{1}{M(R_n)/R_n} \to 0.$$

Thus, $r_n = (R_n - |z_n|)/\rho_n \to \infty$ as $n \to \infty$. Define g_n: $\mathbb{C} \to \mathbb{P}$ by $g_n(\zeta) = f(z_n + \rho_n \zeta)$. Clearly, g_n is meromorphic on $\mathbb{C}$ and $g_n^{\#}(0) = \rho_n f^{\#}(z_n) = 1$. Fix $r > 0$. If $|\zeta| \leq r < r_n$, then

$$|z_n + \rho_n \zeta| \leq |z_n| + \rho_n |\zeta| \leq |z_n| + \rho_n r_n = R_n .$$

Therefore, for $|\zeta| \leq r < r_n$,

$$f^{\#}(z_n + \rho_n \zeta) \leq \frac{R_n M(R_n)}{R_n^2 - |z_n + \rho_n \zeta|^2} \leq \frac{R_n M(R_n)}{R_n^2 - (|z_n| + \rho_n r)^2}$$

and

$$g_n^*(\zeta) = \rho_n f^\#(z_n + \rho_n\zeta) = \frac{R_n^2 - |z_n|^2}{R_n M(R_n)} \frac{R_n M(R_n)}{R_n^2 - (|z_n| + \rho_n r)^2}$$

$$= \frac{R_n + |z_n|}{R_n + |z_n| + \rho_n r} \frac{R_n - |z_n|}{R_n - |z_n| - \rho_n r} \leq \frac{1}{1 - \frac{\rho_n r}{R_n - |z_n|}} .$$

From (1) we conclude that there exists $K(r)$ such that $g_n^\#(\zeta) \leq K(r)$ for all ζ with $|\zeta| \leq r$ and all n. Marty's Theorem implies that $\{g_n\}$ is a normal family on $\mathbb{C}$. Thus, there is a subsequence, which we may assume to be $\{g_n\}$ itself, and a meromorphic function g such that $g_n \to g$ locally uniformly in the spherical metric. Because $g_n^\#(0) = 1$ for all n, we obtain $g^\#(0) = 1$, so g is nonconstant. The above inequality yields $g^\#(\zeta) \leq 1$ for all $\zeta \in \mathbb{C}$. Hence, $g \in \mathscr{Y}$.

Remark. The idea for this theorem stems from work of Lohwater and Pommerenke [3] on normal meromorphic functions in the unit disk $\mathbb{D}$. There is also an analog for Bloch function [7]. Theorem 1 can be stated in more picturesque language as follows. A meromorphic function on $\mathbb{C}$ is not Yosida if and only if its Riemann image surface asymptotically contains the Riemann image surface of a nonconstant Yosida function.

Example. From the definition of the class $\mathscr{Y}$ it is easy to see that $\exp(z^2)$ is not a Yosida function. Nevertheless, it is instructive to use Theorem 1 to establish this fact. Let $c = \exp(i\pi/4)$, $z_n = c(2\pi n)^{1/2}$ and $\rho_n = 1/(2\pi n)^{1/2}$. Clearly, $\rho_n \to 0$ and

$$g_n(\zeta) = f(z_n + \rho_n\zeta) = \exp(2c\zeta + \rho_n^2\zeta^2) \to \exp(2c\zeta) \in \mathscr{Y} .$$

This shows that $f \notin \mathscr{Y}$.

Theorem 1 suggests the following problem which is similar to one raised by Campbell and Wickes [2] for non-normal functions. A meromorphic function f defined on $\mathbb{C}$ is said to contain a Yosida function g if there exist a sequence $\{z_n\}$ of complex numbers and a sequence $\{\rho_n\}$ of positive numbers such that $\rho_n \to 0$ and $g_n\{\zeta\} = f(z_n + \rho_n\zeta) \to g(\zeta)$ locally uniformly on $\mathbb{C}$. Theorem 1 asserts that every non-Yosida meromorphic function on $\mathbb{C}$ contains at least one Yosida function. The problem is to determine all Yosida functions that are contained in a given non-Yosida function. Perhaps a slightly easier problem is to determine the number of distinct Yosida functions that are contained in a given non-Yosida function.

Our characterization of Yosida functions enables us to show that the spherical Bloch constant for the family of nonconstant Yosida functions is equal to the spherical Bloch constant for the full family of nonconstant meromorphic functions on $\mathbb{C}$. We recall the definition of spherical Bloch constants, see [4,5] for more details. Suppose f is meromorphic on a region Ω. We view the Riemann image surface R_f as spread over the Riemann sphere $\mathbb{P}$. For $a \in \Omega$ let $r(a,f)$ denote the spherical radius of the largest unramified spherical disk in R_f with center $f(a)$. Note that $r(a,f) = 0$ precisely when $f^{\#}(a) = 0$. Let $r(f) = \sup\{r(a,f) : a \in \Omega\}$ and $\mathscr{B}_1(\mathbb{C}) = \inf\{r(f)\} : f$ is a nonconstant meromorphic function on $\mathbb{C}$ }. Minda [4] showed that

$$1.0471976\ldots = \frac{\pi}{3} \leq \mathscr{B}_1(c) \leq 2\arctan(1/\sqrt{2}) = 1.22309594\ldots$$

The upper bound is derived from a Weierstrass pe-function and is conjectured to be sharp. This conjectured extremal function is actually a Yosida function since it is doubly periodic. It is discussed in more detail in section 4.

The spherical Bloch constant for $\mathscr{Y}$ is $\mathscr{B}(\mathscr{Y}) = \inf\{r(f) : f$ is a nonconstant Yosida function$\}$. It is clear that $\mathscr{B}_1(\mathbb{C}) \leqslant \mathscr{B}(\mathscr{Y})$. Yamashita [10] obtained the lower bound $\mathscr{B}(\mathscr{Y}) \geqslant 2 \arctan(.24) = .47109\ldots$.

Theorem 2. $\mathscr{B}_1(\mathbb{C}) = \mathscr{B}(\mathscr{Y})$.

Proof. It suffices to show that $r(f) \geqslant \mathscr{B}(\mathscr{Y})$ for any nonconstant meromorphic function f. This is trivial if $f \in \mathscr{Y}$, so we can assume $f \notin \mathscr{Y}$. Then by Theorem 1 there exist sequences $\{z_n\}$ and $\{\rho_n\}$ such that $f(z_n + \rho_n\zeta) = g_n(\zeta) \to g(\zeta)$ locally uniformly, where $g \in \mathscr{Y}$ is nonconstant. It is elementary that $r(g_n) = r(f)$ for all n and straightforward to verify that if $h_n \to h$ locally uniformly, then $\liminf r(h_n) \geqslant r(h)$. Hence, $r(f) = \lim r(g_n) \geqslant r(g) \geqslant \mathscr{B}(\mathscr{Y})$.

Theorem 3. Suppose f is meromorphic in $\mathbb{D}$ and $r(f) < \mathscr{B}_1(\mathbb{C})$. Then f is normal in $\mathbb{D}$.

Proof. Suppose g were not normal in $\mathbb{D}$. Then by a result of Lohwater and Pommerenke [3] there would exist sequences $\{z_n\}$ and $\{\rho_n\}$ with $z_n \in \mathbb{D}$, $\rho_n > 0$ and $\rho_n/(1 - |z_n|) \to 0$ such that $f(z_n + \rho_n\zeta) = g_n(\zeta) \to g(\zeta)$ locally uniformly, where $g \in \mathscr{Y}$ is nonconstant. As in the proof of the preceding theorem we have $r(f) = \lim r(g_n) \geqslant r(g) \geqslant \mathscr{B}_1(\mathbb{C})$. This contradicts $r(f) < \mathscr{B}_1(\mathbb{C})$, so f must be normal in $\mathbb{D}$.

Pommerenke [8] established a version of Theorem 3 with the (probably) smaller constant $\pi/3$ in place of $\mathscr{B}_1(\mathbb{C})$. His technique of proof is quite different.

3. MARDEN CONSTANTS

We define the euclidean Marden constant for a family of holomorphic or meromorphic functions defined on $\mathbb{C}$. Hyperbolic Marden constants were considered for families of holomorphic or meromorphic functions defined on $\mathbb{D}$ in [6]. Let f be holomorphic or meromorphic on $\mathbb{C}$. For $a \in \mathbb{C}$ let $s(a,f) = \sup\{r : f \text{ is univalent in } D(a,r)\}$. Note that $s(a,r) = 0$ if and only if $f^{\#}(a) = 0$. Thus, $D(a,s)$, where $s = s(a,f)$, is the largest disk centered at a in which f is univalent. Define $s(f) = \sup\{s(a,f\} : a \in \mathbb{C}\}$. Roughly speaking, $s(f)$ is the radius of the largest disk in $\mathbb{C}$ in which f is univalent. Note that if $f_n \to f$ locally uniformly, then $\liminf s(f_n) \geqslant s(f)$.

For any positive number m define

$$\mathscr{s}(m) - \inf\{s(f) : f \in \mathscr{Y} \text{ and } \|f\| = m\},$$

$$\mathscr{s}_0(m) = \inf\{s(f) : f \in \mathscr{Y} \text{ is locally schlicht and } \|f\| = m\}.$$

Note that if $f \in \mathscr{Y}$ and $\|f\| = m$, then $g(z) = f(z/m) \in \mathscr{Y}$ and $\|g\| = 1$. Also, $s(f) = s(g)/m$. This observation yields $\mathscr{s}(m) = \mathscr{s}(1)/m$. Similarly, $\mathscr{s}_0(m) = \mathscr{s}_0(1)/m$. Thus, it suffices to investigate $\mathscr{s} = \mathscr{s}(1)$ and $\mathscr{s}_0 = \mathscr{s}_0\ (1)$.

Theorem 4. (a) $\mathscr{s} \geqslant .5610963..$ and (b) $\mathscr{s}_0 \geqslant .7064574..$.

Proof. (a) We will show that $s(f) \geqslant c = .5610963...$ for all $f \in \mathscr{Y}$ with $\|f\| = 1$. We first consider the special case in which $f(0) = 0$ and $f'(0) = 1$. Then $f(z) = z + a_2z^2 + \ldots$ in a neighbourhood of the origin. The condition $f^{\#}(z) \leqslant 1$, or $|f'(z)| \leqslant 1 + |f(z)|^2$, for z near the origin gives

$$1 + \mathrm{Re}\{a_2 z\} + o(|z|^2) \leq 1 + |z|^2 + o(|z|^3) \ .$$

For $z = re^{i\theta}$ this gives $\mathrm{Re}\{a_2 e^{i\theta}\} \leq o(r)$, or $\mathrm{Re}\{a_2 e^{i\theta}\} \leq 0$. Because this holds for all $\theta \in \mathbb{R}$, we have $a_2 = 0$. Now,

$$\arctan |f(z)| \leq \int_0^z f^{\#}(\zeta)|d\zeta| \leq \int_0^z |d\zeta| = |z| \ ,$$

so that $|f(z)| \leq \tan |z|$. Then

$$|f'(z)| \leq 1 + |f(z)|^2 \leq 1 + \tan^2|z| = \sec^2|z| \ .$$

Thus, for $R \leq \pi/2$ we have f holomorphic in $|z| < R$, $f'(0) = 1$, $f''(0) = 0$ and $|f'(z)| \leq \sec^2(R)$. A result of Minda [6] gives $s(0,f) \geq R\cos(R)$. The value $R = .86033326$ yields $s(f) \geq s(0,f) \geq c$.

Now we turn to the general situation: $f \in \mathscr{Y}$ and $\|f\| = 1$. There is a sequence $\{a_n\}$ in $\mathbb{C}$ with $f^{\#}(a_n) \to 1$. Let

$$R_b(z) = \begin{cases} (z - b)/(1 - \bar{b}z) & b \in \mathbb{C} \ , \\ 1/z & b = \infty \ . \end{cases}$$

R_b is a rotation of the sphere $\mathbb{P}$ that sends b to the origin. Let $b_n = f(a_n)$ and let R_n denote R_b with b replaced by b_n. If $g_n(z) = R_n \circ f(z + a_n)$, then g_n is meromorphic on $\mathbb{C}$, $g_n(0) = 0$ and $g_n{}^{\#}(0) = f^{\#}(a_n) \to 1$. Because the spherical derivative is invariant under rotations of $\mathbb{P}$, $g_n{}^{\#}(z) \leq 1$ for all $z \in \mathbb{C}$ and all n. Thus, $\{g_n\}$ is a normal family so there is no harm in assuming that $g_n \to g$ locally uniformly on $\mathbb{C}$. Then $g(0) = 0$ and $g'(0) = g^{\#}(0) = 1$, so g is nonconstant.

Also, $g^{\#}(z) = \lim g_n^{\#}(z) \leq 1$ for all z. This shows that $g \in \mathscr{Y}$, $g(0) = 0$, $g'(0) = 1$ and $\|g\| = 1$. The first part of the proof gives $s(g) \geq c$. Clearly, $s(g_n) = s(f)$ for all n, so $s(f) = \liminf s(g_n) \geq s(g) \geq c$.

(b) The proof of (b) exactly parallels the proof of (a) except that Theorem 1(b) of [6] is used in place of Theorem 1(a) of the same reference. We obtain $s(f) \geq Rh(\cos^2 R)$, where

$$h(t) = \left[\frac{(\pi^2 + 4\log^2 t)^{1/2} + 2\log t}{\pi}\right]^{1/2}$$

The value $R = 1.114$ produces $s(f) \geq .7064574 \ldots$.

Remarks. (a) Yamashita [10] has considered a related constant. Let $s^*(a, f)$ designate the maximum value of r such that f is univalent in $D(a,r)$ and the image $f(D(a,r))$ is starlike with respect to $f(a)$ relative to spherical geometry and $s^*(f) = \sup\{s^*(a,f) : a \in \mathbb{C}\}$. Then $\mathscr{s}^*(m)$ is defined like $\mathscr{s}(m)$ and $\mathscr{s}^* = \mathscr{s}^*(1)$. Yamashita showed that $\mathscr{s}^* \geq .38$.

(b) The exponential function provides an upper bound for $\mathscr{s}_0$. Note that e^{2z} is locally schlicht and $\|e^{2z}\| = 1$. From $s(e^{2z}) = \pi/2$ we get $\mathscr{s}_0 \leq \pi/2$. We present an upper bound for $\mathscr{s}$ in the next section.

4. POSSIBLE EXTREMAL FUNCTION

There is natural Yosida function that is a plausible extremal function for both the Bloch and Marden constants for the family of nonconstant Yosida functions. This function is an analog of the Ahlfors-Grunsky function which is conjectured to extremal for the classical Bloch constant problem. In fact, the construction of our function is just an adaptation to the geometry of the situation at hand of their construction [1]. We refer the reader to [4, 5] for related constructions; we freely employ notation from these references without further mention.

We begin by giving a geometric construction of the function g which will make the values of $r(g)$ and $s(g)$ apparent. Let $\Delta_{1/3}$ be the euclidean equilateral triangle with vertices at 1, ω and ω^2, where $\omega = \exp(2\pi i/3)$. Similarly, $\Delta_{2/3}$ is the regular circular triangle which has all interior angles equal to $2\pi/3$ and vertices at the points $1/\sqrt{2}$, $\omega/\sqrt{2}$ and $\omega^2/\sqrt{2}$. The factor $1/\sqrt{2}$ guarantees that each side of $\Delta_{2/3}$, when viewed on the Riemann sphere $\mathbb{P}$, is an arc of a great circle. The euclidean triangle $\Delta_{1/3}$ leads to a triangulation of $\mathbb{C}$ that is obtained by beginning with $\Delta_{1/3}$ and then reflecting in its sides and the sides of the newly formed triangles. Likewise, the spherical triangle $\Delta_{2/3}$ leads to a triangulation of $\mathbb{P}$ which contains four spherical triangles and four vertices, the extra vertex is at infinity. This triangulation of $\mathbb{P}$ can be viewed as the projection of a regular tetrahedron onto $\mathbb{P}$.

Let g be the unique conformal mapping of $\Delta_{1/3}$ onto $\Delta_{2/3}$ such that $g(\omega^j) = \omega^j/\sqrt{2}$ $(j = 0,1,2)$. Note that $g(0) = 0$. In the notation of [5]

$$g(z) = \frac{1}{\sqrt{2}} f_{2/3} \circ f_{1/3}^{-1}(z),$$

so that

$$g'(0) = \frac{f'_{1/3}(0)}{\sqrt{2}\, f'_{1/3}(0)} = \frac{\Gamma(7/6)\,\Gamma(1/3)}{\sqrt{2}\,\Gamma(1/2)} = .9914928 \ldots .$$

By making use of the Schwarz reflection principle, we can extend g to a doubly periodic meromorphic function defined on $\mathbb{C}$. Hence, $g \in \mathscr{Y}$. It is not difficult to show that $r(g) = 2\arctan(1/\sqrt{2})$ and $s(g) = 1$. Observe that g is locally schlicht on $\mathbb{C}$ except at each vertex of the triangulation of $\mathbb{C}$ that is induced from $\Delta_{1/3}$. At each such vertex g assumes one of the values $1/\sqrt{2}$, $\omega/\sqrt{2}$, $\omega^2/\sqrt{2}$ and ∞ with exact multiplicity 2. This example shows that $\mathscr{B}_1(\mathbb{C}) = \mathscr{B}(\mathscr{Y}) \leq 2\arctan(1/\sqrt{2})$ [4]. We obtain an upper bound for $\mathscr{s}$ after we calculate $\|g\|$.

In order to calculate $\|g\|$, we make use of the fact that $g(z) = p(az + b)$, where $p(z)$ is a Weierstrass pe-function which satisfies the differential equation

$$p'(z)^2 = 4 \left[p(z)^3 - \frac{1}{2\sqrt{2}} \right]$$

We first compute $\|p\|$. By making use of the above differential equation, we find that

$$p^{\#}(z) = \frac{|p'(z)|}{1 + |p(z)|^2} \leq \frac{2(|p(z)|^3 + 1/2\sqrt{2})^{1/2}}{1 + |p(z)|^2} = 2H(|p(z)|)^{1/2},$$

where

$$H(t) = \frac{t^3 + 1/2\sqrt{2}}{(1 + t^2)^2} .$$

Note that equality holds in the preceding inequality at points where $p(z) = 0$. We want to maximize $H(t)$ for $t \in [0, +\infty)$; note that $H(t) \to 0$ as $t \to 0$. Direct calculation of the derivative reveals that the only nonnegative roots of $H'(t) = 0$ are 0, $(\sqrt{3} - 1)/\sqrt{2}$ and $\sqrt{2}$. The maximum value of $H(t)$ on $[0, +\infty)$ is $H(0) = H(\sqrt{2}) = 2^{-3/2}$. The important point is that $H(t)$ is maximized for $t = 0$. Thus, $\|p\| = 2^{1/4}$ and $p^{\#}(z) = 2^{1/4}$ at all points with $p(z) = 0$. Since $g(z) = p(az + b)$, $\|g\| = |a| \, \|p\| = 2^{1/4} \, |a|$ and equality holds at any solution of $g(z) = 0$. Because $g(0) = 0$, we have $\|g\| = g'(0)$. This gives $\mathfrak{s} \leq s(g)\|g\| = g'(0)$, or

$$\mathfrak{s} \leq \frac{\sqrt{2}\ \Gamma(1/2)}{\Gamma(7/6)\ \Gamma(1/3)} = .99144928 \ldots .$$

5. HOLOMORPHIC YOSIDA FUNCTIONS

We now obtain several results for holomorphic Yosida functions. The principle tools for establishing these results are certain distortion theorems for holomorphic normal functions which are due to Pommerenke [9].

Theorem 5. Suppose f is a holomorphic Yosida function. Then

$$|f(z)| \leq \max\{1, |f(z)|\} \leq \max\{1, |f(0)|\}\, e^{2\|f\|\,|z|}.$$

Proof. Pommerenke [9] showed that if g is a holomorphic normal function and $(1 - |w|^2)|g'(w)| \leq \alpha$ for all $w \in \mathbb{D}$, then

$$\log^+|g(w)| \leq \frac{1 + |w|}{1 - |w|} \log^+|g(0)| + \frac{2\alpha|w|}{1 - |w|}. \tag{2}$$

Fix $R > 0$ and consider $g(w) = f(Rw)$. Then $g^\#(w) = Rf^\#(Rw)$, so $(1 - |w|^2)g^\#(w) \leq g^\#(w) \leq R\|f\|$. Then (2) yields

$$\log^+|f(Rw)| \leq \frac{1 + |w|}{1 - |w|} \log^+|f(0)| + \frac{2R\|f\|\,|w|}{1 - |w|}.$$

For $z = Rw$ this becomes

$$\log^+|f(z)| \leq \frac{R + |z|}{R - |z|} \log^+|f(0)| + \frac{2R\ \|f\|\,|z|}{R - |z|}.$$

This holds for all $|z| < R$. Fix z and let $R \to \infty$ to obtain

$$\log^+|f(z)| \leqslant \log^+|f(0)| + 2\|f\|\,|z|.$$

This is equivalent to the inequalities given in the statement of the theorem.

Corollary. If f is a holomorphic Yosida function, then f has order at most one.

Remark. The inequalities in Theorem 5 are sharp. If $f(z) = e^{cz}$, where $c \neq 0$, then $\|f\| = c/2$, $f(0) = 1$ and

$$|f(z)| = e^{\operatorname{Re}\{cz\}} \leqslant e^{|cz|} = e^{2\|f\|\,|z|},$$

with equality whenever $\operatorname{Im}\{cz\} = 0$ and $\operatorname{Re}\{cz\} > 0$. Also, f has order one.

Theorem 6. If f is a holomorphic Yosida function, then

$$|f'(z)| \leqslant 2\|f\| \max\{1, |f(z)|\}.$$

Proof. Pommerenke [9] showed that if g is a holomorphic normal function and $(1 - |w|^2)g^{\#}(w) \leqslant \alpha$ for $w \in \mathbb{D}$, then

$$(1 - |w|^2)|g'(w)| \leqslant 2(\log^+|g(w)| + \alpha) \max\{1, |g(w)|\}.$$

Let $g(w) = f(Rw)$ as before. Then we can take $\alpha = R\|f\|$ and $z = Rw$ to get

$$\frac{(R^2 - |z|^2)|f'(z)|}{R} \leqslant 2(\log^+|f(z)| + R\|f\|) \max\{1, |f(z)|\},$$

or

$$\frac{(R^2 - |z|^2)|f'(z)|}{R^2} \leqslant 2\left[\frac{\log^+|f(z)|}{R} + \|f\|\right] \max\{1, |f(z)|\}.$$

Fix z and let $R \to \infty$ to get the desired result.

The function $f(z) = e^{cz}$ again shows that this result is sharp. Theorem 6 leads to a characterization of nonvanishing holomorphic Yosida functions.

Theorem 7. Let f be a nonvanishing entire function. Then $f \in \mathscr{Y}$ if and only if $f(z) = Ae^{cz}$ for some constants $A \neq 0$ and c.

Proof. The sufficiency is trivial. Now, assume f is a nonvanishing holomorphic Yosida function. For those $z \in \mathbb{C}$ with $|f(z)| \geq 1$ Theorem 6 gives $|f'(z)|/|f(z)| \leq 2\|f\|$. Since $1/f$ is also a nonvanishing holomorphic Yosida function and $\|1/f\| = \|f\|$, $1/|f(z)| \geq 1$ implies

$$\frac{|f'(z)|}{|f(z)|} = \frac{|(1/f)'(z)|}{|1/f(z)|} \leq 2\|f\| .$$

Hence, the holomorphic function f'/f is bounded on $\mathbb{C}$ by $2\|f\|$. Liouville's Theorem implies that $f'/f = c$, a constant. Then $f(z) = Ae^{cz}$, where $A \neq 0$ since f omits the value zero.

Corollary. Let f be a meromorphic function on $\mathbb{C}$ which omits two distinct values. Then $f \in \mathscr{Y}$ if and only if $f(z) = T(e^{cz})$, where $c \in \mathbb{C}$ and T is a Möbius transformation.

REFERENCES

1. L.V. Ahlfors and H. Grunsky, Über die Blochsche Konstant, Math. Z. 42 (1937), 671-673.

2. D.M. Campbell and G. Wickes. Characterizations of normal meromorphic functions, Complex Analysis, Joensuu 1978, pp. 55-72, Lecture Notes in Math. 747, Springer-Verlag, Berlin, 1979.

3. A.J. Lohwater and Ch. Pommerenke, On normal

meromorphic functions, Ann. Acad. Sci. Fenn. Ser. AI, no. 550 (1973), 12 pp.

4. D. Minda, Bloch constants for meromorphic functions, Math. Z. 181 (1982), 83-92.

5. D. Minda, Bloch constants, J. Analyse Math. 41 (1982), 54-84.

6. D. Minda, Marden constants for Bloch and Normal functions, J. Analyse Math. 42 (1982/83), 117-127.

7. D. Minda, Bloch and normal functions on general planar regions, to appear.

8. Ch. Pommerenke, Estimates for normal meromorphic functions, Ann. Acad. Sci. Fenn. Ser. AI, no. 476 (1970), 10 pp.

9. Ch. Pommerenke, Normal functions, Proc. NRL Conf. on Classical Function Theory, Math. Res. Center, Naval Res. Lab., Washington, D.C. 1970.

10. S. Yamashita, Constants for Bloch, normal and Yosida functions, Math. Japonica 30 (1985), 405-418.

11. K. Yosida, On a class of meromorphic functions, Proc. Physico-Math. Soc. Japan 16 (1934), 227-235.

SOME RESULTS ON H^p SPACES AND A^p BERGMAN SPACES

Fuyao Ren
Department of Mathematics
Fudan University
Shanghai, China

In this paper we shall introduce some recent results of our university on H^p spaces and A^p Bergman spaces.

1. ON CORONA PROBLEMS

Let Δ be the unit disc in the complex plane $\mathbb{C}$. Let $H^\infty(\Delta)$ denote the space of bounded analytic functions with supremum norm. In 1962, L. Carleson [4] proved the following corona theorem:

Theorem (Carleson's corona theorem). Let $f_1, f_2, \ldots, f_m \in H^\infty(\Delta)$ satisfy

$$0 < \delta \leq \max_j f_j(z) \leq 1$$

for all $z \in \Delta$. Then there exist $g_1, g_2, \ldots, g_m \in H^\infty(\Delta)$, and there exists a constant $c(m,\delta)$ such that $\sum_{j=1}^{m} f_j g_j = 1$ and $\|g_j\|_\infty \leq c(m,\delta)$ for all $j = 1,2,\ldots,m$.

1.1 Corona Problem on Planer Domains

In 1971, M. Behrens [2] showed that the corona theorem

is valid for some kind of infinitely connected domains D which are obtained from Δ by deleting a sequence of mutually disjoint discs $\Delta_j = \Delta(r_j, c_j)$, where r_j are radius and c_j are centres, $D = \Delta \setminus \bigcup_j \Delta_j(r_j, c_j)$, r_j, c_j converge to zero, and there are mutually disjoint discs $\Delta(R_j, c_j)$ such that $\Sigma r_j / R_j < \infty$.

Recently, Liefen Huang [13] shows that

Theorem 1.1. If the centres of Δ_j are on a common line then the corona theorem is valid on D, and there are corona solutions with bounds depending only on m and δ.

In 1985, Garnett and Jones [7] proved that corona theorem is valid for Denjoy domains $D = C^* \setminus E$, where $E \subset R$.

Recently, Liefen Huang [13] extends their results to some kind of domains. Let $E = \bigcup_j [a_j, b_j)$, let F_j be connected subset of C which meets $[a_j, b_j]$, and let

$$\alpha_j(z) = \arg \frac{z - b_j}{z - a_j} .$$

Theorem 1.2. Let $D = \mathbb{C}^* \setminus \bigcup_j (F_j \cup [a_j, b_j])$, suppose that $\inf_j \alpha_j(z) \geqslant \varepsilon > 0$ for all j. If $f_1, \ldots, f_m \in H^\infty(D)$ satisfy

$$0 < \delta \leqslant \max_j |f_j(z)| \leqslant 1$$

for all z in D, then there exist $g_1, \ldots, g_m \in H^\infty(D)$, and there exists a constant $c(m, \delta)$ such that $\|g_j\|_\infty \leqslant c(m, \delta)$ and

$$\sum_j f_j g_j = 1 .$$

1.2 Case of Higher Dimensions

It is not known whether the corona theorem is true on polydiscs.

In 1975, Arveson [1] and Schuber [17] showed an operator version of corona problem in the plane independently. Recently, Liefen Huang [12] extended their results to higher dimensions.

Let $\alpha = (\alpha_1,\ldots,\alpha_n) \in Z \times \ldots \times Z = Z^n$, we define a partial order on Z^n by $\alpha \geq \beta$ iff $\alpha_j \geq \beta_j$ for all j. Let U^n be the unit polydisc.

Theorem 1.3. If $f_1,\ldots,f_m \in H^\infty(U^n)$, then there exists a constant $M > 0$ such that

$$M^2 \sum_{\substack{\alpha \geq o \\ 1 \leq j \leq m}} \|T_\alpha \otimes L_\alpha f_j x\|^2 \geq \sum_{\alpha \geq o} \|T_\alpha \otimes L_\alpha x\|^2$$

for all $x \in \ell^2 \otimes H^2(U^n)$ and all decreasing projections T_α, i.e. if $\alpha \geq \beta$ and $T_\alpha \leq T_\beta$, if and only if there exist $g_1,\ldots,g_m \in H^\infty(U^n)$ and a constant c such that $\|g_j\|_\infty \leq c$ and $\sum_j f_j g_j = 1$, moreover $M \leq mc$, $c \leq 4M^2$. Where ℓ^2 is the usual Hilbert space $\ell^2 = \{\{x_n\}_1^\infty : \sum_{n=1} |x_n|^2 < \infty\}$, L_α is the projection from $H^2(U^n)$ onto the subspace spanned by single function $z^\alpha = z_1^{\alpha_1} \ldots z_n^{\alpha_n}$.

1.3 Ideals in $H^\infty(\Delta)$

Question: does g^p belong to the ideal $J(f_1,\ldots,f_m)$ which is generated by $f_1,\ldots,f_m$, if $|g| \leq C \max_j |f_j(z)|$ for $z\in\Delta$? where p is integer.

There are examples which say that for $p = 1$, it is not true. It is open for $p = 2$. For $p \geq 3$ Wolff [20] proved that it is true. We consider the case that $p \geq 3$, and there are infinitely many functions f_j.

Theorem 1.4. ([11]). If $\{f_j\}_1^\infty \subset H^\infty(\Delta)$, $\|f\| = \sup_\Delta(\sum_j |f_j(z)|^2)^{1/2}$, $g\in H^\infty(\Delta)$ satisfy

$$|g(z)| \leq M(\sum_j |f_j(z)|^2)^{1/2} ,$$

then there exists $\{h_j\}_1^\infty \subset H^\infty(\Delta)$, $\|h\| \leq C\|f\|M^2$ such that $\sum_j h_j f_j = g^3$.

Naturally we will ask what is the necessary and sufficient condition for $g\in J(f_1,\ldots,f_m)$? Let T_f be the Toeplitz operator on $H^2(\Delta)$, where $f\in L^\infty(\Delta)$.

Theorem 1.5. ([11]). Let $f_1,\ldots,f_m \in H^\infty(\Delta)$. Then there exists $M > 0$ such that

$$M^2 \sum_j \|T_{\bar{f}_j} x\|^2 \geq \|T_{\bar{g}} x\|^2 , \quad x\in H^2(\Delta)$$

if and only if there exists a constant $C > 0$, and there exist $h_1,\ldots,h_m \in H^\infty(\Delta)$ such that $\|h_j\| \leq C$ and $\sum_j f_j h_j = g$.

When $g = 1$, it is shown by Arveson and Schubert in 1975.

Corollary 1.6. If $\Sigma|f_j|^2 = 1$ on $T = \partial\Delta$, $f_j \in VMOA$ for all j, then $g \in J(f_1,\ldots,f_m)$ iff there exists $M > 0$ such that

$$|g(z)| \leqslant M \max_j |f_j(z)|$$

for all $z \in \Delta$.

2. ON MULTIPLIER THEORY

A complex sequence $\{\lambda_n\}$ is called a multiplier from a space A of sequences into a space B of sequences if $\{\lambda_n a_n\} \in B$ whenever $\{a_n\} \in A$. Let (A,B) denote the collection of all multipliers from A into B. A space of analytic functions on the unit disc Δ can be regarded as a space of sequences by identifying each function with the sequence of its Taylor coefficients. In 1984, Campell and Leach [3] asked 22 problems on multipliers. Three of them are the following:

Problem 1. Is $X = \{\{\lambda_n\}: \lambda_n = O(n^\alpha)\ \alpha < 1/p - 3/2\}$ $= (\ell^\infty, B^p)$? where $B^p (0 < p < 1)$ consists of all analytic functions on Δ, satisfying

$$\int_0^1 (1-r)^{1/p-2} M_1(r,f)dr < \infty$$

where

$$M_p(r,f) = \left(\int_0^{2\pi} |f(re^{i\theta})|^p d\theta/2\pi\right)^{1/p}.$$

Problem 2. Is $H^1(\Delta)\subset\ell(2,\infty)$? where $\ell(2,\infty)$ is the space of complex sequences $\{\lambda_n\}$ satisfying

$$\sup_{m\in Z} \sum_{n=2^{m-1}}^{2^m-1} |\lambda_n|^2 < \infty .$$

Problem 3. Is $(H^1,\ell^\infty) = \bigcup_{q<\infty} (H^1,\ell^q)$? $(H^1,H^\infty) = \bigcup_{p<1} (H^p,H^\infty)$? $(H^1,\ell^1) = \bigcup_{p<1} (H^p,\ell^1)$? and if $0 < p < 1$,

$(H^p,\ell^\infty) = \bigcup_{p\leq q<\infty} (H^p,\ell^q)$?

Jianbin Xiao [21] proves that the answers of these three problems are all negative.

Theorem 2.1.

1). For $0 < p < 1$, $X \neq (\ell^\infty,H^p)$.

2). For $0 < p < 2$, $H^p\subset(2,\infty)$.

3). $\bigcup_{q<\infty} (H^1,\ell^q) \neq (H^1,\ell^\infty)$,

$\bigcup_{p<1} (H^p,H^\infty) \neq (H^1,H^\infty)$,

$\bigcup_{p<1} (H^p,\ell^1) \neq (H^1,\ell^1)$,

$\bigcup_{p\leq q<\infty} (H^p,\ell^q) \neq (H^p,\ell^\infty)$ for $0 < p < 1$.

3. ESTIMATES OF BMO NORM ON RIEMANN SURFACE

If $f \in BMOA(\Delta)$, S. Kobayashi showed that [10]

$$\|f\|_{BMO} = \sup_{a\in\Delta} \int_0^{2\pi} |f(e^{i\theta}) - f(a)|^2 P_a(e^{i\theta})\, d\theta/2\pi$$

$$\leq \left(\frac{1}{\pi} \text{Area } f(\Delta)\right)^{1/2},$$

where $P_a(e^{i\theta})$ is the Poisson kernel. If R is a Riemann surface, with a Green function $g(z,a)$ which has logarithm singularity at point a, if f is an analytic function on R, we define

$$\|f\|_{BMO} = \frac{2}{\pi} \sup_{a\in R} \iint_R |f'(z)|^2 g(z,a)\,dxdy$$

$$= \sup H(|f - f(a)|^2)(a),$$

where $H(h)$ is the least harmonic majorent of the subharmonic function h. Kobayashi, in 1985 showed that [9]

$$\|f\|_{BMO} \leq \left(\frac{1}{\pi} \text{Area } f(R)\right)^{1/2} .$$

We extend this result to arbitrary Riemann surface by using the methods of operator theory which is different to Kobayashi. The following theorem is due to Huang [14].

Theorem 3.1. Let R be a Riemann surface, and let f be an analytic function on R, then

$$\|f\|_{BMO} = \sup H(|f - f(a)|^2)(a)$$

$$\leq \left(\frac{1}{\pi} \text{Area } f(R)\right)^{1/2} .$$

4. SOME RESULTS ON BERGMAN SPACES

Let A^p be the Bergman space which consists of analytic functions on Δ, with norm

$$||f||_p = \left(\frac{1}{\pi} \iint_{|z|<1} |f(z)|^p dxdy\right)^{1/p} < \infty.$$

4.1 On Conjugate Functions

Let h^p denote the space consisting of harmonic functions u on Δ satisfying

$$\sup_{0<r<1} M_p(r,u) < \infty .$$

Let a^p denote the space of harmonic functions u which satisfy

$$||u||_p = \left(\frac{1}{\pi} \iint_{|z|<1} |u(z)|^p dxdy\right)^{1/p} < \infty .$$

Let $\tilde{u}(z)$ be the harmonic conjugate of u. The famous Riesz's theorem says that for $1 < p < \infty$, if $u \in h^p$ then $\tilde{u} \in h^p$, in other words, h^p is self-conjugate, but it is false when $p \leqslant 1$, and $p = \infty$. For the space a^p, in 1974, Forelli and Rudin [6] proved that a^p $(1 \leqslant p < \infty)$ is self-conjugate space by using operator theory. In 1985, Axler conjectured that $a^p (0 < p < 1)$ is self-conjugate. Recently, Jianbin Xiao [22] proves this conjecture true.

Theorem 4.1. For $0 < p < 1$, a^p is self-conjugate.

4.2 Derivatives of A^p Functions

Let us consider the question that for functions f of what kind of Bergmann space, we have $f' \in A^p$. We improve some results due to Watanable [19] and Stojan [18], and our results are best possible.

Theorem 4.2. ([23]). If $f' \in A^p$, $0 < p < 2$, then $f \in A^q$, where $q = 2p/(2-p)$. Moreover for each p, the index q is best possible.

Theorem 4.3. ([23]). If $f' \in A^2$, then $f \in VMOA$. Moreover there exists an analytic function f_0 on Δ, such that $f' \in A^2$ but $f_0 \notin H^\infty(\Delta)$.

Theorem 4.4 ([23]). For $2 < p < \infty$, if $f' \in A^p$, then f is continuous on $|z| \leq 1$, but $f(e^{i\theta})$ need not be absolutely continuous on $0 \leq \theta \leq 2\pi$.

5. EXTENSION OF A THEOREM OF CARLESON-DUREN

In 1969, Duren [5] generalized the well-known Carleson measure theorem. In 1975, Hastings [8] proved an analogous result on Bergmann spaces A^p. We generalize both results of Duren and Hastings. The following results are due to Ren and Huang [15].

Theorem 5.1. Let μ be a finite, positive measure on Δ, and suppose that the function $\phi(t)\colon (0,\infty) \to R$ satisfies the following conditions:

i. $\phi(0) = 0$, $\phi(t) > 0$ for $t > 0$,

ii. ϕ is increasing and $\lim (\phi(t)/t) = \infty$ or finite,

iii. ϕ' exists and is increasing in $(0,\infty)$,

iv. $\lim_{t\to o} \phi(ct^{-1})\phi(t^2) = 0$,

v. there exists a constant $B > 0$ such that

$$\sup_{t>o} \frac{t\phi'(t)\ \phi(ct^{-1})}{\phi(t)} = B$$

for all $c > 0$.

Then there exists a constant $C > 0$ depending only on ϕ such that

$$\phi^{-1}\{\int_{U^n} \phi(|f(z)|^p)d\mu(z)\} \leq C\ \|f\|^p_{A^p}$$

for all $f \in H^p$, if and only if there exists a positive constant A depending only on ϕ such that

$$\mu(S_h) \leq \phi(Ah)$$

for every set S_h of the form

$$S_h = \{z = re^{i\theta}: 1 - h \leq r < 1,\ \theta_o \leq \theta \leq \theta_o + h\},$$

where $0 < p < \infty$.

Theorem 5.2. Let μ be a finite, positive measure on U^n, and suppose that the function $\phi(t)$ satisfies the conditions (i) - (iii) in theorem 5.1, and there exists a constant K such that

$$\phi(t_1)\cdot\phi(t_2) \leq K\ \phi(t_1 t_2)$$

for arbitrary $t_1, t_2 > 0$. Then there exists a constant C such that

$$\phi^{-1}\{\int_{U^n}\phi(|f(z)|^p)d\mu(z)\} \leq C\ \|f\|^p_{A^p}$$

for all $f \in A^p$, if and only if there is a constant $A > 0$ such that

$$\mu(S_h) \leq \phi(A \prod_{j=1}^{n} h_j^2)$$

for every set S_h of the form

$$S_h = \{z = (r_1 e^{i\theta}, \ldots, r_n e^{i\theta}): 1-h_j \leq r_j < 1,$$
$$\theta_j^o \leq \theta_j \leq \theta_j^o + h_j,\ 1 \leq j \leq n\},$$

where $h = (h_1, \ldots, h_n)$, $0 < p < \infty$.

If we choose $\phi(t) = t^{p/q}$, $0 \leq p \leq q < \infty$, then we obtain the results of Duren and Hastings. If we choose $\phi(t) = t/(1 + t)$, $\alpha \geq 2$, then we obtain

Corollary 5.3. Let μ be a finite, positive measure on Δ, and suppose $\alpha \geq 2$. Then there is $C > 0$ such that

$$\int_\Delta \frac{|f(z)|^p}{1 + |f(z)|^p}\, d\mu(z) \leq C\ \frac{(\|f\|^p_{H^p})}{1 + \|f\|^p_{H^p}}$$

for all $f \in H^p$, if and only if there is a $A > 0$ such that

$$\mu(S_h) \leq A\ \frac{h^\alpha}{1 + h}$$

for every set S_h of the form in theorem 5.1, where

$0 < p < \infty$.

We also have the same results on the Bergmann space on the unit polydisc U^n [24].

REFERENCES

1. Arveson, W.B., Interpolation problems in Nest algebras, J. of Func. Analysis 20(1975), 208-233.

2. Behrens, M.F., The maximal ideal space of algebras of bounded analytic functions on infinitely connected domains, Trans. Amer. Math. Soc. 161(1971), 359-379.

3. Campbell, D.M. and Leach, R.J., A survey of H^p multipliers as related to classical function theory, Complex Variable, 3(1984), 85-111.

4. Carleson L., Interpolations by bounded analytic functions and the corona problem, Ann. of Math., (2)76(1962), 547-559.

5. Duren, P.L., Extension of a theorem of Carleson, Bull. Amer. Math. Soc., 75(1969), 143-146.

6. Forelli, F. and Rudin, W., Projections on spaces of holomorphic functions in balls, Indiana Univ. Math. J. 24(1974), 593-602.

7. Garnell, J.B. and Jones, P.W., The corona theorem for Denjoy domains, Acta Math. 155(1985), 27-40.

8. Hastings, W.W., A Carleson measure theorem for Bergmann spaces, Proc. Amer. Math. Soc. 52(1975), 237-241.

9. Kobayashi, S., Image areas and BMO norms of analytic functions, Kodai Math. J. 8(1985), 163-170.

10. Kobayashi, S., Range sets and BMO norms of analytic functions, Can. J. Math. 34(1984), 747-755.

11. Huang Le-fen, The ideals in the Banach algebra $H^\infty(D)$, Acta Math. Sinica, 2(1986), 270-279.

12. Huang Le-fen, An operator version of corona problem on polydisks, preprint.

13. Huang Le-fen, The corona theorem for some infinitely connected domains, preprint.

14. Huang Le-fen, Images areas and Garsia's norms of analytic functions, preprint.

15. Ren, Fu-yao and Huang Le-fen, Extension of a theorem of Carleson-Duren, preprint.

16. Rosemblum, M., Corona theorem for infinitely many functions, Integral Eq. and operator theory, 3(1980), 125-137.

17. Schubert C.F., The corona theorem as operator theorem, Proc. Amer. Math. Soc. 69(1978), 73-76.

18. Stojan, D., Some new properties of the space A^p, Mat. Vesnik, 5(18) (33)(1981), 151-157.

19. Watanable, H., On some properties of functions in weighted Bergmann spaces, Pro. Fac. Sci. Tokai Univ. 15(1980), 33-44.

20. Wolff, T., Oral communication.

21. Xian Jian-bin, On the open problems of Campbell and Leach, Kexue Tongbao.

22. Xiao Jian-bin, Some properties of conjugate functions, preprint.

23. Xiao Jian-bin, Derivatives of A^p functions, preprint.

24. Xiao Jian-bin, Carleson measure and multipliers on Bergmann spaces, preprint.

A CLASS OF FUNCTION SPACES

He Yuzan
Inst. of Math., Academia Sinica, Beijing

and

Ouyang Caiheng
Inst. of Math. Sci., Academia Sinica, Wuhan

I. INTRODUCTION

It is well known that the Bloch space $\mathscr{B}$ plays an important role in geometric function theory. And its situation in Bergman space is similar to that of the BMOA in H_p space. In [8], Peterson listed 16 equivalent definitions of BMOA, and recently Axler[2], Yamashita[11] gave several alternative descriptions of $\mathscr{B}$. It has been shown that $\mathscr{B}$ and BMOA can be looked at in many ways that it is of interest and there exist some analogies between $\mathscr{B}$ and BMOA. In this paper we consider a class of function spaces L^{Ba} (or H^{Ba}) where "M function" arises from an entire function with finite order and mean type introduced by Ding and Luo[5] in general version. This class of function spaces is very natural generalization of the classical L_p spaces and includes some important Hardy-Orlicz space, Bergman-Orlicz spaces. We give some equivalent definitions of $\mathscr{B}$ (and BMOA) which shows that there exist some relations between $\mathscr{B}$ (or BMOA) and some class of special L^{Ba} (or H^{Ba}).

II. A CLASS OF BERGMAN-ORLICZ SPACES AND BLOCH SPACE

Let $E(z) = \sum_{n=2}^{\infty} a_n z^n$ be an entire function with finite order $\rho(< +\infty)$, mean type $\sigma(< +\infty)$, and nonnegative coefficients $a_n \geqslant 0$, let $\{p_n\}$ be a sequence $1 \leqslant p_1 < p_2 < \ldots < p_n \nearrow \infty$, and for a fixed $k > 0$ satisfies

$$q_k = \overline{\lim_{n\to\infty}} \, p_n^k / n^{\frac{1}{\rho}} < +\infty \, . \qquad (1)$$

Suppose that $\{B_n\}$ is a sequence of Banach spaces, for $f \in \bigcap_{n=1}^{\infty} B_n$, we form the power series

$$I(\alpha, f) = \sum_{n=2}^{\infty} a_n \|f\|_{B_n}^{n} \alpha^n , \qquad (2)$$

where $\|\cdot\|_{B_n}$ is the norm in B_n. Let d_f denote the radius of convergence of (2) and Ba the set

$$Ba = \{f: f \in \bigcap_{n=1}^{\infty} B_n, \, d_f > 0\}, \qquad (3)$$

the norm of f in Ba is defined by $\|f\|_{Ba}$

$$= \inf_{I(|\alpha|, f) \leqslant 1} \left\{\frac{1}{|\alpha|}\right\} .$$

Ding and Luo [5] proved the following:

<u>Theorem A.</u> If any fundamental sequence of B_n contains a p. p. convergent subsequence, then Ba is also a Banach space.

Let $D = \{z, |z| \leqslant 1\}$, an analytic function $f(z)$ in D is called a Bloch function if

$$\|f\|_{\mathscr{B}} = \sup_{z \in D} \{(1 - |z|^2)|f'(z)|\}$$

is finite. The set of all such $f(z)$ forms a Banach space with the norm $\|f\| = |f(0)| + \|f\|_{\mathscr{B}}$, known as the Bloch space and denoted by $\mathscr{B}$.

For $1 \leqslant p < \infty$, the Bergman space $L_a^p(D, dA)$ is the Banach space of analytic functions $f(z)$ in D such that

$$\|f\|_p = \{\int_D |f(z)|^p \frac{dA(z)}{\pi}\}^{\frac{1}{p}}$$

is finite, where $dA(z)$ denotes the usual element of area on C.

For $\lambda \in D$, let $\phi_\lambda(z) = \dfrac{\lambda - z}{1 - \bar{\lambda}z}$ and $D(\lambda, r) = \{z, |\phi_\lambda(z)| < r\}$. denotes the pseudo-hyperbolic disk with pseudo-hyperbolic centre λ and radius r, and $\pi|D(\lambda, r)|$ the area of $D(\lambda, r)$. Axler[2] and Yamashita[11] proved the following theorem:

<u>Theorem B.</u> Let $1 \leqslant p < \infty$ and $0 < r < 1$, then the following quantities are equivalent:

(A) $\|f\|_{\mathscr{B}}$;

(B) $\text{Sup}\,\{\|f \circ \phi_\lambda - f(\lambda)\|_p, \ \lambda \in D\}$;

(C) $\text{Sup}\,\{[|D(\lambda, r)|^{-1} \int_{D(\lambda,r)} |f(z) - f_{D(\lambda,r)}|^p \dfrac{dA(z)}{\pi}]^{\frac{1}{p}}, \ \lambda \in D\}$;

where $f_{D(\lambda, r)} = |D(\lambda, r)|^{-1} \int_{D(\lambda,r)} f(z) \dfrac{dA(z)}{\pi}$.

(D) $\text{Sup}\ \{[|D(\lambda, r)|^{-1} \int_{D(\lambda,r)} |f(z) - f(\lambda)|^p \frac{dA(z)}{\pi}]^{\frac{1}{p}},\ \lambda \epsilon D\}$;

(E) $\text{Sup}\{\text{dis.}[\bar{f}|_{D(\lambda,r)}, H^\infty(D(\lambda, r)],\ \lambda \epsilon D\}$;

(F) $\text{Sup}\{[\text{area } f(D(\lambda, r))]^{\frac{1}{2}},\ \lambda \epsilon D\}$;

(G) $\text{Sup}\ \{[\int_{D(\lambda,r)} |f'(z)|^2 \frac{dA(z)}{\pi}]^{\frac{1}{2}},\ \lambda \epsilon D\}$.

In (3) we take $B_n = L_a^{p_n}(D, dA)$, $k = 1$, then we get a special Ba space, called a Bergman-Orlicz space and denoted by $L_1^{Ba} = L_1^{Ba}(D, dA)$. For a function defined in D, we set

$$\mathscr{M}(f) = \{g(z),\ g(z) = f \circ \phi_\lambda(z) - f(\lambda),\ \phi_\lambda \in \mathscr{M}\},$$

where

$$\mathscr{M} = \{\ \phi_\lambda(z),\ \ \phi_\lambda(z) = \frac{\lambda - z}{1 - \bar{\lambda} z},\ \ \lambda \epsilon D\}.$$

We have

<u>Theorem 1.</u> There exists a constant c such that

$$c^{-1}\ \|f\|_{\mathscr{B}} \leqslant \ |||f|||\ \leqslant c\,\|f\|_{\mathscr{B}} \tag{4}$$

where $|||f||| = \sup\ \{\ \|g\|_{L_1^{Ba}},\ g \in \mathscr{M}(f)\}$, $\|g\|_{L_1^{Ba}} = \inf_{I(|\alpha|,\ g)\ \leqslant\ 1} \{\frac{1}{|\alpha|}\}$, i.e., $f \in \mathscr{B}$ if and only if $\mathscr{M}(f)$ is bounded in L_1^{Ba}.

<u>Proof</u>. Let $\lambda \in D$ and $g(z) = f \circ \phi_\lambda(z) - f(\lambda)$, noting $\|f\|_{\mathscr{B}} = \|g\|_{\mathscr{B}}$, we have

$$|g(z)| = |\int_0^z g'(\zeta)d\zeta| \leqslant \frac{1}{2} \|f\|_{\mathscr{B}} \log \frac{1 + |z|}{1 - |z|},$$

therefore

$$\int_D |g(z)|^p \frac{dA(z)}{\pi} \leqslant 2^{-p} \|f\|^p_{\mathscr{B}} \int_D (\log \frac{1 + |z|}{1 - |z|})^p \frac{dA(z)}{\pi}$$

$$\leqslant 2^{-(p - 2)} \|f\|^p_{\mathscr{B}} \Gamma(p + 1),$$

thus again

$$\|g\|_{p_n} \leqslant 2^{-(1 - \frac{2}{p_n})} \|f\|_{\mathscr{B}} (\Gamma(p_n + 1))^{\frac{1}{p_n}}.$$

By definition

$$I(\frac{|\alpha|}{\|f\|_{\mathscr{B}}}, g) = \sum_{n=2}^{\infty} a_n \|g\|^n_{p_n} (\frac{|\alpha|}{\|f\|_{\mathscr{B}}})^n$$

$$\leqslant \sum_{n=2}^{\infty} A_n |\alpha|^n, \qquad (5)$$

where $A_n = a_n 2^{-n(1 - \frac{2}{p_n})} (\Gamma(p_n + 1))^{\frac{n}{p_n}}$, we are going to prove that the power series (5) has a positive radius of convergence. In fact, by the Stirling's formula

$$\Gamma(p + 1) = p^p e^{-p} \sqrt{2\pi p}(1 + o(1)), \qquad (6)$$

and the following relations between the order and the Taylor's coefficients of an entire function

$$(\rho e \sigma)^{\frac{1}{\rho}} = \overline{\lim_{n\to\infty}} (n^{\frac{1}{\rho}} \sqrt[n]{a_n}) \qquad (7)$$

and noting the condition (1) with k = 1, we get

$$\overline{\lim_{n\to\infty}} \sqrt[n]{A_n} \leqslant \overline{\lim_{n\to\infty}} \{(\sqrt[n]{a_n}\, n^{\frac{1}{\rho}})(p_n/n^{\frac{1}{\rho}})\, 2^{-(1-2/p_n)}$$

$$(2\pi p_n)^{\frac{1}{p_n}} e^{-1}\} \leqslant (2e)^{-1}\, q_1(\sigma\rho e)^{\frac{1}{\rho}} < \infty,$$

therefore the radius of convergence of (5) is

$$d_f \geqslant \frac{1}{\overline{\lim_{n\to\infty}} \sqrt[n]{A_n}} \geqslant \frac{2}{q_1(\sigma\rho e^{1-\rho})^{\frac{1}{\rho}}} > 0.$$

It shows that there exists $\alpha_0 \in (0,\ 2/q_1(\sigma\rho e^{1-\rho})^{\frac{1}{\rho}})$, say $\alpha_0 = \frac{1}{q_1(\sigma\rho e^{1-\rho})^{\frac{1}{\rho}}}$, such that

$$I\left(\frac{\alpha_0}{\|f\|_{\mathscr{B}}},\ g\right) \leqslant K < \infty,$$

where $K = K(\rho, \sigma, q_1)$ is a constant independent of f. If $K \leqslant 1$, then by the definition

$$\|g\|_{L_1^{Ba}} = \inf_{I(|\alpha|,g)\leqslant 1} \left\{\frac{1}{|\alpha|}\right\} \leqslant \frac{1}{\alpha_0} \|f\|_{\mathscr{B}} = c\, \|f\|_{\mathscr{B}}.$$

If $K > 1$, then

$$1 \geqslant \frac{1}{K} I\left(\frac{\alpha_0}{\|f\|_{\mathscr{B}}},\ g\right) \geqslant I\left(\frac{\alpha_0}{\|f\|_{\mathscr{B}}},\ \frac{g}{K}\right).$$

It shows that $\|\frac{g}{K}\|_{L_1^{Ba}} \leqslant \frac{1}{\alpha_0} \|f\|_{\mathscr{B}}$. Since the norm is homogeneous, we get

$$\|g\|_{L_1^{Ba}} \leqslant \frac{K}{\alpha_0} \|f\|_{\mathscr{B}} = c\, \|f\|_{\mathscr{B}}$$

both of the two cases, we have

$$|||f||| = \sup \{ \|g\|_{L_1^{Ba}},\ g \in \mathcal{M}(f) \} \leqslant c \ \|f\|_{\mathscr{B}}.$$

To prove the first inequality of (4), we take $c = \min \{ \frac{1}{n\sqrt{A_n}} \} = \frac{1}{n_0 \sqrt{a_{n_0}}}$. According to the definition, we have

$$1 \geqslant I\left(\frac{1}{\|g\|_{L_1^{Ba}}},\ g\right) \geqslant a_{n_0} \left(\frac{\|g\|_{p_{n_0}}}{\|g\|_{L_1^{Ba}}}\right)^n,$$

hence

$$\|g\|_{p_{n_0}} \leqslant c \ \ |||f||| .$$

On the other hand, we have

$$\|f\|_{\mathscr{B}} = \sup \{ |g'(0)|,\ g \in \mathcal{M}(f) \}$$

$$= \sup \{ | \lim_{t \to 1} \frac{2}{t^4}$$

$$\times \int_{D(0,t)} \bar{z} \ \ g(z) \frac{dA(z)}{\pi} |,\ g \in \mathcal{M}(f) \}$$

$$= 2 \sup \{ \|g\|_{p_{n_0}},\ g \in \mathcal{M}(f) \},$$

thus again

$$\|f\|_{\mathscr{B}} \leqslant 2c \ \ |||f||| .$$

<u>Corollary.</u> $f \in \mathscr{B}$ if and only if for each $g(z) \in \mathcal{M}(f)$ there exist constants K and α such that

$$\mu(t) = |E_t| = |\{z \in D,\ |g(z)| > t\}| \leqslant K\, e^{-\alpha t} \quad (8)$$

where $|E_t|$ denotes the normalized Lebesgue measure

of E_t, and if $f \in \mathscr{B}$, then $\alpha = 1/\|f\|_{\mathscr{B}}$.

In fact, if $f \in \mathscr{B}$, we take $E(z) = \sum_{n=2}^{\infty} \frac{1}{n!} z^n = e^z - 1 - z$, $p_n = n$, then $\rho = \sigma = 1$ and the condition (1) holds for $k = 1$. Noting that

$$\|g\|_n^n \leq 2^{-(n-2)} n! \ \|f\|_{\mathscr{B}}^n$$

and

$$I\left(\frac{1}{\|f\|_{\mathscr{B}}}, g\right) = \sum_{n=2}^{\infty} \frac{1}{n!} \ \|g\|_n^n \left(\frac{1}{\|f\|_{\mathscr{B}}}\right)^n \leq 2,$$

we have

$$\int_D \exp\left(\frac{|g(z)|}{\|f\|_{\mathscr{B}}}\right) \frac{dA(z)}{\pi} \leq 5,$$

therefore

$$|\{z \in D, \ |g(z)| > t\}| \leq 5e^{-\frac{1}{\|f\|_{\mathscr{B}}} t}.$$

Conversely, if (8) holds, then

$$\|g\|_1 = \int_0^{\infty} \mu(t)dt \leq K/\alpha,$$

thus

$$\|f\|_{\mathscr{B}} \leq 2 \sup \{ \|g\|_1, \ g(z) \in \mathscr{M}(f)\} \leq 2K/\alpha.$$

Now we take $B_n = H_{p_n}(D(0, r))$ and $k = 1/2$, then we obtain another special Ba space, and denote this space by $L_{\frac{1}{2}}^{Ba}(\partial D(0, r), \frac{d\theta}{2\pi})$ or $L_{\frac{1}{2}}^{Ba}$. We have

Theorem 2. For each $r \in (0, 1)$, there exists a constant C_r independent of f such that

$$C_r^{-1} \|f\|_{\mathscr{B}} \leqslant |||f|||_L \leqslant C_r \|f\|_{\mathscr{B}} \tag{9}$$

where

$$|||f|||_L = \sup\{ \|g\|_{L^{Ba}_{1/2}} , g \in \mathscr{M}(f)\},$$

$$\|g\|_{L^{Ba}_{1/2}} = \inf_{I(|\alpha|, g) \leqslant 1} \{\frac{1}{|\alpha|}\}, \; I(\alpha, g)$$

$$= \sum_{n=2}^{\infty} a_n \|g\|^n_{H_{p_n}} \alpha^n,$$

$$\|g\|^{p_n}_{H_{p_n}} = (1/2\pi) \int_0^{2\pi} |g(re^{i\theta})|^{p_n} d\theta .$$

To prove theorem 2, we prove the following lemma.

Lemma 1. $f \in \mathscr{B}$ if and only if for any $\lambda \in D$ and each $r \in (0, 1)$ there exists a positive nonincreasing function $A(r)$ such that for $g(z) \in \mathscr{M}(f)$

$$\hat{\mu}_r(t) = |\hat{E}_t| = |\{\zeta \in \partial D(0, r), |g(\zeta)| > t\}|$$
$$\leqslant e^{- A(r)t^2 + 1}, \tag{10}$$

where $|\hat{E}_t|$ denotes the normalized Lebesgue measure of $\hat{E}_t$ on $\partial D(0, r)$, and if $f \in \mathscr{B}$, then $A(r)^{-1} =$

$$k_r^2 \|f\|^2_{\mathscr{B}} = \frac{r^2}{1 - r^2} \|f\|^2_{\mathscr{B}} .$$

Proof. Let $g(z) = f \circ \phi_\lambda(z) - f(\lambda)$, we have

$$\int_{D(\lambda,r)} |f'(z)|^2 \frac{dA(z)}{\pi} = \int_{D(0,r)} |g'(w)|^2 \frac{dA(w)}{\pi} \leq \|g\|^2_{\mathscr{B}}$$

$$\times \int_{D(0,r)} \frac{1}{(1-|w|^2)^2} \frac{dA(w)}{\pi} = k_r^2 \ \|f\|^2_{\mathscr{B}} ,$$

where $k_r^2 = r^2/1 - r^2$. Setting

$$G(Z) = g(rZ)/k_r \ \|f\|_{\mathscr{B}} , \ |Z| < 1 ,$$

we get

$$\int_D |G'(Z)|^2 \frac{dA(Z)}{\pi} \leq 1 .$$

By Beurling's theorem[4], we have

$$|\{e^{i\theta} \epsilon \partial D, \ |G(e^{i\theta})| > t\}| \leq e^{-t^2+1} .$$

On the other hand,

$$\{e^{i\theta} \in \partial D, \ |G(e^{i\theta})| > t\} = \{\zeta \epsilon D(0, r), \ |g(\zeta)| > k_r \ \|f\|_{\mathscr{B}} t\} ,$$

thus

$$\hat{\mu}_r(t) \leq e^{-\frac{1}{K_r^2 \ \|f\|^2_{\mathscr{B}}} t^2 + 1} .$$

Conversely, if (10) holds, noting $A(r) = \min_{0<\rho\leq r} \{A(\rho)\}$, we have

$$\int_0^{2\pi} |g(\rho e^{i\theta})|^p \frac{d\theta}{2\pi} = p \int_0^{\infty} s^{p-1} \ \hat{\mu}_\rho(s)ds$$

$$\leqslant p \int_0^{\infty} s^{p-1} \ e^{-A(\rho)s^2+1} ds$$

$$\leqslant eA(r)^{-p/2} \ \Gamma(\frac{p}{2} + 1),$$

therefore

$$\int_{D(\lambda,r)} |f(z) - f(\lambda)|^p \frac{dA(z)}{\pi}$$

$$= \int_{D(0,r)} |g(z)|^p | \phi_\lambda'(z) |^2 \frac{dA(z)}{\pi}$$

$$\leqslant (1 - |\lambda|^2)^2 \ (1 - r|\lambda|)^{-4}$$

$$\int_{D(0,r)} |g(z)|^p \frac{dA(z)}{\pi}$$

$$\leqslant e \ r^2(1 - |\lambda|^2)^2 \ (1 - r|\lambda|)^{-4}$$

$$A(r) \ ^{-p/2} \ \Gamma(\frac{p}{2} + 1)$$

Since $D(\lambda, r)$ is also an Euclidean disk with Euclidean centre $\frac{\bar{\lambda}(1 - r^2)}{1 - r^2|\lambda|^2}$ and radius $\frac{r(1 - |\lambda|^2)}{1 - |\lambda|^2 r^2}$, the area of $D(\lambda, r)$ is

$$\pi|D(\lambda, r)| = \pi r^2(1 - |\lambda|^2)^2 \ (1 - |\lambda|^2 r^2)^{-2}.$$

Thus

$$|D(\lambda, r)|^{-1} \int_{D(\lambda,r)} |f(z) - f(\lambda)|^p \frac{dA(z)}{\pi} \leq$$

$$e\, A(r)^{-p/2} \left(\frac{1+r}{1-r}\right)^2 \Gamma\left(\frac{p}{2} + 1\right) .$$

On the other hand, applying the following formula

$$f'(0) = \frac{2}{r^4} \int_{D(0,r)} \overline{z}\, f(z) \frac{dA(z)}{\pi}$$

to $g(z) = f\circ \phi_\lambda(z) - f(\lambda)$, by the Hölder inequality, we get

$$(1 - |\lambda|^2)^p |f'(\lambda)|^p = |g'(0)|^p$$

$$\leq \frac{2^p}{r^{2p+2}} \int_{D(0,r)} |g(z)|^p \frac{dA(z)}{\pi}$$

$$= \frac{2^p (1 - |\lambda|^2)^2}{r^{2p+2}} \int_{D(\lambda,r)} |f(w) - f(\lambda)|^p$$

$$|1 - \overline{\lambda} w|^{-4} \frac{dA(w)}{\pi} \leq \frac{2^{p+4}}{r^{2p}(1 - r^2)^2} |D(\lambda, r)|^{-1}$$

$$\int_{D(\lambda,r)} |f(w) - f(\lambda)|^p \frac{dA(w)}{\pi}$$

$$\leq \frac{2^{p+4} e\, \Gamma(\frac{p}{2} + 1)}{r^{2p} (1 - r)^4 (A(r))^{p/2}}$$

thus $\|f\|_{\mathscr{B}} \leq C_r$.

<u>Proof of Theorem 2.</u> Suppose that $f\in\mathscr{B}$, by lemma 1 we have

$$\int_{\partial D(0,r)} |g(\zeta)|^p \frac{d\theta}{2\pi} = p \int_0^\infty t^{p-1} \hat{\mu}_r(t)dt$$

$$\leq e(k_r \|f\|_{\mathscr{B}})^p \, \Gamma(\frac{p}{2} + 1),$$

thus

$$\|g\|_{H_{p_n}} \leq e^{\frac{1}{p_n}} k_r \|f\|_{\mathscr{B}} (\Gamma(\frac{p_n}{2} + 1))^{\frac{1}{p_n}} .$$

We form the power series

$$I(\frac{|\alpha|}{\|f\|_{\mathscr{B}}}, g) = \sum_{n=2}^\infty a_n \|g\|^n_{H_{p_n}} (\frac{|\alpha|}{\|f\|_{\mathscr{B}}})^n$$

$$\leq \sum_{n=2}^\infty a_n k_r^n e^{\frac{n}{p_n}} (\Gamma(\frac{p_n}{2} + 1))^{\frac{1}{p_n} n} |\alpha|^n$$

$$= \sum_{n=2}^\infty \hat{A}_n |\alpha|^n . \tag{11}$$

By using (6) and (7), and noting the condition (1) with $k = \frac{1}{2}$, we get

$$\overline{\lim} \sqrt[n]{\hat{A}_n} \leq 2k_r e^{-\frac{1}{2}} q_{\frac{1}{2}}(\sigma\rho e)^{\frac{1}{\rho}} < \infty .$$

It shows that the power series (11) has a positive radius of convergence. Similar to the proof of theorem 1, we can complete this proof.

Remark 1. In [6], Gong Sheng pointed out that for $\lambda \in D$, if

$$g(z) - f(\lambda) = f \circ \phi_\lambda(z) = \sum_{n=0}^\infty A_n(\lambda) z^n ,$$

then

$$A_n(\lambda) = \sum_{j=0}^{n-1} (-1)^j \binom{j}{n-1} \frac{\bar{\lambda}^j (1 - |\lambda|^2)^{n-j}}{(n-j)!} f^{(n-j)}(\lambda) .$$

Therefore we have

$$\frac{1}{2} \sum_{n=1}^{\infty} n |A_n(\lambda)|^2 r^{2n} = \int_{D(0,r)} |g'(z)|^2 \frac{dA(z)}{\pi}$$

$$\leq \frac{r^2}{1 - r^2} \|f\|_{\mathscr{B}}^2 . \tag{12}$$

On the other hand, by theorem B

$$\|f\|_{\mathscr{B}}^2 = \|g\|_{\mathscr{B}}^2 \leq \frac{4}{r} \operatorname{dist.}[\bar{f}|_{D(\lambda,r)},$$

$$H^\infty(D(\lambda,r))]^2 \leq \frac{16}{r^2} \sup\{\int_{D(\lambda,r)} |f'(z)|^2$$

$$\frac{dA(z)}{\pi}, \lambda \in D\}$$

$$= \frac{8}{r^2} \sup\{ \sum_{n=1}^{\infty} n |A_n(\lambda)|^2 r^{2n}, \lambda \in D\} . \tag{13}$$

Inequalities (12) and (13) imply that $f \in \mathscr{B}$ if and only if for each $r \in (0, 1)$

$$\sup \{ \sum_{n=1}^{\infty} n |A_n(\lambda)|^2 r^{2n}, \lambda \in D\}$$

is finite.

Especially, f is in the Dirichlet space, i.e.

$$\int_D |f'(z)|^2 \frac{dA(z)}{\pi} < \infty .$$

if and only if

$$\sup \{ \sum_{n=1}^{\infty} n|A_n(\lambda)|^2, \lambda \in D \}$$

is finite.

Remark 2. We can construct other special Ba space, for example, if we take $B_n = H_{p_n}(D)$, $k = \frac{1}{\ell}$, where ℓ is even, then we get a special Ba space $L^{Ba}_{\frac{1}{\ell}}(\partial D, \frac{d\theta}{2\pi})$ and we can prove that $f \in L^{Ba}_{\frac{1}{\ell}}$ if and only if

$$\int_D |f^{\ell-2}(z)||f'(z)|^2 \frac{dA(z)}{\pi} < \infty .$$

III. DISCUSSION ON THE UNIT BALL OF C^n

In this section, we discuss a special Ba space which corresponds to BMOA or $\mathscr{B}$. In [7], we consider BMOA in the unit ball of $\mathbb{C}^n$ which denoted as $B = \{z \in \mathbb{C}^n, \langle z, z\rangle = \sum_{j=1}^{n} z_j \overline{z}_j < 1\}$. Let $d(\zeta, \eta) = |1 - \langle\zeta, \eta\rangle|^{1/2}$, $\zeta, \eta \in \partial B$, denote the quasi-metric and $Q(\zeta, \delta) = \{\eta \in \partial B, d(\zeta, \eta) < \delta\}$ the ball on ∂B, a function f which is integrable on ∂B is said to be in BMO, if

$$\|f\|_* = \sup_{Q \subset \partial B} \frac{1}{\sigma(Q)} \int_Q |f - f_Q| d\sigma < \infty ,$$

where σ is the normalized Lebesgue measure on ∂B, $f_Q = \frac{1}{\sigma(Q)} \int_Q f \, d\sigma$, if in addition its Poisson extension to B, $f(z) = \int_{\partial B} f(\zeta) \frac{(1 - |z|^2)^n}{|1 - \langle z, \zeta\rangle|^{2n}} d\sigma(\zeta)$ is holomorphic, then f is said to be in BMOA. By Rudin's expression[9], the Möbius transformation group of the unit ball is

$$\mathscr{M} = \{\psi = \phi_a U, \quad \phi_a = \frac{a - P_a z - s_a Q_a z}{1 - \langle z, a\rangle}\}$$

where $a \in B$, U is a unitary transformation, $Q_a = I - P_a$, $s_a = (1 - |a|^2)^{1/2}$, and

$$P_a z = \begin{cases} \frac{\langle z, a\rangle}{\langle a, a\rangle} a, & a \neq 0, \\ 0, & a = 0. \end{cases}$$

Set $\mathscr{M}(f) = \{g,\ g(z) = f\circ\psi(z) - f\circ\psi(0),\ \psi \in \mathscr{M}\}$, we have[7].

<u>Theorem C.</u> $f \in$ BMOA if and only if for each $g \in \mathscr{M}(f)$

$$\Lambda(t) = \mu_0(\{\zeta \in \partial B,\ |g(\zeta)| > t\}) \leqslant K e^{-\alpha t} \tag{14}$$

where $\mu_a(E) = \int_E \frac{(1 - |a|^2)^2}{|1 - \langle a, \zeta\rangle|^{2n}} d\sigma(\zeta)$ is the harmonic measure of the set $E(\subset B)$ at $a(\in B)$ with respect to B, and if $f \in$ BMOA, then $\alpha = c/\|f\|_*$, where K and c are two constants independent of f.

Let $H_p(B)$, $1 \leqslant p < \infty$, be the Hardy space[9] in B such that $\|f\|^p_{H_p} = \sup_{0<r<1} \int_{\partial B} |f_r|^p d\sigma$, we take $B_n = H_{p_n}(B)$ and $k = 1$ in (1), we get a special Ba space $H_1^{Ba}(B)$, called a Hardy-Orlicz space, and have

<u>Theorem 3.</u> There exists a constant c such that

$$c^{-1} \|f\|_* \leqslant |||f|||_H \leqslant c \|f\|_* \tag{15}$$

where $|||f|||_H = \sup\{\|g\|_{H_1^{Ba}},\ g \in \mathscr{M}(f)\}$, $\|g\|_{H_1^{Ba}} =$

$$\inf_{I(|\alpha|, g) \leq 1} \{\frac{1}{|\alpha|}\}, \quad I(\alpha, g) = \sum_{n=2}^{\infty} a_n \|g\|_{p_n}^{n} \alpha^n.$$

Proof. In fact, if $f \in BMOA$, by the Theorem C we have

$$\|g\|_{H_p}^{p} \leq \int_{\partial B} |g|^p d\sigma = p \int_0^{\infty} t^{p-1} \Lambda(t) dt$$

$$\leq K p \int_0^{\infty} t^{p-1} \exp(\frac{-c}{\|f\|_*} t) dt$$

$$= \frac{K\Gamma(p+1)}{c^p} \|f\|_*^{p} . \tag{16}$$

Similar to the proof of theorem 1, we can complete the proof of theorem 3.

Bloch functions in $\mathbb{C}^n$ were finely considered by Timoney[10]. Let $f: B \to \mathbb{C}$ be a holomorphic function, $H_z(x, \overline{x})$ denote the Bergman metric in B. For $z \in B$ define

$$Q_f(z) = \sup_{0 \neq x \in \mathbb{C}^n} \{|(\nabla_z f)x| / H_z(x, \overline{x})^{\frac{1}{2}}\} .$$

A holomorphic function f is said to be a Bloch function which is denoted by $\mathscr{B}$, if

$$\|f\|_{\mathscr{B}} = \sup_{z \in B} Q_f(z) < \infty .$$

This space $\mathscr{B}(B)$ of Bloch functions (modulo the constant function) form a Banach space with the Bloch norm $\|\cdot\|_{\mathscr{B}}$. Denote

$$\|f\|_{\chi} = \sup_{z \in B} \{|\nabla_z f|(1 - |z|^2)\} .$$

Timoney proved that $\|\cdot\|_{\mathscr{B}}$ and $\|\cdot\|_{\chi}$ are equivalent, i.e. there exist constants α_n and $\beta_n > 0$ such that

$$\alpha_n \|f\|_X \leq \|f\|_{\mathscr{B}} \leq \beta_n \|f\|_X$$

for each $f \in \mathscr{B}(B)$.

For $1 \leq p < \infty$, the space $L^p \cap H(B)$[9] of all holomorphic functions with the norm

$$\|f\|_p = \{\int_B |f|^p \, dv\}^{\frac{1}{p}} < \infty,$$

is a closed subspace of $L^p(B)$, where dv is the volume element of the normalized Lebesgue measure such that $v(B) = 1$.

If we take $B_n = L^{pn} \cap H(B)$, $k = 1$ and $k = \frac{2n-1}{2n}$, then we get other Ba spaces $L_1^{Ba}(B)$ and $L_{\frac{2n-1}{2n}}^{Ba}(B)$, respectively. We have

<u>Theorem 4.</u> If $f \in \mathscr{B}$, then $\mathscr{M}(f)$ is bounded in $L_1^{Ba}(B)$. Further, if

$$\sup_{g \in \mathscr{M}(f)} |\nabla_z g| \leq (1 - |z|)^{\beta}, \ -1/2n < \beta < 0,$$

then $\mathscr{M}(f)$ is bounded in $L_{\frac{2n-1}{2n}}^{Ba}(B)$.

<u>Proof.</u> Let $z \in B$, then

$$|f(z) - f(0)| = |\int_{t=0}^{1} (\nabla_{tz} f) z \, dt|$$

$$\leq \|f\|_X \int_{t=0}^{1} (1 - |z|t)^{-1} |z| \, dt$$

$$= - \|f\|_X \log (1 - |z|)$$

$$\leqslant - \alpha_n^{-1} \|f\|_{\mathscr{B}} \log (1 - |z|).$$

But since $Q_{f\circ\psi}(z) = Q_f(\psi(z))$, and each $\psi \in \mathscr{M}$ is a bijection, hence $\|f\circ\psi\|_{\mathscr{B}} = \|f\|_{\mathscr{B}}$. Replacing f by $f\circ\psi$ from the above inequality, there leads

$$|g(z)| \leqslant - \alpha_n^{-1} \|f\|_{\mathscr{B}} \log (1 - |z|),$$

therefore, simple computation gets

$$\int_B |g(z)|^p \, dv \leqslant 2n \alpha_n^{-p} \|f\|_{\mathscr{B}}^p \Gamma(p + 1). \qquad (17)$$

A same manipulation with the proof of theorem 1 shows that $\mathscr{M}(f)$ is bounded in $L_1^{Ba}(B)$.

To prove the other assertion in theorem 4, we first prove that for each $g \in \mathscr{M}(f)$, in which

$$V(\{z \in B, |g(z)| > t\}) \leqslant Ke^{-\lambda t^{\frac{2n}{2n-1}}} \qquad (18)$$

where K is an absolute constant and $\lambda = \lambda(n, \beta, \|f\|_{\mathscr{B}})$.

In fact, by theorem 3.4 in [10], the hypothesis implies that f is a Bloch function. Since the unit ball B of $\mathbb{C}^n$ can be regarded as a unit ball B_{R_e} in $\mathbb{R}^{2n}$, $B_{R_e} = \{(x_1, \ldots, x_n, y_1, \ldots, y_n) \in \mathbb{R}^{2n}, \sum_{j=1}^n x_j^2 + y_j^2 < 1\}$, $z_j = x_j + iy_j$, $1 \leqslant j \leqslant n$, $g(z) = g(z_1, \ldots, z_n) = g(x_1, \ldots, x_n, y_1, \ldots, y_n)$ is defined in B_{R_e}. Thus by (17)

$$\int_{B_{Re}} |g(x_1, \dots, x_n, y_1, \dots, y_n)|^{2n} dv$$

$$= \int_B |g(z)|^{2n} dv \leq 2n\alpha_n^{-2n} \|f\|_{\mathscr{B}}^{2n} \Gamma(2n + 1)$$

Noting $\beta > - 1/2n$, we have

$$\int_{B_{Re}} |\frac{\partial g}{\partial x_j}|^{2n} dv = \int_B |\frac{\partial g}{\partial x_j}|^{2n} dv$$

$$\leq 2n \int_0^1 r^{2n-1} dr \int_{\partial B} (\sqrt{2} \mid \nabla_z g \mid)^{2n} d\sigma$$

$$= n2^{n+1} B(2n, 2n\beta + 1),$$

where $B(.,.)$ denotes Beta function. Similarly

$$\int_{B_{Re}} |\frac{\partial g}{\partial y_i}|^{2n} dv \leq n2^{n+1} B(2n, 2n\beta + 1).$$

By Trudinger's theorem[1]

$$\int_{B_{Re}} (\exp(\frac{|g|}{\|g\|_A})^{\frac{2n}{2n-1}} - 1) dv \leq 1,$$

where $\|g\|_A = \inf \{k > 0, \int_{B_{Re}} A(\frac{|g|}{k}) dv \leq 1\}$, $A(t) = \exp t^{\frac{2n}{2n-1}} - 1$, the norm $\|g\|_A$ only depends on n, β and $\|f\|_{\mathscr{B}}$. Thus

$$\exp\left(\|g\|_A^{\frac{2n}{1-2n}}\, t^{\frac{2n}{2n-1}}\right) v(\{z\in B,\ |g(z)| > t\})$$

$$\leq \int_{\{z\in B,\ |g(z)|>t\}} \exp\left(\|g\|_A^{\frac{2n}{1-2n}}\, |g(z)|^{\frac{2n}{2n-1}}\right) dv$$

$$\leq \int_B \exp\left(\frac{|g|}{\|g\|_A}\right)^{\frac{2n}{2n-1}} dv \leq 2,$$

this proves (18). Using (18), we obtain

$$\|g\|_p^p = \int_B |g(z)|^p\, dv$$

$$\leq K\, \lambda^{\frac{(1-2n)p}{2n}}\ \Gamma\left(\frac{2n-1}{2n}\, p + 1\right)$$

$$\leq C(n,\ \beta,\ \|f\|_{\mathscr{B}},\ p)\ \Gamma\left(\frac{2n-1}{2n}\, p + 1\right).$$

Similar to the proof of Theorem 1, we can complete the proof of the second assertion of Theorem 4.

<u>Remark</u>. The above theorems show that the Ba space which corresponds to $\mathscr{B}$ (or BMOA) is a class of special Ba spaces, where the sequence $\{p_n\}$ and the growth of "M function" $E(|z|)$ must satisfies some conditions. Recall that the dual of the Bergman space $L_a^p(D, dA)$ (or $H_p(D)$) $(1 < p < \infty)$ can be identified with $L_a^{p'}(D, dA)$ (or $H_{p'}$) where $\frac{1}{p} + \frac{1}{p'} = 1$, and the pre-dual of H_1(or $L_a^1(D, dA)$) can be identified with VMOA (or

the little Bloch space $\mathscr{B}_0$), a natural question to ask is that what is the dual of L_k^{Ba} (D, dA) (or H_k^{Ba}) and under what conditions the L_k^{Ba} (D, dA) (or H_k^{Ba}) corresponds to $\mathscr{B}_0$ (or VMOA).

REFERENCES

1. Adams, R., Sobolev Spaces, Academic Press, New York, 1975.

2. Axler, S., The Bergman space, the Bloch space and commutators of multiplication operators, Duke Math. J., 53(1986), 315 - 332.

3. Baernstein II. A., Analytic functions of bounded mean oscillation, in "Aspects of Contemporary Complex Analysis", Academic Press, New York, 1980, 3 - 36.

4. Beurling, A., Études sur un Problème de Majoration, thèse, Almquist and Wiksell, Uppsalla, 1933.

5. Ding, X. X. and Lou, P. Z., Ba spaces and some estimates of Laplace operator, J. of Systems Sci. & Math., Sci., 1(1981), 1, 9 - 33.

6. Gong Sheng, A remark on Möbius transformations, Pure & Appl. Math., 1(1985), 1 - 15.

7. Ouyang Caiheng, Baerstein theorem in unit ball of C^n, Kexue Tongbao, 32(1987), 581 - 584.

8. Petersen, K. E., Brownian Motion, Hardy Spaces and BMO, Cambridge Univ. Press, Cambridge, 1977.

9. Rudin, W., Function Theory in the Unit Ball of C^n, Springer-Verlag, New York, 1980.

10. Timoney, R. M., Bloch functions in several complex variables, I, Bull. London Math. Soc., 12(1980), 241-267.

11. Yamashita, S., Criteria for functions to be Bloch, Bull. of Australian Math. Soc., 21(1980), 223 - 227.

APPLICATION OF COMPLEX FUNCTIONS TO CRACK PROBLEMS OF HALF-PLANE WITH DIFFERENT MEDIA*

Lu Jianke (Chien-ke Lu)
Wuhan University
China

The mathematical problems of infinite elastic plane with cracks of different media were discussed in many works, e.g., [1-5]. In [3], the crack problems for a half-plane of single material were also studied. However, in practice, for instance, in geological prospecting, most often occur the similar problems in the half-plane with different media. In this paper, mathematical methods for treating such problems are posed, when the elastic half-plane with cracks consists of two kinds of different media. They are transferred into certain boundary value problems of analytic functions, and then are reduced to singular integral equations along the boundary of the half-plane, interface and cracks, the unique solvability of which is established. It is mentioned that the boundary of the half-plane may be eliminated among the paths of integration, which has obvious advantages. Some other comments are also given in the last section.

---

*Project supported by the National Science Funds of P.R.C.

§1. FORMULATION OF THE PROBLEM AND TRANSFER TO BOUNDARY VALUE PROBLEM

Let the elastic body occupy the lower complex half-plane consisting of two different isotropic media, the upper one of which is a strip S^1: $-h \leqslant \mathrm{Im}z \leqslant 0$ $(h > 0)$, and the lower one, is a half-plane S^2: $\mathrm{Im}z \leqslant -h$, and the interface is denoted by γ: $\mathrm{Im}z = -h$. Both of S^1 and S^2 may contain some cracks. Assume there are p cracks (Liaponov arcs) total in number: $L_j = a_jb_j$, $j=1,\ldots,p$, non-intersecting to each other nor to the interface as well as to the boundary X-axis. Denote $L = \sum_{j=1}^{p} L_j$ and $\Gamma = X + \gamma + L$. The positive sense of Γ is chosen as shown in Fig. The elastic constants of S^k are noted

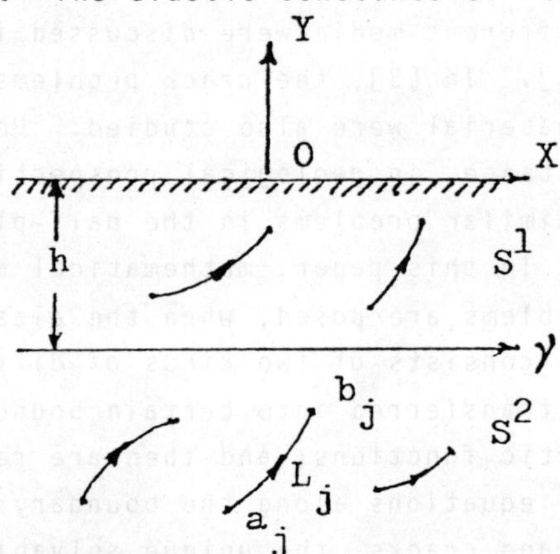

by κ^k, μ^k. Without loss of generality, we assume there is no stress nor rotation at infinity.

The proposed problem may be stated as follows. Given the external stress function $X_n(x)+iY_n(x)$ on X and the external stresses on the positive side and the negative side of L: $X_n^{\pm}(t)+iY_n^{\pm}(t)$, $t \in L$, determine the elastic equilibrium. We always assume they satisfy Hölder conditions (including $x = \pm\infty$ on X).

Denote

$$X+iY = \int_{-\infty}^{+\infty} [X_n(x)+iY_n(x)]dx,$$

$$X_j^{\pm}+iY_j^{\pm} = \int_{L_j} [X_n^{\pm}(t)+iY_n^{\pm}(t)]ds, \ j=1,\ldots,p$$

(s is the arc-length parameter). The external resultant principal stress vector on L_j is

$$X_j+iY_j = (X_j^+ + iY_j^+) + (X_j^- + iY_j^-) .$$

By using the process for separating the multiplicity of Kolosov functions (Cf. [6], §§18,25), we may restrict ourselves to the case

$$X+iY = 0, \quad X_j+iY_j = 0, \ j=1,\ldots,p.$$

As usual, denote

$$f(x) = i \int_{-\infty}^{x} [X_n(\xi)+iY_n(\xi)]d\xi, \qquad x\epsilon X,$$

$$f^{\pm}(t) = \pm i \int_{a_j}^{t} [X_n^{\pm}(\tau)+iY_n^{\pm}(\tau)]ds, \quad t\epsilon L_j,$$

so that $f(\pm\infty)=0$, $f^{\pm}(a_j)=0$, $f^+(b_j)=f^-(b_j)$, $j=1,\ldots,p$.

Hence $f'(x)\epsilon\hat{H}$ on X,[4] $f^{\pm}(t)$, $f'^{\pm}(t)\epsilon H$ on L. Moreover, we assume $f(x)\epsilon\hat{H}$ on X.

Now, the Kolosov functions $\phi(z)$, $\psi(z)$ are sectionally holomorphic functions in $S = S^1+S^2$ with γ as the line of discontinuity. Regarding the external conditions on $X+L$ and the conditions of continuity of stresses and displacements on γ, we have the following boundary value conditons:[6]

$$\phi(x) + x\overline{\phi'(x)} + \overline{\psi(x)} = -f(x) + C_o, \ x\epsilon X, \tag{1.1}$$

$$\phi^+(\sigma) + \sigma\overline{\phi'^+(\sigma)} + \overline{\psi^+(\sigma)} = \phi^-(\sigma) + \sigma\overline{\phi'^-(\sigma)} + \overline{\psi^-(\sigma)}, \ \sigma\epsilon\gamma, \tag{1.2}$$

$$\alpha^1\phi^+(\sigma) - \beta^1[\sigma\overline{\phi'^+(\sigma)} + \overline{\psi^+(\sigma)}] = \alpha^2\phi^-(\sigma) - \beta^2[\sigma\overline{\phi'^-(\sigma)} + \overline{\psi^-(\sigma)}], \ \sigma\epsilon\gamma, \tag{1.3}$$

$$\phi^{\pm}(t) + t\overline{\phi'^{\pm}(t)} + \overline{\psi^{\pm}(t)} = f^{\pm}(t) + C(t), \ t\epsilon L, \tag{1.4}$$

where C_o and $C(t) = C_j$, $t\epsilon L_j$, $j=1,\dots,p$, are undetermined complex constants. Here we have put

$$\alpha^k = \kappa^k/\mu^k, \quad \beta^k = 1/\mu^k, \quad k = 1,2 .$$

The negative sign before $f(x)$ in (1.1) is due to the fact that the elastic region lies in the negative side of X. In general, $C(t)$ may take different constants $C_j^{\pm}$ on different sides of L_j, but, in our case, $C_j^+ = C_j^-$ $(=C_j)$ owing to our definition of $f^{\pm}(t)$ (Cf. [6],§26).

Thus, the problem is now transferred to determine two sectionally holomorphic functions $\phi(z),\psi(z)$ in S satisfying the boundary value conditions (1.1) — (1.4), with finite values $\phi(\infty)$ and $\psi(\infty)$ on $X+\gamma$.

§2. REDUCTION TO SINGULAR INTEGRAL EQUATIONS

In order to solve the boundary value problem (1.1) — (1.4), we introduce a new unknown complex function $\omega(\zeta),\zeta\epsilon\Gamma = X+\gamma+L$, such that, for $z\epsilon S\backslash\gamma$,

$$\phi(z) = \frac{1}{2\pi i}\int_\Gamma \frac{\omega(\zeta)}{\zeta - z}\, d\zeta , \tag{2.1}$$

$$\psi(z) = -\frac{1}{2\pi i}\int_\Gamma \frac{\overline{\omega(\zeta)} + \bar{\zeta}\omega'(\zeta)}{\zeta - z}\, d\zeta + \frac{1}{2\pi i}\int_L \frac{\overline{F(\tau)}}{\tau - z}\, d\tau , \tag{2.2}$$

where

$$F(t) = f^+(t) - f^-(t) .$$

We assume $\omega(x)$, $\omega'(x)\in\hat{H}$ on $X+\gamma$ and $\omega'(t)\in H$ on L with $\omega(a_j) = \omega(b_j) = 0$ $(j=1,\ldots,p)$. The existence of such $\omega(\zeta)$ is to be proved later. We shall ask $\phi(\zeta) = o(1)$, $\phi'(\zeta) = o(1/|\zeta|)$, $\psi(\zeta) = o(1)$ on $X+\gamma$ as $\zeta\to\infty$. Then, if such $\omega(\zeta)$ exists, we must have $\omega(\zeta) = o(1)$, $\omega'(\zeta) = o(1/|\zeta|)$ on $X+\gamma$. Hence $C_0 = 0$ in (1.1).

Substituting (2.1), (2.2) into (1.1) with $z\to x$ from the negative side of X, by Plemelj's formula, we obtain

$$\underline{\underline{K}}_X\omega \equiv \frac{1}{2\pi i}\int_\Gamma \frac{\omega(\zeta)}{\zeta - x}\, d\zeta + \frac{1}{2\pi i}\int_\Gamma \frac{\omega(\zeta)}{\bar{\zeta} - x}\, d\bar{\zeta} - \frac{1}{2\pi i}\int_{L+\gamma} \overline{\omega(\zeta)}\, d\,\frac{\zeta - x}{\bar{\zeta} - x}$$

$$= -f(x) + \frac{1}{2\pi i}\int_L \frac{F(\tau)}{\bar{\tau} - x}\, d\bar{\tau} ,\quad x\in X. \tag{2.3}$$

Substituting them into (1.2), we see that it is identically satisfied.

Substituting them into (1.3), we obtain

$$\underline{\underline{K}}_\gamma \omega \equiv (\alpha^1+\alpha^2+\beta^1+\beta^2)\omega(\sigma) + \frac{\alpha^1-\alpha^2-\beta^1+\beta^2}{\pi i}\int_\Gamma \frac{\omega(\zeta)}{\zeta-\sigma}d\zeta$$

$$+ \frac{\beta^1-\beta^2}{\pi i}\{\int_\Gamma \omega(\zeta)d\log\frac{\zeta-\sigma}{\bar{\zeta}-\bar{\sigma}} + \int_\Gamma \overline{\omega(\zeta)}\, d\frac{\zeta-\sigma}{\bar{\zeta}-\bar{\sigma}}\}$$

$$= -\frac{\beta^1-\beta^2}{\pi i}\int_L \frac{\overline{F(\tau)}}{\bar{\tau}-\bar{\sigma}}d\bar{\tau}, \quad \sigma\in\gamma. \tag{2.4}$$

Finally, substituting them into (1.4) for boundary values on positive or negative side of L, we obtain the same equation

$$\underline{\underline{K}}_L \omega \equiv \frac{1}{\pi i}\int_\Gamma \frac{\omega(\zeta)}{\zeta-t}d\zeta - \frac{1}{2\pi i}\int_\Gamma \omega(\zeta)d\log\frac{\zeta-t}{\bar{\zeta}-\bar{t}}$$

$$-\frac{1}{2\pi i}\int_\Gamma \overline{\omega(\zeta)}\, d\frac{\zeta-t}{\bar{\zeta}-\bar{t}} = \frac{1}{2}G(t) + \frac{1}{2\pi i}\int_L \frac{\overline{F(\tau)}}{\bar{\tau}-\bar{t}}d\bar{\tau}$$

$$+ C_j, \quad t\in L_j, \; j=1,\dots,p, \tag{2.5}$$

where

$$G(t) = f^+(t) + f^-(t), \quad t\in L.$$

(2.3) — (2.5) constitute a singular integral equation along Γ with p undetermined constants $C_1,\dots,C_p$. Its characteristic part is

$$A(\zeta)\omega(\zeta) + \frac{B(\zeta)}{\pi i}\int_\Gamma \frac{\omega(\zeta)}{\zeta-\zeta_0}d\zeta, \quad \zeta_0\in\Gamma,$$

where

$$A(\zeta) = \begin{cases} 0, & X+L, \\ \alpha^1+\alpha^2+\beta^1+\beta^2, & \zeta\in\gamma; \end{cases}$$

$$B(\zeta) = \begin{cases} 1, & X+L, \\ \alpha^1-\alpha^2-\beta^1+\beta^2, & \zeta\in\gamma. \end{cases}$$

We see that $A(\zeta) \pm B(\zeta) \neq 0$ on Γ and so it is of normal type. We should solve it in class h_{2p} (Cf.[7]), i.e., the solution to be sought should be bounded at all the end-points $a_j, b_j (j=1,\ldots,p)$ of the cracks. Then, the solution $\omega(\zeta)$ would actually possess the properties described before if it exists (Cf. [6]).

§3. UNIQUE SOLVABILITY OF THE EQUATION

We shall now prove the obtained singular integral equation on Γ is uniquely solvable in class h_{2p} with constants $C_1,\ldots,C_p$ uniquely suitably chosen.

For simplicity, we rewrite the obtained equations as

$$\underset{=}{K}_{\Gamma}\omega \equiv \underset{=}{K}\omega = H(\zeta) + C(\zeta), \quad \zeta\epsilon\Gamma, \tag{3.1}$$

where $H(\zeta)$ is a given function determined by the boundary value conditions and

$$C(\zeta) = \begin{cases} 0, & \zeta\epsilon X+\gamma, \\ C_j, & \zeta\epsilon L_j,\ j=1,\ldots,p. \end{cases} \tag{3.2}$$

It is easily seen that the index of (3.1) in class h_{2p}[7] is $-p$.

Let us consider first the simplest case $H(\zeta)\equiv 0$, which means there is no external load neither on X nor on each side of L. If $\omega_0(\zeta)$ is a solution of $K\omega = C_0(\zeta)$ in class h_{2p} with $\omega(\zeta) = o(1)$ on $X+\gamma$ as $\zeta \to \infty$, where $C_0(\zeta) = C_j^0$, $\zeta\epsilon L_j (j=1,\ldots,p)$, we shall prove that it must be $C_j^0 = 0$ and $\omega_0(\zeta) \equiv 0$.

Let the functions defined by (2.1), (2.2) be $\phi_0(\zeta)$, $\psi_0(\zeta)$ respectively when $\omega(\zeta) = \omega_0(\zeta)$ is substituted. Repeating the same reasoning as in §2, we then again have (1.1) — (1.4) with replacements ϕ,ψ,C_j by ϕ_0,ψ_0,C_j^0 respectively, which means $\phi_0(z)$, $\psi_0(z)$ are the Kolosov functions in the case considered. Since $\phi_0(z)$

and $\psi_o(z) \to 0$ as $z \to \infty$ with $\omega(\zeta) \to 0$ as $\zeta \to \infty$ on $X+\gamma$, it is then evident, by the uniqueness of the Kolosov functions,

$$\phi_o(z) \equiv 0, \quad \psi_o(z) \equiv 0, \quad z\epsilon S, \tag{3.3}$$

which follows immediately $C_j^o = 0$ $(j=1,\ldots,p)$.

Let

$$\phi_*(z) = \frac{1}{2\pi i}\int_\Gamma \frac{\omega_o(\zeta)}{\zeta-z}\,d\zeta,$$

$$\psi_*(z) = -\frac{1}{2\pi i}\int_\Gamma \frac{\overline{\omega_o(\zeta)}+\bar{\zeta}\omega_o'(\zeta)}{\zeta-z}\,d\zeta, \qquad z\bar{\epsilon}\Gamma. \tag{3.4}$$

They are sectionally holomorphic functions on the whole plane with $\phi_*(z) = \phi_o(z) = 0$, $\psi_*(z) = \psi_o(z) = 0$, $z\epsilon S$. We readily see that $\omega_o(x)$ and $-\overline{\omega_o(x)} - x\omega_o'(x)$ respectively are the boundary values on X of $\phi_*(z)$ and $\psi_*(z)$ in the upper half-plane Z^+ $(\mathrm{Im} z \geqslant 0)$:

$$\phi_*^+(x) = \omega_o(x),\; \psi_*^+(x) = -\overline{\omega_o(x)} - x\omega_o'(x), \tag{3.5}$$

with $\phi_*(\infty) = \psi_*(\infty) = 0$. Eliminating $\omega_o(x)$ in these equalities, we have

$$\phi_*^+(x) + x\overline{\phi_*'^+(x)} + \overline{\psi_*(x)} = 0, \qquad x\epsilon X.$$

This means $\phi_*(z)$, $\psi_*(z)$ are the Kolosov functions in Z^+ with zero boundary condition. Thereby $\phi_*(z) = \psi_*(z) = 0$, $z\epsilon Z^+$. Thus, by (3.5), we have $\omega_o(x) = 0$, $x\epsilon X$.

Then (3.4) becomes

$$\phi_*(z) = \frac{1}{2\pi i}\int_{L+\gamma}\frac{\omega_o(\zeta)}{\zeta - z}\,d\zeta,$$

$$\psi_*(z) = -\frac{1}{2\pi i}\int_{L+\gamma}\frac{\overline{\omega_o(\zeta)}+\bar{\zeta}\omega_o'(\zeta)}{\zeta - z}\,d\zeta, \quad z\bar{\epsilon}L+\gamma. \tag{3.6}$$

If we imagine $S^+ = \bar{Z}^+ + S^1$ is an elastic region with elastic constants κ^1, μ^1 and $S^- = S^2$ as before, and $\phi_*(z)$, $\psi_*(z)$ with $\phi_*(\infty) = \psi_*(\infty) = 0$ represented by (3.6) as the Kolosov functions for the elastic infinite plane with interface γ and cracks L_j, $j=1,\ldots,p$, under zero boundary condition, then $\omega_o(\zeta) = 0$, $\zeta\epsilon L+\gamma$, as shown in [4].

Thus, we have proved $\omega_o(\zeta) \equiv 0$ on Γ as well as $C_j^o = 0$.

Now we consider the general case. We have shown that the homogeneous equation $\underline{\underline{K}}\omega = 0$ has only trivial solution $\omega = 0$. According to the extended Noether's theorem (Cf.[7], §112), the number of linearly independent (in the real coefficient field) complex solutions of its adjoint equation $\underline{\underline{K}}'\chi = 0$ in class h_o is $-2\kappa=2p$, denoted by $\chi_1,\ldots,\chi_{2p}$, and the conditions of (unique) solvability of (3.1) in class h_{2p} are

$$\mathrm{Re}\int_\Gamma[H(\zeta) + C(\zeta)]\chi_j(\zeta)d\zeta = 0, \; j=1,\ldots,2p,$$

or, similarly,

$$\mathrm{Re}\sum_{k=1}^{p} C_k\int_{L_k}\chi_j(t)dt = -\mathrm{Re}\int_\Gamma H(\zeta)\chi_j(\zeta)d\zeta,$$

$$j=1,\ldots,2p.$$

Writing the real and the imaginary parts of $C_1,\ldots,C_p$ as

$\delta_1,\ldots,\delta_{2p}$, we get an equivalent system of real linear equations

$$\sum_{k=1}^{2p} \gamma_{jk}\delta_k = \lambda_j, \quad j=1,\ldots,2p, \tag{3.7}$$

where γ_{jk} are known constants independent of the boundary conditions while $\lambda_j = -\mathrm{Re}\int_\Gamma H(\zeta)\chi_j(\zeta)d\zeta$. In case $H(\zeta) = 0$, $\zeta\epsilon\Gamma$, so that $\lambda_j = 0$ ($j=1,\ldots,2p$), we have already shown that $C_j = C_j^o = 0$, $j=1,\ldots,p$, which are equivalent to $\delta_k = 0$, $k = 1,\ldots,2p$. Thus, the system of homogeneous equations corresponding to (3.7) has only the trivial solution, which asserts $\det(\gamma_{jk}) \neq 0$. Therefore, (3.7) is uniquely solvable for any λ_j's, i.e., for any $H(\zeta)$, and hence (3.1) is also uniquely solvable for uniquely suitably chosen C_j, $j=1,\ldots,p$. This is what we want to prove.

§4. SOME COMMENTS

1° The equation (3.1) may be further reduced by elimination of the path X of integration. In fact, we may rewrite (2.3) as

$$\frac{1}{\pi i}\int_X \frac{\omega(\xi)}{\xi-x}\,d\xi = -\frac{1}{2\pi i}\int_{L+\gamma}\frac{\omega(\zeta)}{\zeta-x}\,d\zeta - \frac{1}{2\pi i}\int_{L+\gamma}\frac{\omega(\zeta)}{\bar{\zeta}-x}\,d\bar{\zeta}$$

$$+\frac{1}{2\pi i}\int_{L+\gamma}\overline{\omega(\zeta)}\,d\,\frac{\zeta-x}{\bar{\zeta}-x} - f(x) + \frac{1}{2\pi i}\int_L \frac{F(\tau)}{\bar{\tau}-x}\,d\bar{\tau}.$$

Since $\omega(\pm\infty) = 0$ on X, by inversion[7], we get

$$\omega(x) = -\frac{1}{2\pi i}\int_{L+\gamma}\omega(\zeta)\mathrm{dlog}\frac{\zeta-x}{\bar{\zeta}-x} - \frac{1}{2\pi i}\int_{L+\gamma}\overline{\omega(\zeta)}\, d\,\frac{\zeta-x}{\bar{\zeta}-x}$$

$$-\frac{1}{\pi i}\int_{X}\frac{f(\xi)}{\xi-x}d\xi - \frac{1}{2\pi i}\int_{L}\frac{F(\tau)}{\bar{\tau}-x}d\bar{\tau};\tag{4.1}$$

here we have changed the order of integration which is allowable because only one in each repeated integrals is singular[7] and have noted that

$$\int_X \frac{d\xi}{\xi-x} = 0,\ x\in X;\quad \int_X \frac{d\xi}{\xi-z} = \begin{cases} i, & \mathrm{Im}z > 0, \\ -i, & \mathrm{Im}z < 0. \end{cases}$$

Substituting (4.1) into (2.4), (2.5), we get a singular integral equation of normal type on $L+\gamma$, which will not be expressed here.

2° In practice, often we should only seek for $\phi'(z) = \Phi(z)$, $\psi'(z) = \Psi(z)$ instead of $\phi(z)$, $\psi(z)$ themselves, which may be represented by $\Omega(\zeta) = \omega'(\zeta)$. By differentiating (3.1) (or that equation mentioned in 1°), we get a singular integral equation for $\Omega(\zeta)$ on Γ (or $L+\gamma$) without any undetermined constants, which should be solved in class h_0, i.e., the solution may have integrable singularity at all the end-points of L, with additional requirements

$$\int_{L_j}\Omega(t)dt = 0,\quad j=1,\ldots,p.\tag{4.2}$$

Equation thus obtained with requirements (4.2) always has unique solution in class h_0 (Cf.[6],§30). Here we have avoided the difficulty for determining the unknown constants, which would greatly facilitate the process of solution.

3° The stress intensity factors depend upon the behaviours of $\Omega(t)$ on L in the neighborhoods of its end-points. Similar formulas of them may be established as in [5].

4° If there is a displacement difference function $g(\sigma)$ to be welded along the two sides of γ, then $g(\sigma)$ should be added to the right-hand member of (1.3). The method given here remains in effect in this case with certain modifications as shown in [6], §31.

5° The method used in this paper with appropriate modifications is also effective for the second fundamental problem, i.e., given the displacements on X and both sides of L (Cf.[8]).

REFERENCES

[1] Cook T.S., Erdogan F., Stresses in bonded materials with a crack perpendicular to the interface, Int. J. Eng. Sci., 10 (1972), 677-697.

[2] Erdogan F., Aksogan O., Bonded half planes containing an arbitrarily oriented crack, Int. J. Solids Structure, 10 (1974), 569-585.

[3] Ioakimidis N.I., Theocaris P.S., A system of curvilinear cracks in an elastic plane, Int. J. of Fracture, 15 (1979), 299-309.

[4] Lu Jianke, The mathematical problems of bonded plane materials with cracks, J. of Wuhan Univ., 1982, No. 2, 1-10 (in Chinese with English abstract).

[5] Lu Chien-ke, Yew Ching H., Bonded half-planes containing two arbitrarily oriented cracks: a study of containment of hydraulically induced fractures, Soc. of Petrol. Eng. J., 1985, Feb., 55-66.

[6] Lu Jianke, Complex Variable Methods in Plane Elasticity, Wuhan Univ. Press, Wuhan, 1986 (in

Chinese).

[7] Muskhelishvili N.I., Singular Integral Equations, 2nd ed., Noordhoff, 1962.

[8] Lu Jianke, On the second fundamental problem of bonded half-planes of different media with cracks, Appl. Math., J. Chinese Univ., 2 (1987), 37-48 (in Chinese with English abstract).

SOME BOUNDARY VALUE PROBLEMS FOR NONLINEAR DEGENERATE ELLIPTIC COMPLEX EQUATIONS

Wen Guo-Chun
Peking University, Yantai University
P.R. China

In this paper, we consider some boundary value problems for nonlinear degenerate elliptic complex equations of first and second order in a multiply connected domain, namely the Riemann-Hilbert problem for degenerate elliptic complex equations of first order, the oblique derivative problem for degenerate elliptic complex equations of second order, etc. In 1984, T. Iwaniec and A. Mamourian discussed the Dirichlet problem for special degenerate elliptic complex equations of first order with many conditions in a simply connected domain (cf. ref. [1]). Now we use another method, cancel some important hypotheses in [1], and obtain the result on solvability of the above boundary value problems for more general degenerate elliptic complex equations.

§1. THE RIEMANN-HILBERT BOUNDARY VALUE PROBLEM FOR DEGENERATE ELLIPTIC COMPLEX EQUATIONS OF FIRST ORDER

From ref. [2] and Chap. 1 in book [3], we can see that the nonlinear elliptic system of two equations

$$\Phi_j(x,y,u,v,u_x,u_y,v_x,v_y) = 0, \quad j = 1,2 \tag{1.1}$$

with some conditions transformed into the complex form

$$w_{\bar{z}} = F(z,w,w_z), \quad F = Q(z,w,w_z)\, w_z + A(z,w) \tag{1.2}$$

We suppose that the complex equation (1.2) satisfies the following degenerate ellipticity conditions:

$$|F(z,w,V_1) - F(z,w,V_2)| = q(z,w,V_1 - V_2)\,|V_1 - V_2|, \tag{1.3}$$

$$0 \leqslant q(z,w,V) \leqslant 1,$$

$$\overline{\lim_{V\to\infty}}\, q(z,w,V) = q_0 < 1, \; z \in D, \; w, \; V \in \mathbb{C}, \tag{1.4}$$

where q_0 is a real constant; $Q(z,w,V)$, $A(z,w)$ are continuous in w, $V \in \mathbb{C}$ (the whole plane) for almost every point $z \in D$, and are measurable in D for all continuous functions $w(z)$ and measurable functions $V(z)$ in D, and satisfies

$$L_p[A(z,w), \bar{D}] \leqslant k < \infty, \; p > 2, \tag{1.5}$$

where p, k are real constants. The above conditions will be called Condition C.

Similarly to Chap. 5 in [3], there is no harm in assuming that D is a circular domain in the unit disc $|z| < 1$, which is bounded by $N + 1$ circles $\Gamma_j = \{|z - z_j| = \gamma_j\}$, $j = 1,\dots,N$, $\Gamma_{N+1} = \Gamma_0 = \{|z| = 1\}$ and $z = 0 \in D$.

The so-called Riemann-Hilbert problem for complex equation (1.2) may be formulated as follows: Find a continuous solution of (1.2) on $\bar{D}$ satisfying the boundary condition

$$\mathrm{Re}\,[\overline{\lambda(t)}\;w(t)] = r(t) + h(t),\ t \in \Gamma, \tag{1.6}$$

where $\lambda(t)$ is the standard form (see Chap. 2 in [3]), $\lambda(t), r(t)$ satisfy

$$|\lambda(t)| = 1,\ t\in\Gamma,\ C_\alpha[\lambda(t),\ \Gamma] \leq 1 < \infty,$$
$$C_\alpha[r(t),\ \Gamma] \leq 1, \tag{1.7}$$

α ($\frac{1}{2} < \alpha < 1$), ℓ are all real constants, and

$$h(t) = \begin{cases} 0,\ t\in\Gamma,\ K = \frac{1}{2\pi}\Delta_\Gamma \arg \lambda(t) \geq N, & \\ \left.\begin{array}{l} h_j,\ t\in\Gamma_j,\ 1 \leq j \leq N - K \\ 0,\ t\in\Gamma_j,\ N - K < j \leq N + 1 \end{array}\right\} & 0 \leq K < N, \\ \left.\begin{array}{l} h_j,\ t\in\Gamma_j,\ 1 \leq j \leq N \\ h_0 + \mathrm{Re} \sum\limits_{m=1}^{-k-1} (H_m^+ + iH_m^-)\,t^m,\ t\in\Gamma_0 \end{array}\right\} & K < 0 \end{cases} \tag{1.8}$$

where $h_j (j = 1,\ldots N)$, $H_m^{\pm}(m = 1,\ldots,|K| - 1)$ are unknown real constants to be determined appropriately. The above boundary value problem will be denoted by <u>Problem A.</u> If the solution $w(z)$ of the Problem A satisfies the point condition:

$$\mathrm{Im}[\overline{\lambda(a_j)}w(a_j)] = b_j,$$
$$j\in\{j\} = \begin{cases} 1,\ \ldots,\ 2K - N + 1,\ K \geq N, \\ N - K + 1,\ \ldots,\ N + 1,\ 0 \leq K < N \end{cases} \tag{1.9}$$

where $a_j\in\Gamma_j$, $j = 1,\ldots,N$, $a_j (j = N + 1,\ldots,2K - N + 1)$ are distinct points on Γ_0, $b_j (j = 1,\ldots,2K - N + 1)$ are all real constants, and $|b_j| \leq 1$, then the Problem A with the condition (1.9) is called <u>Problem B.</u>

In ref. [1], the authors discussed the case of $F(z,w,w_z) = h(w_z)$, $q(z,w,|V|) = q(|V|)$, $A(z,w) = 0$, and suppose that $q(t)$ is continuous on $[0, \infty)$; $\overline{\lim_{t\to\infty}}\, q(t) = q_0 < 1$; $q(t) < 1$, $t \in (0,\infty)$ and $tq^2(\sqrt{t})$ is a concave and increasing function. Under these hypotheses, they considered the Dirichlet Problem for the complex equation $w_{\bar{z}} = h(w_z)$ in $D = \{|z| < 1\}$. In this paper, we do not assume that $tq^2(\sqrt{t})$ is concave and increasing in $(0, \infty)$ and the condition $q(t) < 1$, $t\in(0, \infty)$ in ref. [1] is replaced by $q(z,w,t) \leq 1$, $t\in(0,\infty)$. For instance, the functions

$$F(z,w,V) = \begin{cases} V^2/2, & \text{for } |V| \leq 1, \\ \overline{V}^{-2}/2, & \text{for } |V| > 1 \end{cases}$$

and

$$F(z,w,V) = \begin{cases} |V|, & \text{for } |V| \leq 1, \\ (|z| + |w| + 2)|V|/(|z| + |w| \\ \quad + 1 + |V|), & \text{for } |V| > 1 \end{cases}$$

satisfy the Condition C.

In order to discuss solvability of the Problems A and B for the complex equation (1.2), we introduce the function

$$F^{(m)}(z,w,V) = \begin{cases} F(z,w,V), \ q(z,w,V) \leq 1 - \frac{1}{m}, \\ \quad m \text{ is a positive integer}, \\ Q_m(z,w,V)V + A(z,w), \\ \quad Q(z,w,V) > 1 - \frac{1}{m}, \\ Q_m(z,w,V) = (1 - \frac{1}{m})e^{i\arg Q(z,w,V)}, \end{cases} \tag{1.10}$$

and consider the uniformly elliptic complex equation

$$w_{\bar{z}} = F^{(m)}(z,w,w_z). \tag{1.11}$$

By Chap. 5 in book [3], we know that the Problems A and B are solvable, and if the complex equation (1.2) satisfies the Condition C, then any solution $w(z)$ of the Problem B for (1.2) satisfies some estimates as a lemma.

<u>Lemma 1.1.</u> Let the complex equation (1.2) satisfy the Condition C. Then the Problem A and the Problem B for the corresponding complex equation (1.11) are solvable, and any solution $w(z)$ of the Problem B($K \leq 0$) for the complex equation (1.2) satisfies the estimate:

$$L_{p_0}[|w_{\bar{z}}| + |w_z|, \bar{D}] \leq M_1,$$
$$C_\beta [w(z), \bar{D}] \leq M_2, \tag{1.12}$$

where $M_j = M_j(q_0, p_0, k, \alpha, \ell, D, M)(j = 1, 2)$ are constants, $\beta = 1 - 2/p$, and M, $p_0(2 < p_0 < \min(p, \frac{1}{1-\alpha}))$ are constants satisfying (1.13) and (1.16) respectively. In the case of $K > 0$, if we assume that the constant q_0 in (1.4) is small enough, then the estimate (1.12) remains true.

Proof. By using the condition (1.4), there exists a positive constant M, such that

$$q(z,w,V) \leq \delta = (1 + q_0)/2 < 1, \text{ when } |V|>M. \tag{1.13}$$

We first discuss the case of $K \leq 0$. The solution $w(z)$ of the Problem B for the complex equation $w_{\bar{z}} = \omega(z)$ ($\omega(z) \in L_{p_0}(\bar{D})$) can be expressed as

$$w(z) = \Phi(z) + \hat{T}\omega, \tag{1.14}$$

where $\Phi(z)$ is a unique solution of the Problem B for analytic functions, $\hat{T}\omega$ is a double integral as follows

$$\left.\begin{aligned}\hat{T}\omega &= -\frac{1}{\pi}\iint_D [P(z,\zeta)\omega(\zeta) + Q(z,\zeta)\overline{\omega(\zeta)}] \\ &\qquad d\sigma_\zeta + \Psi(z), \\ P(z,\zeta) &= \frac{1}{2}[G_1(z,\zeta) + G_2(z,\zeta) + H_1(z,\zeta) \\ &\qquad - H_2(z,\zeta)],\ z\in\bar{D}, \\ Q(z,\zeta) &= \frac{1}{2}[G_1(z,\zeta) - G_2(z,\zeta) + H_1(z,\zeta) \\ &\qquad + H_2(z,\zeta)],\ \zeta\in D,\end{aligned}\right\} \tag{1.15}$$

where $\Psi(z)$ is an analytic function in D, and $\hat{S}\omega = (\hat{T}\omega)_z$ possesses the following property

$$\delta\Lambda_{p_o} < 1,\ \Lambda_{p_o} = \sup_{\omega(z)\in L_{p_o}(\bar{D})} \frac{L_{p_o}(\hat{S}\omega,\bar{D})}{L_{p_o}(\omega,\bar{D})}, \tag{1.16}$$

in which p_o is a proper constant satisfying $2 < p_o < p$ (cf. §2, Chap 2 in ref. [3]). Substituting $w(z) = \Phi(z) + \hat{T}\omega$ into the complex equation (1.11), we obtain the integral equation

$$\begin{aligned}\omega(z) &= F^{(m)}(z,w,w_z) = F^{(m)}(z,w,w_z) - \\ &\quad - F^{(m)}(z,w,\Phi') + F^{(m)}(z,w,\Phi') \\ &= Q^{(m)}(z,w,S\omega)S\omega + F^{(m)}(z,w,\Phi')\ .\end{aligned} \tag{1.17}$$

Taking

$$L_{p_o}[\Phi',\ \bar{D}] \leqslant k' = k'(p_o,\ \alpha,\ \ell,\ D) < \infty \tag{1.18}$$

into account (cf. Chap. 2 in [3]), we have

$$|\omega(z)| \leq |Q^{(m)}|\ |\hat{S}\omega| + |Q^{(m)}|\ |\Phi'| + |A|,$$

$$L_{p_0}[\omega, \bar{D}] \leq \delta\, L_{p_0}[\hat{S}\omega, \bar{D}] \tag{1.19}$$

$$+ \left(\iint_{|\hat{S}\omega| \leq M} |Q^{(m)}\ \hat{S}\omega|^{p_0} d\sigma_\zeta\right)^{1/p_0}$$

$$+ L_{p_0}(\Phi', \bar{D}) + L_{p_0}(A, \bar{D}) \leq \delta\, \Lambda_{p_0} L_{p_0}(\omega, \bar{D})$$

$$+ k + k' + \pi M,$$

consequently

$$L_{p_0}(\omega, \bar{D}) \leq (k + k' + \pi M)/(1 - \delta\Lambda_{p_0}) = M_1. \tag{1.20}$$

From (1.20) and the result of Chap. 5 in [3], the estimate (1.12) can be derived.

Next we discuss the case of $K > 0$ and suppose that the constant q_0 in (1.13) is small enough, such that

$$\delta\, \Lambda_{p_0} < 1. \tag{1.21}$$

Thus similarly to the case of $K \leq 0$, the estimate (1.12) can be obtained.

Now, we shall prove the result of solvability of the Problems A and B for the complex equation (1.2), namely

<u>Theorem 1.1.</u> Under the hypothesis in Lemma 1.1, the Problem A and the Problem B for (1.2) have a solution.

Proof. Because the solution of the Problem B is also a solution of the Problem A for the complex equation (1.2), we only need to prove the solvability of the Problem B for (1.2).

According to Lemma 1.1, the Problem B for the complex equation (1.11) has a solution $w_m(z)$ and the solution $w_m(z)$ satisfy the estimate (1.12), i.e.

$$L_{p_0}[|w_{m\bar{z}}| + |w_{mz}|, \bar{D}] \leq M_1,$$
$$C_\beta[w_m(z), \bar{D}] \leq M_2. \tag{1.22}$$

Hence we can select a subsequence of $\{w_m(z)\}$, which uniformly converges to $w_0(z)$ on $\bar{D}$, and $w_0(z)$ satisfies the boundary condition (1.6), the point condition (1.9) and the second estimate in (1.22). Moreover, from $\{w_{m\bar{z}}\} = \{\omega_m(z)\}$ and $\{w_{mz}\}$, we can choose subsequences $\{\omega_{m_k}(z)\}$ and $\{w_{m_k z}\}$, which weakly converge to $\omega_0(z)$ and $Y_0(z)$ on $\bar{D}$ respectively, and $w_m(z) = \Phi(z) + \hat{T}\omega_m$, where $\Phi(z)$, $\hat{T}\omega_m$ are similar to (1.14), (1.15), and $\{\hat{T}\omega_m\}$ uniformly converges to $\hat{T}\omega_0$ on $\bar{D}$. For convenience, we denote still $\{w_{m_k}(z)\}$, $\{\omega_{m_k}(z)\}$ and $\{w_{m_k z}\}$ by $\{w_m(z)\}$, $\{\omega_m(z)\}$ and $\{w_{mz}\}$, respectively. On account of the definition of generalized derivative, for any function $\phi(z) \in D_0^1(D)$, the following formula holds:

$$\iint_D [w_m(z)\,\phi_z + w_{mz}\,\phi(z)]\,d\sigma_z = 0. \tag{1.23}$$

When $m \to \infty$, we obtain

$$\iint_D [w_0(z)\phi_z + Y_0(z)\phi(z)]d\sigma_z = 0. \tag{1.24}$$

So $[w_0(z)]_z = Y_0(z) = \Phi'(z) + \hat{S}\omega_0$, $z \in D$.

In the following, we prove that $c_m(z) = F^{(m)}(z, w_m, w_{0z}) - F(z, w_0, w_{0z})$ possesses the property:

$$L_{p_0}[c_m(z), \bar{D}] \to 0, \text{ as } m\to\infty. \tag{1.25}$$

In fact, $F^{(m)}(z,w_m,w_{oz}) - F^{(m)}(z,w_o,w_{oz})$ and $F^{(m)}(z,w_o,w_{oz}) - F(z,w_o,w_{oz})$ converge to 0 for almost every point $z \in D$, hence $c_m(z)$ possesses the same property. Thus for two arbitrary sufficiently small positive constants ε_1 and ε_2, there exists a subset D_* in D and a positive integer N, so that $\text{mes}D_* < \varepsilon_1$ and $|c_m| < \varepsilon_2$ on $\bar{D}\backslash D_*$ for $n > N$. By the Hölder inequality and the Minkowski inequality, we have

$$\begin{aligned} L_{p_0}[c_m,\bar{D}] &\leq L_{p_0}[c_m,D_*] + L_{p_0}[c_m,\bar{D}\backslash D_*] \\ &\leq L_{p_0}\{Q_m(z,w_m,w_{oz}) - Q(z,w_o,w_{oz})\, w_{oz} \\ &\quad + A(z,w_m) - A(z,w_o),\, D_*\} + \varepsilon_2(\text{mes}D)^{1/p_0} \\ &\leq 2L_{p_0}[\Phi' + \hat{S}\omega_o,\, D_*] + 2k \\ &\quad + \varepsilon_2\pi^{1/p_0} \leq 2\varepsilon_1^{1/p_2} L_{p_1}[\Phi' + \hat{S}\omega_o,\, D_*] \\ &\quad + 2k + \varepsilon_2\pi^{1/p_0} = \varepsilon, \end{aligned} \tag{1.26}$$

where p_1 is a constant such that $2 < p_0 < p_1 < \min(p, 1/(1-\alpha))$, $\delta\Lambda_{p1} < 1$ and $p_2 = p_0p_1/(p_1 - p_0)$. Therefore the formula (1.25) holds.

It remains to prove that $F^{(m)}(z,w_m,w_{mz}) - F^{(m)}(z,w_m,w_{oz})$ weakly converges to 0 on $\bar{D}$. Due to

$$\begin{aligned} &|F^{(m)}(z,w_m,w_{mz}) - F^{(m)}(z,w_m,w_{oz})| \\ &\leq |(w_m - w_o)_z|, \end{aligned} \tag{1.27}$$

we only prove that $g_m(z) = |(w_m - w_o)_z|$ weakly converges to 0 on $\bar{D}$. Otherwise, we may choose a subsequence of $\{g_m(z)\}$ which weakly converges to $g_o(z)$ on $\bar{D}$ and $g_o(z)$

> 0 on a positive measurable set $E \subset D$. However $(w_m - w_0)_z$ weakly converges to 0 on $\bar{D}$. This contradiction proves that $|F^{(m)}(z,w_m,w_{mz}) - F^{(m)}(z,w_m,w_{oz})|$ and $F^{(m)}(z,w_m,w_{mz}) - F^{(m)}(z,w_m,w_{oz})$ weakly converge to 0 on $\bar{D}$, i.e. $F^{(m)}(z,w_m,w_{mz})$ weakly converges to $F(z,w_o,w_{oz})$ on $\bar{D}$. On the basis of the definition of the generalized derivative $w_{m\bar{z}}$, when $m \to \infty$, for any function $\phi(z) \in D_o^1(D)$, we have

$$\iint_D [w_o(z)\ \phi_{\bar{z}} + F(z,w_o,w_{oz})\ \phi(z)]d\sigma_z = 0. \tag{1.28}$$

Hence $w_{o\bar{z}} = F(z,w_o,w_{oz})$. This shows that $w_o(z)$ is a solution of the complex equation (1.2) on D. This completes the proof of Theorem 1.1.

If $A(z,w)$ in the complex equation (1.2) is replaced by $A_1(z,w)w + A_2(z,w)\bar{w} + A_3(z,w)$, in this case, the complex equation (1.2) is represented by

$$\begin{aligned} w_{\bar{z}} &= F(z,w,w_z), \quad F = Q(z,w,w_z)w_z \\ &+ A_1(z,w)w + A_2(z,w)\bar{w} + A_3(z,w), \end{aligned} \tag{1.29}$$

where $A_j(z,w)\ (j = 1,2,3)$ satisfy

$$\begin{aligned} &L_p[A_j(z,w), \bar{D}] \leq \varepsilon k, \\ &j = 1,2, \quad L_p[A_3(z,w), \bar{D}] \leq k, \end{aligned} \tag{1.30}$$

in which $\varepsilon (0 < \varepsilon \leq 1)$ is a constant, then using the similar method, we can obtain the result of solvability of the Problems A and B for (1.29), as a corollary.

<u>Corollary 1.1.</u> Suppose that the complex equation (1.29) satisfies the conditions (1.3), (1.4) and (1.30), the

constant ε in (1.30) is small enough, and the constant q_0 in (1.4) is sufficiently small if $K > 0$. Then the Problem A and the Problem B for (1.29) are solvable.

§2. THE OBLIQUE DERIVATIVE PROBLEM FOR DEGENERATE ELLIPTIC COMPLEX EQUATIONS OF SECOND ORDER

In this section, we discuss the complex form of degenerate elliptic complex equations of second order:

$$\left.\begin{aligned} &u_{z\bar{z}} = F(z,u,u_z,u_{zz}), \\ &\quad F = \mathrm{Re}[Q(z,u,u_z,u_{zz})u_{zz}] + A(z,u,u_z), \\ &\quad A = \mathrm{Re}[A_1u_z] + A_2u + A_3, \quad A_j = A_j(z,u,u_z), \\ &\quad j = 1,2,3, \end{aligned}\right\} \tag{2.1}$$

and assume that (2.1) satisfies <u>Condition C</u>, namely functions $Q(z,u,u_z,V)$ and $A_j(z,u,u_z)$ $(j = 1,2,3)$ are measurable for any continuous differentiable function $u(z)$ and measurable function $V(z)$ in D, and satisfy the condition

$$\begin{aligned} &L_p[A_j(z,u,u_z), \bar{D}] \leqslant \varepsilon k, \; j = 1,2, \\ &\quad L_p[A_3(z,u,u_z), \bar{D}] \leqslant k, \end{aligned} \tag{2.2}$$

where $p(> 2)$, $\varepsilon(0 < \varepsilon \leqslant 1)$, $k(0 \leqslant k < \infty)$ are constants. Moreover, the above functions are continuous in $u \in \mathbb{R}$ (the real axis), u_z, $V \in \mathbb{C}$ for almost every point $z \in D$, and satisfy

$$\begin{aligned} &|F(z,u,u_z,V_1) - F(z,u,u_z,V_2)| \\ &\quad = q(z,u,u_z,V_1 - V_2)|V_1 - V_2|, \; z \in D, \end{aligned} \tag{2.3}$$

$$0 \leq q(z,u,u_z,V) \leq 1, \ \overline{\lim_{V\to\infty}}\ q(z,u,u_z,V)$$
$$= q_o < 1, \ u \in \mathbb{R}, \ u_z, \ V\in\mathbb{C}. \qquad (2.4)$$

Without loss of generality, we assume that D is an $(N + 1)$ - connected circular domain as stated in §1.

The so-called oblique derivative problem P for the second order equation (2.1) may be formulated as follows: Find a continuous differentiable solution $u(z)$ for (2.1) on $\bar{D}$ satisfying the boundary condition

$$\mathrm{Re}[\overline{\lambda(t)}u_t] = \sigma(t)u(t) + r(t) + h(t), \ t\in\Gamma, \qquad (2.5)$$

where $\lambda(t)$, $r(t)$, $h(t)$ are stated as in §1, and $\sigma(t)$ satisfies the condition

$$C_\alpha[\sigma(t), \Gamma] \leq \varepsilon\ell, \ \frac{1}{2} < \alpha < 1, \ 0 \leq \ell < \infty. \qquad (2.6)$$

In order to discuss the solvability of the above Problem P, we introduce the complex equation of first order

$$w_{\bar{z}} = \mathrm{Re}[Q(z,u,w,w_z)\, w_z] + A(z,u,w) \qquad (2.7)$$

and the boundary condition and the point condition

$$\mathrm{Re}[\overline{\lambda(t)}w(t)] = \sigma(t)u(t) + r(t) + h(t), \ t\in\Gamma, \qquad (2.8)$$

$$\mathrm{Im}[\overline{\lambda(a_j)}w(a_j)] = b_j, \ j\in\{j\}, \qquad (2.9)$$

where a_j, b_j as stated in §1, and $u(z)$, $w(z)$ satisfy the following relation

$$u(z) = \mathrm{Re}\int_o^z [2w(z) + \sum_{j=1}^{N} \frac{id_j}{z - z_j}]\, dz + b_o, \qquad (2.10)$$

where $d_j (j = 1,\dots,N)$ are appropriate real constants, such that the function determined by the integral in (2.10) is single-valued in D, and b_0 is a real constant satisfying the condition: $|b_0| \leq \ell$. We denote by Problem Q the above boundary value problem (2.7) - (2.10).

Lemma 2.1. Let the second order equation (2.1) satisfy the Condition C, and the constant ε in (2.2.), (2.6) is small enough.

Then (1) When $K = \frac{1}{2\pi}\Delta_\Gamma \arg\lambda(t) \leq 0$, the Problem Q for (2.1) has a solution.

(2) When $K > 0$ and the constant q_0 is sufficiently small, the Problem Q for (2.1) is solvable.

Proof. We first consider the following complex equation of first order

$$\left.\begin{aligned} w_{\bar z} &= F^{(m)}(z,u,w,w_z), \\ F^{(m)} &= Q_m(z,u,w,w_z)w_z + A(z,u,w), \end{aligned}\right\} \tag{2.11}$$

where $Q_m(z,u,w,w_z)$ is defined as follows:

$$Q_m(z,u,w,w_z) = \begin{cases} Q(z,u,w,w_z), & \text{if } q(z,u,w,V) < 1 - \frac{1}{m}, \\ (1 - \frac{1}{m})\exp i\arg Q(z,u,w,V), & \text{if } q(z,u,w,V) \geq 1 - \frac{1}{m}, \end{cases} \tag{2.12}$$

where m is a positive integer. From the result of Chap. 5 in [3], we know that under the hypothesis in this theorem, the Problem Q for the complex equation (2.11) has a solution $[w_m(z), u_m(z)]$. According to the method of the proof in Lemma 1.1 and using the following representation of the solution $w_m(z)$ of the Problem Q

$$w_m(z) = \Phi_m(z) + \hat{T}\omega_m, \tag{2.13}$$

where $\Phi_m(z)$ is an analytic function which can be expressed as

$$\Phi_m(z) = \frac{1}{2\pi}\int_\Gamma T(z,t)[-\sigma(t)u_m(t) + r(t)]d\theta + \Phi_0(z), \tag{2.14}$$

$T(z,t)$ is the Schwarz kernel of the Problem B for analytic functions, $\Phi_0(z)$ is an analytic function in D (cf. Ref. [4]), and

$$u_m(z) = \mathrm{Re}\int_0^z [w_m(z) + \sum_{j=1}^N \frac{id_{jm}}{z - z_j}]dz + b_0, \tag{2.15}$$

we can prove that $[w_m(z), u_m(z)]$ satisfies the estimates

$$L_{p_0}[|w_{m\bar{z}}| + |w_{mz}|, \bar{D}] \leq M_1,$$
$$C_\beta[w_m(z), \bar{D}] \leq M_2, \tag{2.16}$$

$$L_{p_0}[|u_{mz\bar{z}}| + |u_{mzz}|, \bar{D}] \leq M_3,$$
$$C'_\beta[u_m(z), \bar{D}] \leq M_4, \tag{2.17}$$

where $2 < P_0 < \min(p, \frac{1}{1-\alpha})$, $\beta = 1 - 2/p_0$, $M_j = M_j(q_0, P_0, k, \alpha, \ell, D, M)$, $j = 1,\ldots,4$ in which M is a sufficiently large positive constant, such that

$$q(z,u,w,V) \leq \delta = (1 + q_0)/2 < 1, \text{ for } |V| > M. \tag{2.18}$$

If $K > 0$, then we choose δ in (2.18) to be an appropriate small positive constant. Later similarly to the proof of Theorem 1.1, we can select subsequences of $\{w_m(z)\}$ and $\{u_m(z)\}$, which uniformly converge to

$w_0(z)$ and $u_0(z)$ respectively, and $[w_0(z), u_0(z)]$ is just a solution of the Problem Q for the first order complex equation (2.7).

<u>Theorem 2.1.</u> Under the hypothesis in Lemma 2.1, the Problem P for the second order equation (2.1) has N solvability conditions.

Proof. Let us substitute the solution $[w(z), u(z)]$ of the Problem Q for the complex equation (2.7) into (2.10). If $d_j = 0 (j = 1,\ldots,N)$, i.e. when N equalities:

$$\mathrm{Re} \int_{\Gamma j} w(z)dz = 0, \; j = 1,\ldots,N \tag{2.19}$$

hold, we may choose $d_j = 0$ $(j = 1,\ldots,N)$, then the function $u(z)$ is just a solution of the Problem P for the second order equation (2.1). This shows that the Problem P for (2.1) possesses N solvability conditions as stated in (2.19).

Finally, we introduce simply the oblique derivative problem for the complex form of degenerate elliptic system of second order equations, i.e. the system of second order complex equations

$$\left.\begin{aligned} &u_{jz\bar{z}} = F_j(z,u_1,u_2,u_{1z},u_{2z},u_{1zz},u_{2zz}),\\ &\quad F_j = \mathrm{Re}\,[Q_{j1}\,u_{1zz} + Q_{j2}\,u_{2zz}] + A_j\,,\\ &\quad A_j = \mathrm{Re}\,[A_{j1}\,u_{1z} + A_{j2}\,u_{2z}] + A_{j3}\,u_1\\ &\quad + A_{j4}\,u_2 + A_{j5}\,,\ Q_{jm} = Q_{jm}(z,\,u_1,\,u_2,\\ &\quad u_{1z},\,u_{2z},\,u_{1zz},\,u_{2zz})\,,\ m = 1,\,2,\\ &\quad A_{jm} = A_{jm}(z,\,u_1,\,u_2,\,u_{1z},\,u_{2z})\,,\\ &\quad m = 1,\ldots,5,\ j = 1,\,2. \end{aligned}\right\} \tag{2.20}$$

We suppose that (2.20) satisfies the so-called Condition C,

in which the main conditions are as follows:

$$L_p[A_{jm}(z, u_1, u_2, u_{1z}, u_{2z}), \bar{D}] \leq \varepsilon k,$$
$$m = 1, \ldots, 4. \quad L_p[A_{j5}, \bar{D}] \leq k, \tag{2.21}$$
$$j = 1, 2$$

$$\sum_{m=1}^{2} |Q_{jm}(z, u_1, u_2, u_{1z}, u_{2z}, V_1, V_2)| \leq \frac{1}{2},$$
$$j = 1, 2, \overline{\lim_{|Vm| \to \infty}} |Q_{jm}| \leq q_{jm}, \; m = 1, 2, \tag{2.22}$$
$$q_{j1} + q_{j2} < \frac{1}{2}, \; j = 1, 2$$

where $q_{jm}(j, m = 1, 2)$, $\varepsilon(\leq 1)$ are non-negative constants.

The oblique derivative problem for complex system (2.20) in the $(N + 1)$ - connected circular domain D is defined to be the Problem of finding a continuous differentiable solution $[u_1(z), u_2(z)]$ for (2.20) on $\bar{D}$, which satisfies the boundary condition

$$\operatorname{Re}[\overline{\lambda_j(t)} u_{jt} + \sigma_{j1}(t) u_1(t) + \sigma_{j2}(t) u_2(t)] =$$
$$r_j(t) + h_j(t), \; t \in \Gamma, \; j = 1, 2, \tag{2.23}$$

where $|\lambda_j(t)| = 1$, $\lambda_j(t)$, $\sigma_{jm}(t)$, $r_j(t)$ satisfy

$$C_\alpha[\lambda_j(t), \Gamma] \leq 1, \; C_\alpha[\sigma_{jm}(t), \Gamma] \leq \varepsilon \ell,$$
$$m = 1, 2, \; C_\alpha[r_j(t), \Gamma] \leq 1, \; j = 1, 2, \tag{2.24}$$

in which α $(\frac{1}{2} < \alpha < 1)$, $\varepsilon(0 < \varepsilon \leq 1)$, $\ell(0 \leq \ell < \infty)$ are real constants.

By using the methods stated as before, we transform the second order complex system (2.20) into the corresponding system of first order complex equations, and apply the results of Chap. 6 and Chap. 7 in ref. [3], the solvability theorem of the Problem P for (2.20) with some conditions can be obtained.

REFERENCES

1. Iwaniec T., Mamourian A., On the first order nonlinear differential systems with degeneration of ellipticity. Proc. of the second Finnish-Polish Summer School in Complex Analysis at Jyväskylä, 1984, 41-52.

2. Fang Ai-nong, Quasiconformal mappings and the theory of functions for systems of nonlinear elliptic differential equations of first order. Acta Math. Sinica 23 (1980), 280-292.

3. Wen Guo-chun, Linear and nonlinear elliptic complex equations. Shanghai Science and Technology Publishing House, 1986.

4. Wen Guo-chun, Conformal mappings and boundary value problems. Higher Education Press, 1985.

THE HADAMARD PRODUCT OF LEGENDRE SERIES

Mo Yeh
Department of Mathematics
Shandong University
People's Republic of China

1. INTRODUCTION

Let

$$\varphi(z) = \sum_{n=0}^{\infty} a_n z^n, \quad \psi(z) = \sum_{n=0}^{\infty} b_n z^n \tag{1.1}$$

be two power series with positive radii of convergence R_1, R_2 respectively. By use of (1.1), Hadamard constructed the power series

$$\sum_{n=0}^{\infty} a_n b_n z^n. \tag{1.2}$$

Obviously

$$\overline{\lim_{n\to\infty}} \sqrt[n]{|a_n||b_n|} \leqslant (\overline{\lim_{n\to\infty}} \sqrt[n]{|a_n|})(\overline{\lim_{n\to\infty}} \sqrt[n]{|b_n|}) = \frac{1}{R_1 R_2}.$$

Hence the radius of convergence of the power series (1.2) is $R \geqslant R_1 R_2$. Therefore the power series (1.2) represents an analytic function in $|z|<R$. We denote this analytic function by $\Phi(z)$, then

$$\Phi(z) = \sum_{n=0}^{\infty} a_n b_n z^n. \tag{1.3}$$

We call $\Phi(z)$ the Hadamard product (Hp product) of $\varphi(z)$ and $\psi(z)$, and denote

$$\Phi(z) = \varphi(z)*_p\psi(z). \tag{1.4}$$

At the same time, the power series (1.2) is called the Hp product of the power series (1.1). We have the following famous Hadamard's multiplication theorem [1].

Theorem A. If $\varphi(z)$ has singularities at $\zeta_1, \zeta_2, \ldots$ and $\psi(z)$ at $\eta_1, \eta_2, \ldots,$ then the singularities of $\Phi(z)$ are to be found among the points $\zeta_m \eta_n$.

The proof of the theorem A depends on the following representation of $\Phi(z)$ as integral

$$\Phi(z) = \frac{1}{2\pi i}\int_C \Phi(w)\psi(\frac{z}{w})\frac{dw}{w} \tag{1.5}$$

where C is a contour including the origin, on which $|W|<R_1$, $|z/w|<R_2$.

There are some Chinese authors who discuss the geometric properties of the Hadamard products [2,3].

In this paper instead of power series we consider the Hadamard product of the Legendre series.

2. DEFINITION

Let

$$f(z) = \sum_{n=0}^{\infty} a_n P_n(z), \quad g(z) = \sum_{n=0}^{\infty} b_n P_n(z), \tag{2.1}$$

where $P_0(z)$, $P_1(z)$, $P_2(z)$, ..., are Legendre polynomials; and the ellipses of convergence for them are E_α, E_β respectively, where α, β are positive numbers [4]. We construct the Legendre series

$$\sum_{n=0}^{\infty} a_n b_n P_n(z). \tag{2.2}$$

We know that

$$\overline{\lim_{n\to\infty}} \sqrt[n]{|a_n|} = e^{-\alpha}, \quad \overline{\lim_{n\to\infty}} \sqrt[n]{|b_n|} = e^{-\beta}. \tag{2.3}$$

Therefore

$$\overline{\lim_{n\to\infty}} \sqrt[n]{|a_n b_n|} \leqslant \left(\overline{\lim_{n\to\infty}} \sqrt[n]{|a_n|}\right)\left(\overline{\lim_{n\to\infty}} \sqrt[n]{|b_n|}\right)$$

$$= e^{-(\alpha+\beta)}. \tag{2.4}$$

Let

$$\overline{\lim_{n\to\infty}} \sqrt[n]{|a_n b_n|} = e^{-\gamma},$$

then the ellipse of convergence of the Legendre series (2.2) is E_γ; and from (2.4) we know that

$$e^{-\gamma} \leqslant e^{-(\alpha+\beta)}.$$

Therefore

$$\gamma \geqslant \alpha + \beta. \tag{2.5}$$

Obviously the Legendre series (2.2) represents an analytic function inside E_γ; and we denote it by $F(z)$; then

$$F(z) = \sum_{n=0}^{\infty} a_n b_n P_n(z). \tag{2.6}$$

We call $F(z)$ the Hadamard product (H product) of $f(z)$ and $g(z)$ and denote it by

$$F(z) = f(z)*g(z). \tag{2.7}$$

We also use $F(z)$, $f(z)$, $g(z)$ to denote the complete analytic functions generated by their Legendre series.

When at least one of upper limits in (2.3) is a limit, then

$$\gamma = \alpha + \beta. \tag{2.8}$$

3. THE SINGULARITIES OF F(z)

For power series we can represent $\Phi(z) = \varphi(z)*_p\psi(z)$ simply as (1.5); so we can easily prove theorem A. But for Legendre series we cannot easily find the representation of $F(z) = f(z)*g(z)$ by $f(z)$, $g(z)$, it is not so easy to prove a theorem for the Legendre series similar to theorem A.

Let a be a constant. We call a the algebraic branch-point of second degree of the square root $\sqrt{a-z}$. For Legendre series we have the following

Theorem 1. Suppose that

$$f(z) = -\sum_{k=1}^{m} \frac{A_k}{\sqrt{1-2z\alpha_k+\alpha_k^2}}, \quad g(z) = -\sum_{j=1}^{\ell} \frac{B_j}{\sqrt{1-2z\beta_j+\beta_j^2}}, \tag{3.1}$$

where A_k, $\alpha_k(k=1,2,\ldots,m)$; B_j, $\beta_j(j=1,2,\ldots,\ell)$ are non-zero constants and the moduli of α_k, β_j are either all greater than 1, or all less than 1. Then the singularities of $F(z) = f(z)*g(z)$ are all algebraic branch-points of second degree and can be found from the branch-points of $f(z)$, $g(z)$.

Proof. Obviously all the singularities of $f(z)$, $g(z)$ are algebraic branch-points of second degree, and they are

$$\left.\begin{aligned}\zeta_k &= \frac{1}{2}(\alpha_k + \frac{1}{\alpha_k}) \quad (k=1,2,\ldots,m), \\ \eta_j &= \frac{1}{2}(\beta_j + \frac{1}{\beta_j}) \quad (j=1,2,\ldots,\ell).\end{aligned}\right\} \tag{3.2}$$

Corresponding to Legendre series (2.1) we have the power series (1.1). From (3.1) we know that [5]

$$\varphi(z) = \sum_{k=0}^{m} \frac{A_k}{z-\alpha_k}, \quad \psi(z) = \sum_{j=1}^{\ell} \frac{\beta_j}{z-\beta_j}. \tag{3.3}$$

We consider the case that all the moduli of $\alpha_k(k=1,2,\ldots, m)$, $\beta_j(j=1,2,\ldots,\ell)$ are greater than 1. From (3.3) we have

$$a_n = -\sum_{k=1}^{m} \frac{A_k}{\alpha_k^{n+1}}, \quad b_n = -\sum_{j=1}^{\ell} \frac{B_j}{\beta_j^{n+1}},$$

where $n=0,1,2,\ldots$. Therefore

$$\Phi(z) = \varphi(z)*_p\psi(z) = \sum_{n=0}^{\infty} \{(\sum_{k=1}^{m} \frac{A_k}{\alpha_k^{n+1}})(\sum_{j=1}^{\ell} \frac{B_j}{\beta_j^{n+1}})\}z^n$$

$$= \sum_{n=0}^{\infty} \{\sum_{k=1}^{m} \sum_{j=1}^{\ell} \frac{A_k B_j}{(\alpha_k \beta_j)^{n+1}}\}z^n$$

$$= -\sum_{k=1}^{m} \sum_{j=1}^{\ell} \frac{A_k B_j}{z-\alpha_k \beta_j}.$$

From this we know that

$$F(z) = \sum_{k=1}^{m} \sum_{j=1}^{\ell} \frac{A_k B_j}{\sqrt{1-2z\alpha_k\beta_j+\alpha_k^2\beta_j^2}}. \tag{3.4}$$

Then we see that all the singularities of $F(z)$ are algebraic branch-points of second degree with representations

$$\xi_{k,j} = \frac{1}{2}\left(\alpha_k\beta_j + \frac{1}{\alpha_k\beta_j}\right)(k=1,2,\ldots,m;\ j=1,2,\ldots,\ell). \tag{3.5}$$

From (3.2) we know that

$$\alpha_k = \zeta_k + \sqrt{\zeta_k^2-1},\ \frac{1}{\alpha_k} = \zeta_k - \sqrt{\zeta_k^2-1}\ (k=1,2,\ldots,m);$$

$$\beta_j = \eta_j + \sqrt{\eta_j^2-1},\ \frac{1}{\beta_j} = \eta_j - \sqrt{\eta_j^2-1}\ (j=1,2,\ldots,\ell).$$

Then from (3.5) we know that

$$\begin{aligned}\xi_{k,j} &= \frac{1}{2}\{(\zeta_k+\sqrt{\zeta_k^2-1})(\eta_j+\sqrt{\eta_j^2-1})\\ &\quad + (\zeta_k-\sqrt{\zeta_k^2-1})(\eta_j-\sqrt{\eta_j^2-1})\}\\ &= \zeta_k\eta_j + \sqrt{(\zeta_k^2-1)(\eta_j^2-1)}\end{aligned}$$

$$(k=1,2,\ldots,m;\ j=1,2,\ldots,\ell). \tag{3.6}$$

Hence we have proved that all the singularities of $F(z)$ can be found from the singularities of $f(z)$ and $g(z)$.

If all the moduli of $\alpha_k(k=1,2,\ldots,m)$, $\beta_j(j=1,2,\ldots,\ell)$ are less than 1; we can write (3.1) in the form

$$f(z) = -\sum_{k=1}^{m} \frac{A_k}{\alpha_k\sqrt{1-\frac{2z}{\alpha_k}+\frac{1}{\alpha_k^2}}},$$

$$g(z) = -\sum_{j=1}^{\ell} \frac{\beta_j}{\beta_j \sqrt{1 - \frac{2z}{\beta_j} + \frac{1}{\beta_j^2}}}.$$

Since all the moduli of $\frac{1}{\alpha_k}$ $(k=1,2,\ldots,m)$, $\frac{1}{\beta_j}$ $(j=1,2,\ldots,\ell)$ are greater than 1. Using the proved result we have

$$F(z) = \sum_{k=1}^{m} \sum_{j=1}^{\ell} \frac{A_k B_j}{\alpha_k \beta_j \sqrt{1 - \frac{2z}{\alpha_k \beta_j} + \frac{1}{(\alpha_k \beta_j)^2}}}.$$

Hence (3.4) is still true, and we complete the proof of theorem 1.

By use of the method in the proof of theorem 1, we can study the case that $\varphi(z)$, $\psi(z)$ corresponding to $f(z)$, $g(z)$ have poles of order higher than 1.

From the gap theorem of the Legendre series [6,7] we deduce the

Gap theorem. Let (n_k) be a subsequence of non-negative integers such that

$$n_{k+1} > (1+\theta)n_k \quad (k=1,2,\ldots),$$

where θ is a fixed positive number. Suppose that the ellipses of convergence of $f(z)$, $g(z)$ are respectively E_α, E_β where α, β are finite positive numbers. If $a_n=0$ except a_{n_k} and the second upper limit in (2.3) is a limit, then the ellipse $E_{\alpha+\beta}$ is the natural boundary of $F(z) = f(z)*g(z)$.

4. THE CASE OF INTEGRAL FUNCTIONS

Suppose that $f(z)$, $g(z)$ represented by (2.1) both are integral functions of order ρ_1, ρ_2 respectively. Then their ellipses of convergence both are E_∞. We know that [4]

$$\sqrt[n]{|a_n|} \to 0, \quad \sqrt[n]{|b_n|} \to 0 \tag{4.1}$$

as $n \to \infty$, and

$$\rho_1 = \overline{\lim_{n\to\infty}} \frac{n\log n}{\log\frac{1}{|a_n|}}, \quad \rho_2 = \overline{\lim_{n\to\infty}} \frac{n\log n}{\log\frac{1}{|b_n|}}.$$

When n is large enough, both $\log\frac{1}{|a_n|}$ and $\log\frac{1}{|b_n|}$ are positive, then from the relation between the upper and lower limits we get

$$\frac{1}{\rho_1} = \underline{\lim_{n\to\infty}} \frac{\log\frac{1}{|a_n|}}{n\log n}, \quad \frac{1}{\rho_2} = \underline{\lim_{n\to\infty}} \frac{\log\frac{1}{|b_n|}}{n\log n}. \tag{4.2}$$

From (4.1) we have

$$\sqrt[n]{|a_n||b_n|} \to 0$$

as $n \to \infty$. Therefore $F(z) = f(z)*g(z)$ is also an integral function. Suppose that the order of $F(z)$ is ρ; then when ρ_1, ρ_2 are both finite positive numbers, we have

$$\frac{1}{\rho} = \varliminf_{n\to\infty} \frac{\log\frac{1}{|a_n b_n|}}{n\log n} = \varliminf_{n\to\infty}\left\{\frac{\log\frac{1}{|a_n|}}{n\log n} + \frac{\log\frac{1}{|b_n|}}{n\log n}\right\}$$

$$\geq \frac{1}{\rho_1} + \frac{1}{\rho_2} = \frac{\rho_1+\rho_2}{\rho_1\rho_2}.$$

Hence the relation between ρ and ρ_1, ρ_2 is

$$\rho \leq \frac{\rho_1\rho_2}{\rho_1+\rho_2}\ . \tag{4.3}$$

If at least one of the lower limits in (4.2) is a limit, then

$$\rho = \frac{\rho_1\rho_2}{\rho_1+\rho_2}. \tag{4.4}$$

If at least one of the numbers ρ_1,ρ_2 is 0 or ∞, then $\rho=0$ or $\rho\leq\min(\rho_1,\rho_2)$.

Suppose that ρ_1, ρ_2 are finite positive numbers and the types of $f(z)$, $g(z)$ are σ_1, σ_2. We know that [4]

$$\left.\begin{aligned} (e\rho_1\sigma_1)^{\frac{1}{\rho_1}} &= 2\varlimsup_{n\to\infty} n^{\frac{1}{\rho_1}}\sqrt[n]{|a_n|}, \\ (e\rho_2\sigma_2)^{\frac{1}{\rho_2}} &= 2\varlimsup_{n\to\infty} n^{\frac{1}{\rho_2}}\sqrt[n]{|b_n|}. \end{aligned}\right\} \tag{4.5}$$

If at least one of the lower limits in (4.2) is a limit, then the order ρ of $F(z)$ given by (4.4) is also a finite positive number. Let the type of $F(z)$ be σ, then

$$(e\rho\sigma)^{\frac{1}{\rho}} = 2 \overline{\lim_{n\to\infty}} n^{\frac{1}{\rho}} \sqrt[n]{|a_n||b_n|}. \tag{4.6}$$

From (4.4) we know that

$$\frac{1}{\rho} = \frac{1}{\rho_1} + \frac{1}{\rho_2} .$$

From this and (4.5), (4.6) we have

$$(e\rho\sigma)^{\frac{1}{\rho}} = 2 \overline{\lim_{n\to\infty}} \{(n^{\frac{1}{\rho_1}} \sqrt[n]{|a_n|})(n^{\frac{1}{\rho_2}} \sqrt[n]{|b_n|})\}$$

$$\leqslant \frac{1}{2}(e\rho_1\sigma_1)^{\frac{1}{\rho_1}} (e\rho_2\sigma_2)^{\frac{1}{\rho_2}}$$

$$= \frac{1}{2} e^{\frac{1}{\rho_1} + \frac{1}{\rho_2}} (\rho_1\sigma_1)^{\frac{1}{\rho_1}} (\rho_2\sigma_2)^{\frac{1}{\rho_2}} .$$

From this we get

$$2(\rho\sigma)^{\frac{1}{\rho}} \leqslant (\rho_1\sigma_1)^{\frac{1}{\rho_1}} (\rho_2\sigma_2)^{\frac{1}{\rho_2}} . \tag{4.7}$$

If at least one of the upper limits in (4.5) is a limit, then

$$2(\rho\sigma)^{\frac{1}{\rho}} = (\rho_1\sigma_1)^{\frac{1}{\rho_1}} (\rho_2\sigma_2)^{\frac{1}{\rho_2}} . \tag{4.8}$$

Combining the results mentioned above we get the following

Theorem 2. Let $f(z)$, $g(z)$ represented by (2.1) be integral functions of order ρ_1, ρ_2 respectively. If ρ_1, ρ_2 are all finite positive numbers, and at least one of the lower limits in (4.2) is a limit, then $F(z) = f(z)*g(z)$ is an integral function with finite positive order

$$\rho = \frac{\rho_1\rho_2}{\rho_1+\rho_2} .$$

If at least one of the numbers ρ_1, ρ_2 is zero or ∞, then $\rho=0$ or $\rho\leqslant\min(\rho_1,\rho_2)$. Suppose that ρ_1, ρ_2 are finite positive numbers, and $\rho = \frac{\rho_1\rho_2}{\rho_1+\rho_2}$ and the type of $f(z)$, $g(z)$ are respectively σ_1, σ_2. If at least one of the upper limits in (4.5) is a limit, then the type σ of $F(z)$ is determined by the equation

$$2(\rho\sigma)^{\frac{1}{\rho}} = (\rho_1\sigma_1)^{\frac{1}{\rho_1}}\ (\rho_2\sigma_2)^{\frac{1}{\rho_2}} .$$

5. THE GENERALIZED HADAMARD PRODUCT

Let $\Omega(x,y)$ be a function of two complex variables. When $\varphi(z)$, $\psi(z)$ are defined by power series, we call

$$\Phi(z) = \sum_{n=0}^{\infty}\Omega(a_n,b_n)z^n$$

the generalized Hadamard product (G.Hp product) of $\varphi(z)$ and $\psi(z)$ and denote it by

$$\Phi(z) = \varphi(z)\Omega_p\psi(z).$$

When M is a positive integer, we call

$$\Phi(z) = \sum_{n=0}^{\infty} a_n^M z^n$$

the M_{Hp} power of $\varphi(z)$, and denote it by $\Phi(z) = \{\varphi(z)\}^{\langle M \rangle_p}$. Obviously

$$\{\varphi(z)\}^{\langle M \rangle_p} = \varphi(z) *_p \varphi(z) *_p \cdots *_p \varphi(z)$$

where the operation $*_p$ is repeated M-1 times.

Instead of power series, when $f(z)$, $g(z)$, are represented by Legendre series (2.1), we call

$$F(z) = \sum_{n=0}^{\infty} \Omega(a_n, b_n) P_n(z)$$

the generalized Hadamard product (G.H product) and denote it by

$$F(z) = f(z) \Omega g(z).$$

Similarly we call

$$F(z) = \sum_{n=0}^{\infty} a_n^M P_n(z)$$

the M_H power of $f(z)$, and denote it by $F(z) = \{f(z)\}^{\langle M \rangle}$. Obviously

$$\{f(z)\}^{\langle M \rangle} = f(z) * f(z) * \cdots * f(z)$$

where the operation $*$ is repeated M-1 times.

We consider a special case. Suppose that M, L are both positive integers and $ML \geq 2$. Let

$$\Omega(x,y) = x^M y^L,$$

then

$$F(z) = f(z)\Omega g(z) = \{f(z)\}^{\langle M \rangle} * \{g(z)\}^{\langle L \rangle}.$$

Then for theorem 1 we have the following

Corollary 1. Suppose that M, L are both positive integers and $ML \geq 2$ and $f(z)$, $g(z)$ satisfy all the conditions in the theorem 1. Then all the singularities of $F(z) = \{f(z)\}^{\langle M \rangle} * \{g(z)\}^{\langle L \rangle}$ are algebraic branch-points of second degree and can be found from the singularities of $f(z)$, $g(z)$.

Proof. Consider $f(z)*f(z)$. From theorem 1 we know that

$$f(z)*f(z) = \sum_{k=1}^{m} \sum_{j=1}^{m} \frac{A_k A_j}{\sqrt{1-2z\alpha_k\alpha_j+\alpha_k^2\alpha_j^2}} .$$

Therefore all the singularities of $f(z)*f(z) = \{f(z)\}^{\langle 2 \rangle}$ are algebraic branch-points of second degree with representations similar to (3.6). Similarly all the singularities of $\{f(z)\}^{\langle M \rangle}$ and $\{g(z)\}^{\langle L \rangle}$ are also algebraic branch-points of second degree with representation similar to (3.6). Then from theorem 1 we obtain the corollary 1.

For theorem 2 we have the following

Corollary 2. Suppose that M, L are both positive

integers and $ML \geq 2$ and $f(z)$, $g(z)$ represented by (2.1) are integral functions with finite positive order ρ_1, ρ_2 respectively. If at least one of two lower limits in (4.2) is a limit, then

$$F(z) = \{f(z)\}^{\langle M\rangle} * \{g(z)\}^{\langle L\rangle}$$

is an integral function of order

$$\rho = \frac{\rho_1\rho_2}{L\rho_1 + M\rho_2}.$$

If at least one of the numbers ρ_1, ρ_2 is zero, then $\rho=0$. Suppose that ρ_1, ρ_2 are finite positive numbers and $\rho = \rho_1\rho_2/(L\rho_1+M\rho_2)$, and the types of $f(z)$, $g(z)$ are respectively σ_1, σ_2. If at least one of the upper limits in (4.5) is a limit, then the type of $F(z)$ is determined by the equation

$$2^{M+L-1}(\rho\sigma)^{\frac{1}{\rho}} = (\rho_1\sigma_1)^{\frac{M}{\rho_1}}(\rho_2\sigma_2)^{\frac{L}{\rho_2}}.$$

Proof. Obviously $\{f(z)\}^{\langle M\rangle}$ is an integral function with order

$$\overline{\lim_{n\to\infty}} \frac{n\log n}{\log\frac{1}{|a_n|^M}} = \frac{1}{M}\overline{\lim_{n\to\infty}} \frac{n\log n}{\log\frac{1}{|a_n|}} = \frac{\rho_1}{M}.$$

Similarly $\{g(z)\}^{\langle L\rangle}$ is also an integral function with order $\frac{\rho_2}{L}$.

If ρ_1, ρ_2 are both finite positive numbers, then $\frac{\rho_1}{M}, \frac{\rho_2}{L}$ are also both finite positive numbers. Then if at least one of the lower limits in (4.2) is a limit from theorem 2 we know that $F(z) = \{f(z)\}^{\langle M\rangle} * \{g(z)\}^{\langle L\rangle}$ is an integral function with order

$$\rho = \frac{\frac{\rho_1}{M} \cdot \frac{\rho_2}{L}}{\frac{\rho_1}{M} + \frac{\rho_2}{L}} = \frac{\rho_1 \rho_2}{L\rho_1 + M\rho_2} .$$

If at least one of the numbers ρ_1, ρ_2 is zero, then at least one of the numbers $\rho_1/M, \rho_2/L$ is also zero, therefore $\rho=0$.

When ρ_1, ρ_2 are finite positive numbers, obviously the order of $\{f(z)\}^{\langle M\rangle}$, $\{g(z)\}^{\langle L\rangle}$ are also finite positive numbers. Suppose that their type are $\tilde{\sigma}, \hat{\sigma}$ respectively. Then from (4.5) we know that

$$\left(e\frac{\rho_1}{M}\,\tilde{\sigma}\right)^{\frac{M}{\rho_1}} = 2\ \overline{\lim_{n\to\infty}}\ n^{\frac{M}{\rho_1}}\ \sqrt[n]{|a_n|}^{M}$$

$$= 2\left\{\frac{1}{2}(e\rho_1\sigma_1)^{\frac{1}{\rho_1}}\right\}^{M} = \frac{1}{2^{M-1}}(e\rho_1\sigma_1)^{\frac{M}{\rho_1}} .$$

From this we obtain

$$\tilde{\sigma} = M2^{(1-M)\rho_1/M}\sigma_1 .$$

Similarly we have

$$\hat{\sigma} = L2^{(1-L)\rho_2/M}\sigma_2.$$

When at least one of the lower limits in (4.2) is a limit and at least one of the upper limits in (4.5) is a limit, from theorem 2 we know that the type of $F(z)$ is determined by the equation

$$2(\rho\sigma)^{\frac{1}{\rho}} = \{\frac{\rho_1}{M}\cdot M\cdot 2^{\rho_1(1-M)/M}\sigma_1\}^{\frac{M}{\rho_1}}\{\frac{\rho_2}{L}\cdot L\cdot 2^{\rho_2(1-L)/L}\sigma_2\}^{\frac{L}{\rho_2}},$$

i.e.

$$2^{M+L-1}(\rho\sigma)^{\frac{1}{\rho}} = (\rho_1\sigma_1)^{\frac{M}{\rho_1}}(\rho_2\sigma_2)^{\frac{L}{\rho_2}}.$$

The proof is completed.

REFERENCES

1. Titchmarsh, E.C., The Theory of Functions, Oxford University Press (1932).

2. Yang Dinggong, Hadamard products of certain p-valent functions, Journal of Mathematics (PRC), No. 3 (1983) (in Chinese).

3. Wu Zhuo Ren, On Sakaguchi's function family and Hadamard product, Scientia Sinica, 2 (1985), 104-110.

4. Mo Yeh, Legendre series, Journal of Shandong University, No. 7 (1956) (in Chinese).

5. Mo Yeh, Integrability theorems for series of variable terms, Journal of Shandong University, No. 3 (1981) (in Chinese).

6. Mo Yeh, The gap theorem of the Legendre series, Journal of Shandong University, No. 4 (1979) (in Chinese).

7. Mo Yeh, The Theory of Legendre Functions, Shandong University Press (1987) (in Chinese).

ON PRIMALITY OF THE BESSEL FUNCTIONS

Zhong Chang-Yong and Song Guo-Dong
Department of Mathematics
East China Normal University
Shanghai
The People's Republic of China

It is proved in this paper that the Bessel functions $J_\nu(z)$ are prime when ν is odd, and z^2 is the only right factor of $J_\nu(z)$ when ν is even.

1. INTRODUCTION

Steinmetz [1] proved that if $F(z)$ is a transcendental meromorphic function satisfying the linear differential equation

$$w^{(n)}+a_{n-1}(z)w^{(n-1)}+\dots+a_0(z)w+a(z) = 0$$

with $a(z)$, $a_0(z)$, ..., $a_{n-1}(z)$ being rational functions, then F is pseudo-prime. That is, every factorization of the form

$$F = f\circ g \tag{1}$$

with f being meromorphic and g entire (g may be meromorphic when f is rational) implies that either f is rational or g is a polymonial. As an application, Bessel functions $J_\nu(z), \nu = 0,\pm1,\pm2,\dots,$ which satisfy the second order differential equations

$$w''+\frac{1}{z}w'+(1-\frac{\nu^2}{z^2})w = 0,$$

are pseudo-prime. A question arises naturally that if $J_\nu(z)$ is prime for each ν, that is, every factorization of the form (1) with F being J_ν implies that either f is fractional linear or g is linear. And the answer is contained in the following.

Theorem $J_\nu(z)$ is prime when ν is odd; and it has the only factorization

$$J_\nu(z) = \left(\sum_{n=0}^{\infty}(-1)^n \frac{2^{-\nu-2n}\zeta^{\nu/2+n}}{n!\Gamma(n+\nu+1)}\right)\circ z^2 \tag{2}$$

in the sense of equivalence (Two factorizations $F = f\circ g$ and $F = \tilde{f}\circ\tilde{g}$ are called to be equivalent if there is a linear transformation T such that $\tilde{f} = f\circ T$, $\tilde{g} = T^{-1}\circ g$) when ν is even.

2. PRELIMINARIES

It is well-known that $J_\nu(z)$ where ν is an integer is entire, and it has the expansion at $z=0$:

$$J_\nu(z) = \sum_{n=0}^{\infty}(-1)^n \frac{(\frac{z}{2})^{\nu+2n}}{n!\Gamma(n+\nu+1)}, \quad \nu = 0,1,2,... \tag{3}$$

and

$$J_{-\nu}(z) = (-1)^\nu J_\nu(z). \tag{4}$$

Also, it is easily seen that $J_\nu(z)$ is non-periodic and of order 1.

In proving our theorem we need the following lemmas.

Lemma 1 [2]. Let F be a transcendental entire function which is not periodic. Then F is prime if and only if F is E-prime (prime in entire sense).

Lemma 2 [3]. $J_\nu(z)$ has infinitely many zeros, all of

which are real and for the large $z>0$, we have the asymptotic formula

$$J_\nu(z) \sim \sqrt{\frac{2}{\pi z}} \cos\left(z - \frac{\pi}{4} - \frac{\nu\pi}{2}\right) \tag{5}$$

Lemma 3 [4]. Let $f(z)$ be an entire function of order less than 2. If $f(z)$ is real for real z and the zeros of $f(z)$ are real, then the zeros of $f'(z)$ are also real and separate those of $f(z)$ from each other.

Lemma 4 [5]. Let $f(z)$ be an entire function. If there exists an unbounded sequence $\{a_n\}$ such that almost all the roots of $f(z) = a_n$ lie on one straight line, then $f(z)$ is a polynomial of degree at most 2.

Lemma 5 [6]. Let $F(z)$ be an entire function of finite order, and let $F'(z)$ have infinitely many zeros. If for any complex number a, the simultaneous equations

$$\begin{cases} F(z) = a \\ F'(z) = 0 \end{cases}$$

have only finitely many solutions, then $F(z)$ is E-left-prime, that is, every factorization of the form(1) with f being entire and g transcendental entire implies that f is linear.

3. PROOF OF THE THEOREM

By Lemma 1, we need only to show that J_ν is E-prime. Also, according to Steinmetz theorem just mentioned in the very beginning of §1, J_ν is pseudo-prime. Thus, if we write $J_\nu(z) = f \circ g(z)$, then the only cases to deal with are: (a) f is transcendental entire, and g is a polynomial; (b) f is a polynomial, and g

entire.

Case (a). Since $J_\nu = f \circ g$ has infinitely many zeros, so does f. Assume that all zeros of f are $a_1, a_2, \ldots$. By Lemma 2, all the roots of $g(z) = a_n (n=1,2,\ldots)$ lie on the real axis. Lemma 4 implies that g is a polynomial of degree at most 2. Suppose that g is of degree 2 and write (without loss of generality) $g(z) = (z+a)^2$. We discuss the following two subcases

(i) ν is even. Then expansion (3) implies that

$$J_\nu(z) = f((z+a)^2) \tag{6}$$

is an even function, from which we derive

$$J_\nu(z+2a) = J_\nu(-z-2a) = J_\nu(z).$$

Since J_ν is not periodic, we must have $a=0$, i.e. $J_\nu(z) = f(z^2)$ with $f(\zeta)$ being as in formula (2).

(ii) ν is odd. By (4) J_ν is an odd function. On the other hand, from (6) we obtain

$$J_\nu(z+2a) = -J_\nu(-z-2a) = -J_\nu(z),$$

$$J_\nu(z+4a) = -J_\nu(z+2a) = J_\nu(z),$$

which implies $a=0$ as in (i). Hence $J_\nu(z) = f(z^2)$ is also an even function. This contradiction shows that g must be linear.

Case (b). For any complex number a, consider the simultaneous equations

$$\begin{cases} J_\nu(z) = a \\ J'_\nu(z) = 0 \end{cases} \tag{7}$$

If $a=0$, then (7) has at most one solution $z=0$ by the theory of ordinary differential equations. If $a \neq 0$, then by Lemma 3, the roots of $J'(z)=0$ are real, so are solutions of (7). The asymptotic formula (5) combining (4) implies that if z is real and $|z|$ sufficiently large, then $|J_\nu(z)| < |a|$.

Therefore, (7) has only finitely many solutions. By Lemma 3, J' has infinitely many zeros, and by Lemma 5, f must be linear. And the proof is completed.

REFERENCES

1. Steinmetz, N., Uber die faktorisierbaren Lösungen gewöhnlicher differentialgleichungen, Math. Zeit., 170 (1980), 169-180.

2. Gross, F., Factorization of entire functions which are periodic mod g, Indian J. Pure and Appl. Math. Vol 2(3), p.568(1971).

3. Spiegel, M.R., Theory and problems of advanced mathematics for engineers and scientists (Chapter 10), McGraw-Hill Book Company, 1971.

4. Titchmarsh, E.C., The theory of functions, Oxford Univ. Press, 1957.

5. Edrei, A., Meromorphic functions with three radically distributed values, Trans. Amer. Math. Soc. 78 (1955), 276-293.

6. Ozawa, M., On certain criteria for the left-primeness of entire functions, Kodai Math. Sem. Rep. 26 (1975), 304-317.

MEROMORPHIC FUNCTION WHOSE DERIVATIVE HAS LARGEST SUM OF DEFICIENCIES

Dai Chongji and Jin Lu
Department of Mathematics
East China Normal University
Shanghai, P.R. China

In [1] we proved

THEOREM A. Let $f(z)$ be a meromorphic function of lower order $\mu < \infty$. If $\sum\limits_{a} \delta(a,f') = 2$, then

$$P_o + P_1 \leqslant \mu,$$

where P_o, P_1 are the numbers of non-zero finite deficient values of $f(z)$, $f'(z)$ respectively.

In this paper, we shall improve this result, by proving the following.

THEOREM 1. Let $f(z)$ be a meromorphic function of lower order $\mu < \infty$, k be a positive integer, if

$$\sum_{a} \delta(a, f^{(k)}) = 2, \tag{1}$$

then (a)

$$\sum_{i=-\infty}^{\infty} P_i \leqslant \mu, \tag{2}$$

where P_i $(i = 0, \pm1, \pm2, \ldots)$ are the numbers of non-zero finite deficient values of $f^{(i)}(z)$, $f^{(o)}(z) = f(z)$ and when i is negative, $f^{(i)}(z)$ is defined as $|i|$-th primitive function of $f(z)$ (if existent).

(b) Especially, $P_i = 0$ for $i > k$ and

$$\delta(\infty, f^{(i)}) = \delta(0, f^{(i)}) = 1 \tag{3}$$

Furthermore, we shall prove

THEOREM 2. Let $f(z)$ be a meromorphic function of lower order $\mu < \infty$. If (1) is satisfied, then

$$\sum_{i=-\infty}^{\infty} \sum_{a \neq 0,\infty} \delta(a, f^{(i)}) \leqslant 1, \tag{4}$$

and if the equality in (4) holds, then each deficiency is an integral multiple of $\frac{1}{\mu}$.

1. LEMMAS

LEMMA 1. Let $g(z)$ be a meromorphic function in $|z| < R(0 < R < \infty)$, then for $z = re^{i\theta}$ and ρ, $r < \rho < R$ and for any positive integer k, we have

$$\log^+\left|\frac{g^{(k)}(z)}{g(z)}\right| \leqslant C\{1 + \log^+\rho + \log^+\frac{1}{\rho-r} + \log^+ R$$

$$+ \log^+\frac{1}{\delta(z)} + \log^+\frac{1}{r} + \log^+ T(\rho,g) + \log^+\log^+\frac{1}{|g(0)|}\}$$

where $R = n(\rho,g) + n(\rho, \frac{1}{g})$, $\delta(z) = \min_a \{|z-a|\}$ with a being taken at all zeros and poles of $g(z)$ in $|z| \leqslant \rho$, and C is a constant depending only on k.

Proof. As in [1, Lemma 2], this lemma can be proved by induction.

LEMMA 2. Let $f(z)$ be a meromorphic function of lower order $\mu < \infty$. If (1) is satisfied, then for an arbitrary set of finite deficient values $\{a_{i\ell}\}_{\ell=1}^{\tau_i}$ of $f^{(i)}(z)$ $(i = 0, 1, \ldots, k)$, there exists a sequence of positive numbers $\{\rho_j\}_{j=1}^{\infty}$, $\alpha^j \leqslant \rho_j < \alpha^{j+1}$, $\alpha = \exp(\mu+1)$ and sets $\{E_{i\ell}(j)\}_{\ell=1}^{\tau_i}$ which are on $|z| = \rho_j$, so that meas $E_{i\ell}^{*}(j) = \text{meas}\,\{\theta,\ 0 \leqslant \theta < 2\pi,\ \rho_j e^{i\theta} \in E_{i\ell}(j)\} \geqslant \eta > 0\{\ell = 1, 2, \ldots, \tau_i;\ j = 1, 2, \ldots\}$, and for $z \in E_{i\ell}(j)$ we have

$$\log|f^{(i)}(z) - a_{i\ell}| < - KT(\rho_j, f^{(k+1)}), \quad \ell = 1,2,\ldots,\tau_i,$$

and

$$\log|f^{(m)}(z)| < - KT(\rho_j, f^{(k+1)}),$$

for $m > i$, where η, K are absolute positive constants independent on j.

Proof. Using Lemma 1, as in [1, Lemma 1], only changing $f'(z)$ to $f^{(k)}(z)$, $f''(z)$ to $f^{(k+1)}(z)$, and then taking

$$\delta_0 = \min_{0 \leqslant i \leqslant k} \{\delta(a_{i1}, f^{(i)}), \ldots, \delta(a_{i\tau_i}, f^{(i)})\},$$

we can prove Lemma 2.

2. PROOF OF THEOREM 1

First, similar to the proof of [2, Theorem 1], we have

$$2 = \sum_a \delta(a,f^{(k)}) = \sum_{a\neq\infty} \delta(a,f^{(k)}) + \delta(\infty,f^{(k)})$$

$$\leqslant \sum_{a\neq\infty} \delta(a,f^{(k)}) + \Theta(\infty,f^{(k)}) \leqslant \sum_a \Theta(a,f^{(k)}) \leqslant 2.$$

hence $\delta(\infty,f^{(k)}) = \Theta(\infty,f^{(k)})$. Since

$$N(r,f^{(k)}) \geqslant 2\bar{N}(r,f^{(k-1)}) = 2\bar{N}(r,f^{(k)}),$$

hence $2\Theta(\infty,f^{(k)}) > 1 + \delta(\infty,f^{(k)})$,

$$\delta(\infty,f^{(k)}) = \Theta(\infty,f^{(k)}) = 1.$$

Using [3] or [2, Lemma 1] we have

$$\delta(0,f^{(k+1)}) = \delta(\infty,f^{(k+1)}) = 1. \tag{5}$$

Again from $\sum_a \delta(a,f^{(k+1)}) = 2$, we have

$$\delta(0,f^{(k+2)}) = \delta(\infty, f^{(k+2)}) = 1,$$

and so on. Thus we deduce (3) and $P_i = 0$ $(i > k)$.

Set (see [4, Theorem 1 and Lemma 5])

$$C_j = C(\alpha^j) = |C(\alpha^j)|e^{i\omega_j}. \tag{6}$$

From [2], [5] we know that $f^{(k)}(z)$ is of regular growth and μ is a positive integer. We divide $|z| = \rho_j$ into 2μ arcs $\alpha_1(j), \ldots, \alpha_\mu(j)$; $\beta_1(j), \ldots, \beta_\mu(j)$, so that $\cos(\mu\phi + \omega_j) \geqslant 0$, $\cos(\mu\phi + \omega_j) \leqslant 0$, on $\alpha_\nu(j)$, $\beta_\nu(j)$ $(\nu = 1, 2, \ldots, \mu)$ respectively. Obviously, $\alpha_\nu(j)$ and

$\beta_\nu(j)$ are apart from each other and the angular measure of each $\alpha_\nu(j)$ and $\beta_\nu(j)$ $(\nu = 1, 2, \ldots, \mu)$ is $\frac{\pi}{\mu}$. Again set

$$\alpha^*_\nu(j) = \{\phi;\ 0 \leqslant \phi < 2\pi,\ \rho_j e^{i\phi} \in \alpha_\nu(j)\};$$
$$\beta^*_\nu(j) = \{\phi,\ 0 \leqslant \phi < 2\pi,\ \rho_j e^{i\phi} \in \beta_\nu(j)\}. \tag{7}$$

Without loss of generality we assume,

$$\alpha^*_\nu(j) = [\psi_{\nu j},\ \psi_{\nu j} + \frac{\pi}{\mu}];\ \beta^*_\nu(j) = [\phi_{\nu j},\ \phi_{\nu j} + \frac{\pi}{\mu}],$$
$$\nu = 1, 2, \ldots, \mu.$$

Take $\phi_o(> 0)$ sufficiently small and set

$$\widetilde{\alpha^*_\nu}(j) = [\psi_{\nu j} + \phi_o,\ \psi_{\nu j} + \frac{\pi}{\mu} - \phi_o];$$
$$\widetilde{\beta^*_\nu}(j) = [\phi_{\nu j} + \phi_o,\ \phi_{\nu j} + \frac{\pi}{\mu} - \phi_o],\ \nu = 1, 2, \ldots, \mu. \tag{8}$$

Using the same method as in the proof of [1, Theorem 1] and using Lemma 1, we can prove that for each $a_{i\ell}(\neq 0, \infty)$ (see Lemma 2) there exists a $\widetilde{\beta^*_\nu}(j)$ $(j > j_o)$, such that for $|z| = \rho_j$, $\arg z \in \widetilde{\beta^*_\nu}(j)$ we have,

$$|f^{(i)}(z) - a_{i\ell}| < e^{-KT(\rho_j,\ f^{(k+1)})};$$
$$|f^{(m)}(z)| < e^{-KT(\rho_j, f^{(k+1)})} \quad (m > i). \tag{9}$$

From (9) we can prove that each $a_{i\ell}(\neq 0, \infty)$ $(i = 0, 1, \ldots, k;\ \ell = 1, 2, \ldots, \tau_i)$ is only correspondent with one $\widetilde{\beta^*_\nu}(j)$, and if $(i,\ell) \neq (i',\ell')$, then $a_{i\ell}$ and $a_{i'\ell'}$

are correspondent to different $\widetilde{\beta^*_\nu}(j)$. But the total number of $\widetilde{\beta^*_\nu}(j)$ is μ, so $\sum_{i=0}^{k} P_i \leqslant \mu$. Combining this with (b) we have

$$\sum_{i=0}^{\infty} P_i \leqslant \mu. \tag{10}$$

If $f(z)$ has N-th primitive function $(f^{(-N)}(z))$, then $f^{(k)}(z)$ is $(k+N)$-th derivative of $f^{(-N)}(z)$. From (10), we have

$$\sum_{i=-N}^{\infty} P_i \leqslant \mu. \tag{11}$$

If N is an arbitrary positive integer and $f^{(-N)}(z)$ exists, noticing the right side of (11) is independent of N, (2) follows.

3. PROOF OF THEOREM 2

We have

$$K(f^{(k+1)}) = \overline{\lim_{r\to\infty}} \frac{N(r,f^{(k+1)}) + N(r,\frac{1}{f^{(k+1)}})}{T(r,f^{(k+1)})}. \tag{12}$$

From [4, Theorem 1 and Lemma 5], setting $p = \mu$, for $\varepsilon > 0$, $j > j_o$, $z \in \Gamma_j - E_j$ we have

$$\left|\log|f^{(k+1)}(j)| - \mathrm{Re}\{C_j z^\mu\}\right| < 4\varepsilon|C_j|r^\mu, \tag{13}$$

where C_j occurs in the proof of Theorem 1, $\Gamma_j = \{z;\ \alpha^j \leqslant |z| < \alpha^{j+\frac{3}{2}}\}$, E_j consists of a finite number of circles (Γ_j) and the sum of radius of these cicles is no more than $4e\delta\alpha^{j+2}$, where δ is a sufficiently small positive number. And we have

$$T(\sigma r, f^{(k+1)}) = \sigma^{\mu} T(r, f^{(k+1)})(1 + O(1))$$

$$T(r, f^{(k+1)}) = (1 + O(1)) \frac{C(r)\, r^{\mu}}{\pi}$$

$$|C(\sigma r) - C(r)| = O\{|C(r)]\}, \ z = re^{i\phi}. \tag{14}$$

Taking ϕ_o of (8) sufficiently small and $\varepsilon < \frac{\sin(\mu\phi_o)}{10}$, by the same reasoning as in the proof of [1, Theorem 1], we may choose $\{\rho_j\}$ in Lemma 2 such that $|z| = \rho_j \subset \Gamma_j - E_j$ and $(|z| = \rho_j) \cap (\Gamma_j) = \phi$. Thus for $|z| = \rho_j$, $\arg z \in \widetilde{\alpha_\nu^*}(j)$, from (13) we have

$$\log|f^{(i)}(z) - a_{i\ell}| + \log \frac{|f^{(k+1)}(z)|}{|f^{(i)}(z) - a_{i\ell}|} = \log|f^{(k+1)}(z)|$$

$$\geq |C_j|\rho_j^{\mu} \cos(\mu\phi + \omega_j) - \varepsilon\{|C_j|\rho_j^{\mu}\}$$

$$\geq \frac{1}{2}\sin(\mu\phi_o)\, \rho_j^{\mu}\, |C_j| \quad (> 0).$$

Therefore, for $|z| = \rho_j$, $\arg z \in \widetilde{\alpha_\nu^*}(j)$, we have

$$\log^+ \frac{1}{|f^{(i)}(z) - a_{i\ell}|} \leq \log^+ \frac{|f^{(k+1)}(z)|}{|f^{(i)}(z) - a_{i\ell}|}. \tag{15}$$

Since the order of $f(z)$ is finite, using Nevanlinna's well known Lemma of logarithmic derivative, we have

$$\varlimsup_{j\to\infty} \frac{1}{2\pi\, T(\rho_j, f^{(i)})} \int_{\widetilde{\alpha_\nu^*}(j)} \log^+ \frac{1}{|f^{(i)}(\rho_j e^{i\phi}) - a_{i\ell}|}\, d\phi = 0. \tag{16}$$

For $i = 0, 1, \ldots, k$; $\ell = 1, 2, \ldots, P_i$; $\nu = 1, 2, \ldots, \mu$. (for P_i see Theorem 1), set

$$d^{\nu}_{i\ell}(j) = \frac{1}{2\pi\ T(\rho_j, f^{(i)})} \int_{\widetilde{\beta^*_\nu(j)}} \log^+ \frac{1}{|f^{(i)}(\rho_j e^{i\phi}) - a_{i\ell}|}\, d\phi. \tag{17}$$

Fix i, for $\ell = 1$, for each j, we rearrange $\{d^1_{i1}(j), d^2_{i1}(j), \ldots, d^{\mu}_{i1}(j)\}$ according to increasing order. Since the number of permutations of $\{d^{\nu}_{i1}(j)\}^{\mu}_{\nu=1}$ is finite, there exists an infinite subsequence $\{j_1\}$ of $\{j\}$, which corresponds to a permutation $\{\nu_{11}, \nu_{12}, \ldots, \nu_{1\mu}\}$ of $\{1, 2, \ldots, \mu\}$, so that

$$d^{\nu_{11}}_{i1}(j_1) \geqslant d^{\nu_{12}}_{i1}(j_1) \geqslant \ldots \geqslant d^{\nu_{1\mu}}_{i1}(j_1).$$

In the same way, we can obtain an infinite subsequence $\{j_2\}$ of $\{j_1\}$ and a corresponding permutation $\{\nu_{21}, \ldots, \nu_{2\mu}\}$ of $\{1, 2, \ldots, \mu\}$, so that

$$d^{\nu_{21}}_{i2}(j_2) \geqslant d^{\nu_{22}}_{i2}(j_2) \geqslant \ldots \geqslant d^{\nu_{2\mu}}_{i2}(j_2),$$

and so on. Thus we have an infinite subsequence $\{j_o\}$ of $\{j\}$, with no loss of generality, we may assume $\{j_o\} = \{j\}$ and permutations $\{\nu_{\ell 1}, \nu_{\ell 2}, \ldots, \nu_{\ell\mu}\}^{P_i}_{\ell=1}$ of $\{1, 2, \ldots, \mu\}$, so that for $i = 0, 1, \ldots, k$,

$$d_{i1}^{\nu_{11}}(j) \geq d_{i1}^{\nu_{12}}(j) \geq \dots \geq d_{i1}^{\nu_{1\mu}}(j)$$

$$d_{i2}^{\nu_{21}}(j) \geq d_{i2}^{\nu_{22}}(j) \geq \dots \geq d_{i2}^{\nu_{2\mu}}(j)$$

....

$$d_{iP_i}^{\nu_{P_i1}}(j) \geq d_{iP_i}^{\nu_{P_i2}}(j) \geq \dots \geq d_{iP_i}^{\nu_{P_i\mu}}(j). \tag{18}$$

For $i = 0$, $\ell = 1$, we have an infinite subsequence $\{j_{01}\}$ of $\{j\}$ so that

$$\lim_{j_{01}\to\infty} d_{01}^{\nu_{11}}(j_{01}) = \varliminf_{j\to\infty} d_{01}^{\nu_{11}}(j),$$

and for $i = 0$, $\ell = 2$, we have infinite subsequence $\{j_{02}\}$ of $\{j_{01}\}$ so that

$$\lim_{j_{02}\to\infty} d_{02}^{\nu_{21}}(j_{02}) = \varliminf_{j_{01}\to\infty} d_{02}^{\nu_{21}}(j_{01}),$$

and so on. Since $\{d_{i\ell}^{\nu_{\ell t}}(j)\}$ are finite, we have an infinite subsequence $\{j_o\}$ of $\{j\}$, and with no loss of generality, we may also assume $\{j_o\} = \{j\}$, so that for i = 0, 1, ..., k; $\ell = 1, 2, \dots, P_i$; $t = 1, 2, \dots, \mu$, $\lim_{j\to\infty} d_{i\ell}^{\nu_{\ell t}}(j)$ exists which is denoted by

$$D_{i\ell}^{\nu_{\ell t}} = \lim_{j\to\infty} d_{i\ell}^{\nu_{\ell t}}(j).$$

From (18) we have

$$D_{i\ell}^{\nu_{\ell 1}} \geq D_{i\ell}^{\nu_{\ell 2}} \geq \dots \geq D_{i\ell}^{\nu_{\ell\mu}} \geq 0.$$

Assume that for fix i, ℓ, $D_{i\ell}^{\nu_{\ell N(i,\ell)}}$ is the smallest non-zero one of $\{D_{i\ell}^{\nu_{\ell t}}\}_{t=1}^{\mu}$, $N(i,\ell) \leq \mu$. From (9) we know that if $(i,\ell) \neq (i',\ell')$, then $D_{i\ell}^{\nu_{\ell t}}$ $(1 \leq t \leq N(i,\ell))$ and $D_{i'\ell'}^{\nu_{it'}}$ $(1 \leq t' \leq N(i',\ell'))$ correspond to different elements of $\{\widetilde{\beta_\nu^*(j)}\}_{\nu=1}^{\mu}$ $(j > j_o)$. But the total number of $\widetilde{\beta_\nu^*(j)}$ is μ, so

$$\sum_{i=0}^{k} \sum_{\ell=1}^{P_i} N(i,\ell) \leq \mu. \tag{19}$$

Next we will prove $\delta(a_{i\ell}, f^{(i)}) \leq \frac{N(i,\ell)}{\mu}$, $(i, = 0, 1, \dots, k;\ \ell = 1, 2, \dots, P_i)$. From Theorem 1, $\delta(\infty, f^{(k+1)}) = 1$, so for $0 \leq i \leq k$, we have

$$\begin{aligned} T(r,f^{(k+1)}) &= m(r,f^{(k+1)}) + N(r,f^{(k+1)}) \\ &\leq m(r,\frac{f^{(k+1)}}{f^{(i)}}) + m(r,f^{(i)}) + o\{T(r,f^{(k+1)})\} \\ &\leq T(r,f^{(i)}) + o\{T(r,f^{(i)})\} + o\{T(r,f^{(k+1)})\}. \end{aligned}$$

Hence

$$\overline{\lim_{r\to\infty}} \frac{T(r,f^{(k+1)})}{T(r,f^{(i)})} \leq 1, \tag{20}$$

From (14), (20) we have

$$\overline{\lim_{j\to\infty}} \frac{|C_j|\rho_j^{\mu}}{T(\rho_j, f^{(i)})} \leq \lim_{j\to\infty} \frac{|C_j|}{|C(\rho_j)|} \cdot \lim_{j\to\infty} \frac{|C(\rho_j)\rho_j^{\mu}|}{T(\rho_j, f^{(k+1)})} \cdot$$

$$\overline{\lim_{j\to\infty}} \frac{T(\rho_j, f^{(k+1)})}{T(\rho_j, f^{(i)})} \leq 1 \,.\, \pi \,.\, 1 = \pi. \tag{21}$$

Set $e(j) = (0, 2\pi) - \sum_{\nu=1}^{\mu} \widetilde{\alpha_j^*}(j) - \sum_{\nu=1}^{\mu} \widetilde{\beta_\nu^*}(j)$, then meas $e(j) = 4\mu\phi_0$. From (16) and the definition of $N(i, \ell)$, we have

$$\delta(a_{i\ell}, f^{(i)}) = \overline{\lim_{r\to\infty}} \frac{1}{2\pi\, T(r, f^{(i)})} \int_0^{2\pi} \log^+ \frac{1}{|f^{(i)}(re^{i\phi}) - a_{i\ell}|}\, d\phi$$

$$\leq \overline{\lim_{j\to\infty}} \frac{1}{2\pi\, T(r, f^{(i)})} \int_0^{2\pi} \log^+ \frac{1}{|f^{(i)}(\rho_j e^{i\phi}) - a_{i\ell}|}\, d\phi$$

$$\leq \overline{\lim_{j\to\infty}} \frac{1}{2\pi T(\rho_j, f^{(i)})} \Big(\sum_{\nu=1}^{\mu} \int_{\widetilde{\beta_\nu^*}(j)} + \int_{e(j)} \Big) \log^+ \frac{1}{|f^{(i)}(\rho_j e^{i\phi}) - a_{i\ell}|}\, d\phi$$

$$\leq \overline{\lim_{j\to\infty}} \frac{1}{2\pi T(\rho_j, f^{(i)})} \sum_{\nu=1}^{N(i,\ell)} \int_{\widetilde{\beta_{i\ell}^{\nu\ell t}}} \log^+ \frac{1}{|f^{(i)}(\rho_j e^{i\phi}) - a_{i\ell}|}\, d\phi$$

$$+ \overline{\lim_{j\to\infty}} \frac{1}{2\pi T(\rho_j, f^{(i)})} \int_{e(j)} \log^+ \frac{1}{|f^{(i)}(\rho_j e^{i\phi}) - a_{i\ell}|}\, d\phi = I + II \text{ (say)}. \tag{22}$$

For $|z| = \rho_j$, $\arg z \in \widetilde{\beta_{i\ell}^{\nu\ell t}}(j)$, from (14), we have

$$\log^+ \frac{1}{|f^{(k+1)}(z)|} \leq -|C_j|\rho_j^{\mu} \cos(\mu\phi + \omega_j) + 4\,\varepsilon |C_j|\rho_j^{\mu} \,. \tag{23}$$

For $1 \leqslant t \leqslant N(i,\ell)$, from (23) we have

$$\frac{1}{2\pi}\int_{\widetilde{\beta_{i\ell}(j)}^{\nu_{\ell t}}} \log^+ \frac{1}{f^{(i)}(\rho_j e^{i\phi}) - a_{i\ell}}\, d\phi$$

$$\leqslant \frac{1}{2\pi}\int_{\widetilde{\beta_{i\ell}^{\nu_{\ell t}}(j)}} \log^+ \frac{|f^{(k+1)}(\rho_j e^{i\phi})|}{|f^{(i)}(\rho_j e^{i\phi}) - a_{i\ell}|}\, d\phi$$

$$+ \frac{1}{2\pi}\int_{\widetilde{\beta_{i\ell}^{\nu_{\ell t}}(j)}} \log^+ \frac{1}{|f^{(k+1)}(\rho_j e^{i\phi})|}\, d\phi$$

$$\leqslant o\{T(\rho_j, f^{(i)})\} - \frac{|C_j|\rho_j^{\mu}}{2\pi}\int_{\widetilde{\beta_{i\ell}^{\nu_{\ell t}}(j)}} \cos(\mu\phi + \omega_j)\, d\phi + 4\varepsilon|C_j|\rho_i^{\mu}$$

$$= \frac{\cos(\mu\phi_o)}{\mu\pi}|C_j|\rho_j^{\mu} + o\{T(\rho_j, f^{(i)})\} + 4\varepsilon|C_j|\rho_j^{\mu}. \tag{24}$$

From (21), (24) and the fact that ε, ϕ_o can be chosen arbitrarily small, we have

$$I \leqslant \frac{\cos(\mu\phi_o)}{\mu} N(i,\ell) \leqslant \frac{N(i,\ell)}{\mu}. \tag{25}$$

For $\phi \in e(j)$, from the definition of $e(j)$ and (14), and taking ϕ_o sufficiently small, we have

$$\log^+ \frac{1}{|f^{(k+1)}(z)|} \leqslant |C_j|\rho_j^{\mu}(|\cos(\mu\phi + \omega_j)| + 4\varepsilon)$$

$$\leqslant (\sin \mu\phi_o + 4\varepsilon)|C_j|P_j^{\mu}. \tag{26}$$

Hence from (26), we have

$$\frac{1}{2\pi}\int_{e(j)} \log^+ \frac{1}{|f^{(i)}(\rho_j e^{i\phi}) - a_{i\ell}|} d\phi$$

$$\leq \frac{1}{2\pi}\int_{e(j)} \log^+ \frac{|f^{(k+1)}(\rho_j e^{i\phi})|}{|f^{(i)}(\rho_j e^{i\phi}) - a_{i\ell}|} d\phi$$

$$+ \frac{1}{2\pi}\int_{e(j)} \log^+ \frac{1}{|f^{(k+1)}(\rho_j e^{i\phi})|} d\phi$$

$$\leq o\{T(\rho_j, f^{(i)})\} + \frac{\mu\phi_o}{\pi}(\sin \mu\phi_o + 4\varepsilon)\,|C_j|\,\rho_j^{\mu}. \tag{27}$$

From (21), (22) we have

$$\text{II} = O(\phi_o). \tag{28}$$

From (25), (28) and the fact that ϕ_o can be chosen arbitrarily small, we obtain

$$\delta(a_{i\ell}, f^{(i)}) \leq \frac{N(i,\ell)}{\mu}. \tag{29}$$

Using (19), for $a_{i\ell} \neq 0, \infty$ ($i = 0, 1, \ldots, k$; $\ell = 1, 2, \ldots, P_i$) we have

$$\sum_{i=0}^{k} \sum_{\ell=1}^{P_i} \delta(a_{i\ell}, f^{(i)}) \leq \sum_{i=0}^{k} \sum_{\ell=1}^{P_i} \frac{N(i,\ell)}{\mu} \leq 1.$$

Combining with (b) of Theorem 1, we deduce

$$\sum_{i=0}^{\infty} \sum_{a\neq 0,\infty} \delta(a, f^{(i)}) \leq 1.$$

If $f(z)$ has N-th primitive function, then $f^{(k)}(z)$ is (k+N)-th derivative $f^{(-N)}(z)$, so that

$$\sum_{i=-N}^{\infty} \sum_{a\neq 0,\infty} \delta(a,f^{(i)}) \leqslant 1. \tag{30}$$

Since the right side of (30) is independent of N, if $f(z)$ has primitive function of arbitrary order, then (4) follows.

If the equality of (4) holds, then from

$$1 = \sum_{i=-\infty}^{\infty} \sum_{\ell=1}^{P_i} \delta(a_{i\ell}, f^{(i)}) \leqslant \sum_{i=-\infty}^{\infty} \sum_{\ell=1}^{P_i} \frac{N(i,\ell)}{\mu} \leqslant 1, \ (a_{i\ell}\neq 0,\infty)$$

we derive $\delta(a_{i\ell}, f^{(i)}) = \frac{N(i,\ell)}{\mu}$ $(i = 0, \pm 1, \pm 2, \ldots;\ \ell = 1, 2, \ldots, P_i)$. In this way the Theorem 2 is proved.

REFERENCES

1. Dai, C.J. and L. Jin. Number of deficient values of a class of meromorphic functions. Kodai Math. J. Vol. 10, No. 1, (1987) pp. 74-82.

2. Ozawa, M. On the deficiencies of meromorphic functions. Kodai, Math. Sem. Rep. 20(1968) 385-388.

3. Edrei, A. and W.H. J. Fuchs. On the growth of meromorphic functions with several deficient values. Trans. Amer. Math. Soc. 93(1959), 293-328.

4. Ederi, A. and W.H.J. Fuchs. Valeurs deficientes et valeurs asymptotiques des fonctions meromorphes, Comment. Math. Helv. 33(1959) 258-295.

5. Drasin, D. Proof of a conjecture of F. Nevanlinna concerning functions which have deficiency sum two.

BOREL SETS OF ENTIRE FUNCTIONS

Hai-long Ao
Department of Mathematics
Peking University
Beijing, P.R. China

The notion of Picard sets of entire functions was first introduced by Lehto [1]. It is defined as follows: A set $E \subset \mathbb{C}$ is called a Picard set of entire functions, if each non-constant entire function $f(z)$ takes every finite value a in $\mathbb{C}\setminus E$, except at most one finite value a. In other words, Picard's theorem on entire functions still holds in $\mathbb{C}\setminus E$. Since then some works have been done on such sets, for instance the works [2], [3], [4]. It is natural to extend the notion of Picard sets to that of Borel sets which corresponds to a well known theorem of Borel. This is what we are going to do. In the present work we shall confine ourselves to the case of entire functions and show that certain sets are Borel sets of such functions. The case of meromorphic functions will be treated in a subsequent work.

1. NOTATIONS AND DEFINITIONS

Consider a meromorphic function $f(z)$ in the complex plane and a non-empty set $E \subset \mathbb{C}$. We denote by $n(r,f;E)$ the number of poles of $f(z)$ on $\{|z| < r\} \cap E$, each pole being counted according to its order of multiplicity. Next we define

$$N(r,f;E) = \int_0^r \frac{n(t,f;E)-n(0,f;E)}{t}\,dt + n(0,f;E)\log r.$$

Similarly we define $\bar{n}(r,f;E)$ and $\bar{N}(r,f;E)$ for which each pole of $f(z)$ is counted only once.

Definition 1. Consider a number $0 < \rho < \infty$ and a set $S \subset \mathbb{C}$. Let $S^c = \mathbb{C} \setminus S$. If for each entire function $f(z)$ of order ρ, the equality.

$$\overline{\lim_{r\to\infty}} \frac{\log n(r, \frac{1}{f-a}; S^c)}{\log r} = \rho$$

holds for every finite value a, except at most one finite value a, then S is called a Borel set of the entire functions of order ρ.

Definition 2. A set $S \subset \mathbb{C}$ is called a Borel set of the entire functions of finite positive order, if for each number $0 < \rho < \infty$, S is a Borel set of the entire functions of order ρ.

2. MAIN RESULTS

THEOREM 1. Let $0 < \rho < \infty$ be a number, $\{a_n\}$ $(|a_n| > 1)$ a sequence of complex numbers and $\{\rho_n\}$ $(0 < \rho_n < 1)$ a sequence of positive numbers. Assume that the following conditions are satisfied.

(1) The exponent of convergence of the sequence $\{a_n\}$ is zero, and there is a constant $\eta' > 0$ such that

$$|a_{n+1}| - |a_n| > \max\{\eta', |a_n|^{1+\rho+\eta'}\}. \quad (1)$$

(2) We have

$$\lim_{n\to\infty} \frac{\log\log\frac{1}{\rho_n}}{\log|a_n|} \geq \rho. \qquad (2)$$

Then the set $S = \bigcup_{n=1}^{\infty} D_n$, $D_n = (|z-a_n| < \rho_n)$ is a Borel set of the entire functions of order ρ.

THEOREM 2. Let $\{a_n\}$ $(|a_n| > 1)$ be a sequence of complex numbers and $\{\rho_n\}$ $(0 < \rho_n < 1)$ a sequence of positive numbers. Assume that the following conditions are satisfied:

(1) There is a constant $\alpha \geq 0$ such that

$$|a_{n+1}| - |a_n| \geq \frac{|a_n|}{(\log|a_n|)^{\alpha}}. \qquad (3)$$

(2) We have

$$\lim_{n\to\infty} \frac{\log\log\frac{1}{\rho_n}}{\log|a_n|} = \infty. \qquad (4)$$

Then the set $S = \bigcup_{n=1}^{\infty} D_n$, $D_n = (|z-a_n| < \rho_n)$ is a Borel set of the entire functions of finite positive order.

3. PROOF OF THEOREM 1

We need the following two lemmas [4]:

LEMMA 1. Let $h(z)$ be a function holomorphic for $|z| \leq R$, having no zero and $|\log|h(z)|| \leq M$ for $|z| = R$. Then for $|z| = r < R$ we have

$$\left|\frac{h'(z)}{h(z)}\right| \leq \frac{2MR}{(R-r)^2}.$$

LEMMA 2. Let $P(z)$ be a polynomial of degree k

whose zeros are all in the circle $|z| < R_o$. Then for $|z| = R \geq R_o$ we have

$$\left|\frac{P'(z)}{P(z)}\right| > \frac{k}{2R} .$$

To prove the Theorem 1, we consider first the case $1 < \rho < \infty$, and set $\eta' = 10\eta$. Now suppose that the conclusion of Theorem 1 is untrue, then there is an entire function $f(z)$ of order ρ such that

$$\varliminf_{r\to\infty} \frac{\log\{n(r,\frac{1}{f};\ S^c) + n(r,\frac{1}{f-1};\ S^c)\}}{\log r} = \lambda_1 < \rho.$$

Take a positive number λ such that $\lambda_1 < \lambda < \rho$. Then for $r \geq r_o$ we have

$$n(r,\frac{1}{f};\ S^c) + n(r,\frac{1}{f-1};\ S^c) + N(r,\frac{1}{f};\ S^c)$$

$$+ N(r,\frac{1}{f-1};\ S^c) < r^{\lambda}$$

set

$$D_n^{(k)} = \{|z-a_n| < 2^k \rho_n\}, \quad k = 0,1,2,..., \quad n=1,2,...$$

n being a positive integer, define k_n to be the smallest of the integers k such that $f(z)$ and $f(z) - 1$ have no zero in $D_n^{(k+1)} - D_n^{(k)}$. We are going to show that k_n exists, when n is sufficiently large, and we have

$$2^{k_n}\rho_n < 5\eta. \tag{5}$$

Suppose that this is untrue. Take k_o such that $2\eta \leq 2^{k_o}\rho_n < 5\eta$. Then the total number of the zeros of

$f(z)$ and $f(z) - 1$ in $D_n^{(k_o)} - D_n$ is at least k_o. By condition (1),

$$|a_n| - |a_{n-1}| \geq 10\eta, \quad |a_{n+1}| - |a_n| \geq 10\eta.$$

So when n is sufficiently large, $D_n^{(k_o)}$ and D_m $(m \neq n)$ are disjoint, and we have

$$D_n^{(k_o)} - D_n \subset S^c,$$

$$k_o \leq n\{|a_n| + 5\eta, \tfrac{1}{f}; S^c\}$$

$$+ n\{|a_n| + 5\eta, \tfrac{1}{f-1}; S^c\} = O(|a_n|^\lambda). \qquad (6)$$

On the other hand, from $2\eta \leq 2^{k_o}\rho_n$, we have

$$k_o \geq \frac{1}{\log 2} \log\frac{2\eta}{\rho_n} > |a_n|^{\rho-\varepsilon}.$$

This inequality is incompatible with (6), if we take $\varepsilon < \rho - \lambda$.

Set $D_n^* = D_n^{(k_n)}$, $\tilde{D}_n = \{|z-a_n| < 4\eta\}$. From the proof given above we see that

$$k_n = O(|a_n|^\lambda) \qquad (7)$$

and $\rho_n^* = 2^{k_n}\rho_n$ also satisfies the inequality (2). On the other hand, without loss of generality, we may assume that for any n, k, on the circle $|z-a_n| = 2^k\rho_n$, $f(z)$ does not take the values 0 and 1. In fact, if necessary, it is sufficient to use $\rho_n' = t_n\rho_n$ instead of ρ_n, where $1 \leq t_n \leq 2$ is suitably chosen.

Now we are going to find an upper bound of $T(r,f)$ in basing upon the second fundamental theorem:

$$T(r,f) < N(r,\tfrac{1}{f}) + N(r,\tfrac{1}{f-1}) - N(r,\tfrac{1}{f'}) + O(\log r). \quad (8)$$

Let p_n, q_n and α_n be respectively the number of zeros in D_n^* of $f(z)$, $f(z) - 1$ and $f'(z)$, all zeros being counted according to their orders of multiplicity. Set

$$\nu_n = p_n + q_n - \alpha_n, \quad \mu_n = \max(\nu_1, \dots, \nu_n).$$

From (8), for

$$|a_N| + \rho_N^* \leqslant r \leqslant |a_{N+1}| - \rho_{N+1}^*, \quad (9)$$

we have

$$\begin{aligned} T(r,f) &< N(r,\tfrac{1}{f}) + N(r,\tfrac{1}{f-1}) - N(r,\tfrac{1}{f'}) + O(\log r) \\ &\leqslant O(r^\lambda) + N(r,\tfrac{1}{f}; S^*) + N(r,\tfrac{1}{f-1}; S^*) \\ &\quad - N(r,\tfrac{1}{f'}; S^*) \\ &\leqslant O(r^\lambda) + \sum_{n=1}^{N} \{(p_n+q_n)\log\frac{r}{|a_n|-\rho_n^*} \\ &\quad - \alpha_n \log\frac{r}{|a_n| + \rho_n^*}\} \\ &\leqslant O(r^\lambda) + \sum_{\substack{|a_n|<r \\ \nu_n>0}} \nu_n \log\frac{r}{|a_n|} \\ &\quad + \sum_{n=1}^{N} (p_n+q_n+\alpha_n)\log\frac{|a_n|+\rho_n^*}{|a_n|-\rho_n^*}. \end{aligned}$$

It is easy to see that $\sum_{n=1}^{\infty} (p_n+q_n+\alpha_n)\log\{(|a_n|+\rho_n^*)/(|a_n|-\rho_n^*)\}$ converges, hence

$$T(r,f) \leq O(r^{\lambda}) + N\mu_N \log r \qquad (10)$$

where

$$S^* = \bigcup_{n=1}^{\infty} D_n^* .$$

By hypothesis, for any $\varepsilon > 0$, $A(\varepsilon) = \sum_{n=1}^{\infty} 1/|a_n|^2$ converges, hence

$$\frac{N}{|a_N|^{\varepsilon}} < \sum_{n=1}^{N} \frac{1}{|a_n|^{\varepsilon}} < A(\varepsilon), \quad N < A(\varepsilon)|a_N|^{\varepsilon}$$

and then by (10), we get

$$T(r,f) \leq O(r^{\lambda}) + A(\varepsilon)r^{\varepsilon}\mu_N \log r \qquad (11)$$

where r satisfies the inequality (9).

On the other hand, there is a sequence r_k tending to ∞, such that

$$T(r_k,f) > r_k^{\rho-\varepsilon}. \qquad (12)$$

We may assume that $|a_{N_k}| + \rho_{N_k}^* \leq r_k \leq |a_{N_k+1}| - \rho_{N_k+1}^*$. In fact, if for a term r_k we have $|a_{N_k}| - \rho_{N_k}^* < r_k < |a_{N_k}| + \rho_{N_k}^*$, then for $r_k' = |a_{N_k}| + \rho_{N_k}^*$ we have

$$T(r_k',f) \geq T(r_k,f) > (|a_{N_k}| - \rho_{N_k}^*)^{\rho-\varepsilon} > \frac{1}{2}r_k'^{\,\rho-\varepsilon},$$

and it is sufficient to replace r_k by r_k'. From (11) and (12), we get

$$\mu_{N_k} > r_k^{\rho-3\varepsilon} \qquad (13)$$

when k is large enough.

Let $\mu_{N_k} = \nu_{M_k}$, $M_k \leqslant N_k$. Evidently $M_k \to \infty$, as $k \to \infty$. From (11) and (13), we see that for $|a_{M_k}| + \rho^*_{M_k} \leqslant R \leqslant |a_{M_k}| + 9\eta$, we have

$$T(R,f) \leqslant 2A(\varepsilon)R^{\varepsilon}\mu_{M_k}\log R = 2A(\varepsilon)R^{\varepsilon}\nu_{M_k}\log R. \tag{14}$$

For the sake of simplicity write $M = M_k$. We are going to estimate ν_M.

Assume $p_M \geqslant q_M$. Then $p_M \geqslant \nu_M/2$ and (14) becomes

$$T(R,f) \leqslant 4A(\varepsilon)R^{\varepsilon}p_M\log R. \tag{15}$$

Let ζ_s $(s=1,\ldots,p_M)$ be the zeros of $f(z)$ in D^*_M and set $P(z) = \prod_{s=1}^{p_M}(z-\zeta_s)$. Then

$$f(z) = P(z)h_1(z)$$

where $h_1(z)$ is an entire function having no zero in the circle $2D^*_M = \{|z-a_n| < 2\rho^*_M\}$. On the circle $\partial\tilde{D}_M = \{|z-a_M| = 4\eta\}$ we have

$$\log|P(z)| > p_M\log 3\eta$$

and in the cirle $|z-a_M| \leqslant 4\eta$,

$$\log|f(z)| \leqslant \log M(|a_M| + 4\eta, f)$$
$$\leqslant \eta^{-1}(1+o(1))|a_M|T(|a_M|+6\eta,f). \tag{16}$$

Consequently on $\partial\tilde{D}_M$, we have

$$\log|h_1(z)| \leq \eta^{-1}(1+o(1))|a_M|T(|a_M|+6\eta,f)$$

$$- p_M \log 3\eta . \qquad (17)$$

By the principle of maximum modulus, (17) also holds in $\tilde{D}_M$. So in $D_M^* \subset \tilde{D}_M$ we have

$$\log|f(z)| = \log|P(z)| + \log|h_1(z)|$$

$$\leq p_M \log(2\rho_M^*) + \eta^{-1}(1+o(1))|a_M|T(|a_M|+6\eta,f)$$

$$- p_M \log 3\eta$$

$$< -p_M \log\frac{1}{\rho_M^*} - p_M \log\frac{3\eta}{2}$$

$$+ \eta^{-1}(1+o(1))|a_M|4A(\varepsilon)|a_M|^{\varepsilon} p_M \log|a_M|$$

$$= p_M\{4(1+o(1))\frac{A(\varepsilon)}{\eta}|a_M|^{1+\varepsilon}\log|a_M| - \log\frac{1}{\rho_M^*}\}.$$

Noting that ρ_n^* satisfies the inequality (2) and $\rho > 1$ and taking $\varepsilon < \rho - 1$, we see that when M is sufficiently large, we have

$$\log|f(z)| < 0 \qquad (18)$$

for $z \in D_M^*$. It follows that $|f(z)| < 1$ and $f(z)$ does not take the value 1 in D_M^*. So we have $q_M = 0$ and

$$\nu_M = p_M - \alpha_M . \qquad (19)$$

We may assume that there is an infinite number of integers M such that $p_M > q_M$. For such an integer M, let ζ_t' $(t=1,\ldots,\beta_M)$ be the zeros of $f(z)$ in $\bar{\bar{D}}_M - D_M^*$, and define $Q(z) = \prod_{t=1}^{\beta_M} (z-\zeta_t')$. Since $f(z)$ has no zero in $2D_M^* - D_M^*$, we have

$$|\zeta_t' - a_M| \geq 2\rho_M^*, \tag{20}$$

$$f(z) = P(z)h_1(z) = P(z)Q(z)h(z)$$

where $h(z)$ is an entire function having no zero in $\tilde{\tilde{D}}_M$. By Poisson-Jensen formula, for $|z| = R' < R$, $R = |a_M| + 6\eta$, we have

$$\log|h(z)| = \frac{1}{2\pi}\int_0^{2\pi} \log|h(Re^{i\theta})| \operatorname{Re}\left(\frac{Re^{i\theta}+z}{Re^{i\theta}-z}\right)d\theta$$

$$- \sum_{|\xi|<R, h(\xi)=0} \log\left|\frac{R^2-\bar{\xi}z}{R(z-\bar{\xi})}\right|. \tag{21}$$

Consider the zeros ξ of $h(z)$ in the circle $|z| < R$. If $\xi \in S^*$, then $\xi \in \bigcup_{n=1}^{M-1} D_n^*$ and hence $|\xi| < |a_{M-1}| + \eta$. On the other hand the number of the zeros ξ which do not belong to S^* is less than $n(R, \frac{1}{f}, S^c) \leq O(|a_M|^\lambda)$. Noting also that the zeros ξ are exterior to $\tilde{D}_M$ and hence $|\xi - a_M| > 4\eta$, hence when $|z-a_M| \leq 2\eta$, we have $|z-\xi| \geq 2\eta$ and

$$1 < \left|\frac{R^2-\bar{\xi}z}{R(z-\xi)}\right| \leqslant \frac{2R^2}{2\eta R} = \frac{R}{\eta},$$

$$\begin{aligned}|\log|h(z)|| &\leqslant \frac{R+R'}{R-R'}\{m(R,h) + m(R,\tfrac{1}{h})\}\\ &\quad + n(|a_{M-1}|+\eta,\tfrac{1}{f})\log\tfrac{R}{\eta} + O(|a_M|^{\lambda})\log\tfrac{R}{\eta}\\ &= |a_M|O\{m(R,h) + m(R,\tfrac{1}{h})\\ &\quad + T(R,f) + |a_M|^{\lambda}\}. \qquad (22)\end{aligned}$$

We have

$$\begin{aligned}m(R,h) = T(R,h) &\leqslant T(R,f) + T(R,\tfrac{1}{P}) + T(R,\tfrac{1}{Q})\\ &\leqslant T(R,f) + T(R,P) + T(R,Q) + \log\left|\frac{1}{P(0)Q(0)}\right|\\ &\leqslant T(R,f) + m(R,P) + m(R,Q)\\ &\leqslant T(R,f) + (p_M+\beta_M)\log 2R,\end{aligned}$$

and similarly

$$\begin{aligned}m(R,\tfrac{1}{h}) &\leqslant T(R,\tfrac{1}{f}) + T(R,P) + T(R,Q)\\ &\leqslant T(R,f) + (p_M+\beta_M)\log 2R + O(1).\end{aligned}$$

On the other hand, from (13) and (19),

$$O(|a_M|^{\lambda}) = o(p_M), \quad \beta_M \leqslant O(|a_M|^{\lambda}).$$

Finally combining these results with (15), we get

$$|\log|h(z)|| \leqslant p_M|a_M|O\{|a_M|^{c}\log|a_M|\}. \qquad (23)$$

By Lemma 1, for $z\in\bar{D}_M^*$ we have

$$\left|\frac{h'(z)}{h(z)}\right| = O\{p_M|a_M|^{1+\varepsilon}\log|a_M|\} \tag{24}$$

and for $|z-a_M| = \rho_M^*$, by Lemma 2, we have

$$\left|\frac{P'(z)}{P(z)}\right| \geq \frac{p_M}{2\rho_M^*}. \tag{25}$$

From (20), for $|z-a_M| = \rho_M^*$, we have

$$\left|\frac{\zeta_t'-a_M}{z-a_M}\right| \geq 2, \quad t=1,2,\ldots,\beta_M.$$

Hence

$$\left|\frac{Q'(z)}{Q(z)}\right| = \left|\sum_{t=1}^{\beta_m}\frac{1}{z-\zeta_t'}\right| \leq \frac{1}{|z-a_M|}\sum_{t=1}^{\beta_m}\frac{1}{\left|\frac{\zeta_t'-a_M}{z-a_M}\right|-1}$$

$$\leq \frac{\beta_M}{\rho_M^*} \tag{26}$$

and from the relation $\beta_M = O(|a_M|^{\lambda}) = o(p_M)$, we see that

$$\left|\frac{Q'(z)}{Q(z)}\right| = o\left(\left|\frac{P'(z)}{P(z)}\right|\right) \tag{27}$$

uniformly for $|z-a_M| = \rho_M^*$. Since ρ_M^* also satisfies (2), we have

$$\frac{1}{\rho_M^*} > \exp(|a_M|^{\rho-\varepsilon})$$

when M is sufficiently large. Then by (24), (25), we have also

$$\left|\frac{h'(z)}{h(z)}\right| = o\left(\left|\frac{P'(z)}{P(z)}\right|\right) \tag{28}$$

uniformly for $|z-a_M| = \rho_M^*$. Consequently when M is sufficiently large, for $|z-a_M| = \rho_M^*$, we have

$$\left|\frac{P'(z)}{P(z)}\right| > \left|\frac{h'(z)}{h(z)}\right| + \left|\frac{Q'(z)}{Q(z)}\right| \geq \left|\frac{h'(z)}{h(z)} + \frac{Q'(z)}{Q(z)}\right|$$

and

$$|Q(z)h(z)P'(z)| > |P(z)Q(z)h'(z) + P(z)h(z)Q'(z)|.$$

BY Rouche's theorem, the functions

$$f'(z) = Q(z)h(z)P'(z) + \{P(z)h(z)Q'(z) + P(z)Q(z)h'(z)\}$$

and $Q(z)h(z)P'(z)$ have the same number of zeros in $D_M^* = \{|z-a_M| < \rho_M^*\}$. But $Q(z)$ and $h(z)$ have no zero in D_M^*, so $f'(z)$ and $P'(z)$ have the same number of zeros in D_M^*. By a known result, the $p_M - 1$ zeros of $P'(z)$ are in the convex hull of the zeros of $P(z)$. But the zeros of $P(z)$ are in the set D_M^*, hence $P'(z)$ has $p_M - 1$ zeros in D_M^*, and

$$\alpha_M = p_M - 1.$$

Then by (19), we get $\nu_M = p_M - \alpha_M = 1$, which is incompatible with (13). So for the case $\rho > 1$, Theorem 1 is proved.

To prove Theorem 1 for the case $0 < \rho \leq 1$, we write condition (1) in the form

$$|a_{n+1}| - |a_n| \geq |a_n|^{1-\rho+\eta'}. \tag{1'}$$

Then the proof can be carried on as above, except (16),

(17), (18). We are going to give the proof of these three inequalities as follows: We may assume $\eta' < \rho$. On $\partial\tilde{D}_M$ (here we take $\tilde{D}_m = \{|z-a_M| < 4\}$, we have

$$\log|P(z)| \geq p_M \log 3,$$

$$\log|f(z)| \leq \log M(|a_M|+4,f)$$

$$\leq \frac{2|a_M|+|a_M|^{1-\rho+\frac{\eta'}{2}}+4}{|a_M|^{1-\rho+\frac{\eta'}{2}}-4} T(|a_M|+|a_M|^{1-\rho+\frac{\eta'}{2}},f)$$

$$= 2(1+o(1))|a_M|^{\rho-\frac{\eta'}{2}} T(|a_M| + |a_M|^{1-\rho+\frac{\eta'}{2}}, f) \qquad (16')$$

$$\log|h_1(z)| < (2+o(1))|a_M|^{\rho-\frac{\eta'}{2}} T(|a_M|+|a_M|^{1-\rho+\frac{\eta'}{2}},f) - p_M \log 3. \qquad (17')$$

By the principle of maximum modulus, (17') also holds in $\tilde{D}_M$, then from (11) and (13), taking $\varepsilon = \eta'/z$, we have for $z \in D_M^*$,

$$\log|f(z)| = \log|P(z)| + \log|h_1(z)|$$

$$\leqslant p_M \log 2\rho_M^*$$

$$+ (2+o(1))|a_M|^{\rho - \frac{\eta'}{2}} T(|a_M| + |a_M|^{1-\rho+\frac{\eta'}{2}}, f)$$

$$- p_M \log 3$$

$$< |a_M|^{\rho - \frac{\eta'}{2}} O(|a_M|^{\lambda} + |a_M|^2 p_M \log|a_M|\}$$

$$- p_M \log 3 - p_M \log\frac{1}{\rho_M^*}$$

$$\leqslant p_M \{O(|a_M|^{\rho - \frac{\eta'}{2} + \varepsilon} \log|a_M| + |a_M|^{\lambda})$$

$$- \log\frac{1}{\rho_M^*}\}.$$

Then by condition (2) we get (18). Theorem 1 is now proved.

4. PROOF OF THEOREM 2

In view of Theorem 1, it is sufficient to show that the exponent of convergence of the sequence $\{a_n\}$ is 0. This follows from the following lemma:

LEMMA 3. Assume that $\{a_n\}$ satisfies the condition (3). If we denote by $n(r)$ the number of the points a_n in the circle $|z|\leqslant r$, then

$$n(r) \leqslant \frac{4}{1+\alpha}\log^{1+\alpha} r + O(1).$$

Proof. By the condition (3), we have

$$n(r + \frac{1}{2}\frac{r}{\log^{\alpha} r}) - n(r) \leqslant 1 = \frac{2\log^{\alpha} r}{r}\frac{r}{2\log^{\alpha} r}.$$

For $r \geqslant e^{\alpha}$, $\log^{\alpha} r/r$ is decreasing. Set

$$r_0 = e^{\alpha}, \quad r_n = r_{n-1}\left(1 + \frac{1}{2}\,\frac{1}{\log^{\alpha} r_{n-1}}\right), \quad n=1,2,\ldots$$

Then

$$n(r_n) - n(r_{n-1}) \leq \frac{2\log^{\alpha} r_{n-1}}{r_{n-1}}\,\frac{r_{n-1}}{2\log^{\alpha} r_{n-1}}$$

$$\leq \int_{r_{n-1}}^{r_n} \frac{4\log^{\alpha} t}{t}\,dt$$

$$= \frac{4}{1+\alpha}\left(\log^{1+\alpha} r_n - \log^{1+\alpha} r_{n-1}\right)$$

where we make use of the inequality $r_{n-1} > r_n/2$ for $r_{n-1} > \exp((\frac{1}{2})^{\frac{1}{\alpha}})$. It follows that

$$n(r_n) - n(r_o) \leq \frac{4}{1+\alpha}\left(\log^{1+\alpha} r_n - \log^{1+\alpha} r_o\right).$$

Since $r_n \to \infty$, given r sufficiently large, we can get r_n such that $r_{n-1} < r \leq r_n$. Then

$$n(r) \leq n(r_n) \leq \frac{4}{1+\alpha}\log^{1+\alpha} r_n + O(1)$$

$$\leq \frac{4}{1+\alpha}\log^{1+\alpha}\left(r + \frac{1}{2}\,\frac{r}{\log^{\alpha} r}\right) + O(1)$$

$$\leq \frac{4}{1+\alpha}\log^{1+\alpha} r + O(1).$$

5. OTHER RESULTS

THEOREM 3. Let $a>0$ be a positive number. Then the set $A = \{\mathrm{Re}z \geq 0, |\mathrm{Im}z| \leq a\}$ is a Borel set of the entire functions whose order is finite and greater than 1/2.

The proof of this theorem bases upon the following

lemma [6]:

LEMMA 4. Let $f(z)$ be a meromorphic function of order ρ $(0 < \rho < \infty)$ of which the total number of deficient values is p and the total number of Borel directions is q. Then $p<q$ and when $\rho > p/2$, we have $p \leq q - 1$.

To prove Theorem 3, suppose on the contrary that there is an entire function $f(z)$ of finite order $\rho>1/2$ such that

$$\overline{\lim_{r\to\infty}} \frac{\log\{n(r,\frac{1}{f}; A^C) + n(r,\frac{1}{f-1}; A^C)\}}{\log r} < \rho. \qquad (29)$$

Consider a number θ_0, $0 < \theta_0 < 2\pi$, and the angular domain

$$\Omega(\theta_0,\eta) = (|\arg z - \theta_0| < \eta)$$

where $\eta < \min(\pi/2, \theta_0, 2\pi-\theta_0)$. Then the set $A \cap \Omega(\theta_0, \eta)$ is bounded and hence

$$n(r,\frac{1}{f}; \Omega(\theta_0,\eta)) \leq O(1) + n(r,\frac{1}{f}; A^C),$$

$$n(r,\frac{1}{f-1}; \Omega(\theta_0,\eta)) \leq O(1) + n(r,\frac{1}{f-1}; A^C).$$

Then by (29), we have

$$\overline{\lim_{r\to\infty}} \frac{\log n(r,\frac{1}{f}; \Omega(\theta_0,\eta))}{\log r} < \rho,$$

$$\overline{\lim_{r\to\infty}} \frac{\log n(r,\frac{1}{f-1}; \Omega(\theta_0,\eta))}{\log r} < \rho.$$

Consequently $\arg z = \theta_0$ is not a Borel direction of $f(z)$. It follows that $f(z)$ can have only one Borel direction $\arg z = 0$. So we have $q = 1$. By Lemma 4, $p \leqslant 1$. Then we have $\rho > 1/2 \geqslant p/2$, and again by Lemma 4, $p \leqslant q - 1 = 0$. This contradicts the fact that $f(z)$ is an entire function.

It should be pointed out that the conclusion of Theorem 3 is no longer valid when $\rho \leqslant 1/2$. For example the function $f(z) = \cos\sqrt{z}$ has order 1/2, but the zeros of $f(z)$ and $f(z) - 1$ are all on the positive real axis.

So far we have considered only Borel sets of entire functions of finite positive order. Now we are going to give an extension of Theorem 1 to the case of entire functions of infinite order.

First let us give some definitions.

DEFINITION 2 [7], [8]. Let $f(z)$ be an entire function of infinite order and $\rho(r)$ $(r \geqslant r_o)$ a function satisfying the following conditions:

(1) $\rho(r)$ is positive, continuous and non-decreasing for $r \geqslant r_o$ and tends to ∞ with r.

(2) Set $U(r) = r^{\rho(r)}$ $(r \geqslant r_o)$. We have

$$\lim_{r\to\infty} \frac{\log U(R)}{\log U(r)} = 1, \quad R = r + \frac{r}{\log U(r)}.$$

Such a function $\rho(r)$ is called an order of $f(z)$, if

$$\overline{\lim_{r\to\infty}} \frac{\log T(r,f)}{\log U(r)} = 1.$$

DEFINITION 3 Consider a function $\rho(r)$ satisfying the conditions (1) and (2) in Definition 2 and a set $S \subset \mathbb{C}$. If for each entire function $f(z)$ of infinite order with $\rho(r)$ as an order, the equality

$$\overline{\lim_{r\to\infty}} \frac{\log n(r,\frac{1}{f-a};\ S^c)}{\rho(r)\log r} = 1$$

holds for every finite value a, except at most one finite value a, then S is called a Borel set of the entire functions of infinite order with $\rho(r)$ as an order, where $S^c = \mathbb{C}\backslash S$.

THEOREM 4. Let $\rho(r)$ be a function satisfying the conditions (1) and (2) in Definition 2. Let $\{a_n\}$ ($|a_n| > \max(1,r_o)$) be a sequence of complex numbers and $\{\rho_n\}$ $(0 < \rho_n < 1)$ be a sequence of positive numbers. Assume that the following conditions are satisfied:

1° There is a number $\eta > 0$ such that

$$|a_{n+1}| - |a_n| \geqslant \min(\eta, \frac{|a_n|}{\rho(|a_n|)\log|a_n|}). \qquad (1")$$

2° We have

$$\lim_{n\to\infty} \frac{\log\log\frac{1}{\rho_n}}{\rho(|a_n|)\log|a_n|} \geqslant 1. \qquad (2")$$

Then the set $S = \bigcup_{n=1}^{\infty} D_n$, $D_n = (|z-a_n| < \rho_n)$ is a Borel set of the entire functions of infinite order with $\rho(r)$ as an order.

This theorem is proved in a similar manner as for Theorem 1. Suppose that the conclusion of Theorem 4 is untrue. Then there is an entire function $f(z)$ of infinite order with $\rho(r)$ as an order such that

$$\overline{\lim_{r\to\infty}} \frac{\log\{n(r,\frac{1}{f};\ S^c) + n(r,\frac{1}{f-1};\ S^c)}{\rho(r)\log r} = \lambda_1 < 1.$$

Set $U(r) = r^{\rho(r)}$. As in the proof of Theorem 1, we can find a sequence of circles $D_n^* = \{|z-a_n| < \rho_n^*\}$ such that $f(z)$ and $f(z)-1$ have no zero in $\{\rho_n^* \leqslant |z-a_n| \leqslant 2\rho_n^*\}$, and that ρ_n^* also satisfies (2"). Set $S^* = \bigcup_{n=1}^{\infty} D_n^*$.

In the second fundamental theorem

$$T(r,f) < N(r,f) + N(r,\frac{1}{f}) + N(r,\frac{1}{f-1}) - N(r,\frac{1}{f'})$$

$$+ 12\log^+T(R,f) + 9\log^+\frac{1}{R-r} + 12\log R + O(1),$$

take $R = r+r/\log U(r)$ and make use of the condition (2) in Definition 2, we find

$$T(r,f) \leqslant O(r^{\lambda\rho(r)}) + N\mu_N \log r \quad (\lambda_1<\lambda<1) \tag{9"}$$

for

$$|a_N| + \rho_N^* \leqslant r \leqslant |a_{N+1}| - \rho_{N+1}^* \tag{10"}$$

where μ_N is defined as in the proof of Theorem 1. When N is sufficiently large, we have

$$|a_N| > \sum_{n=1}^{N-1} (|a_{n+1}| - |a_n|) \geqslant \sum_{n=1}^{N-1} \min\{\eta, \frac{|a_n|}{\log U(|a_n|)}\}$$

$$\geqslant (N-1)/\rho(|a_N|),$$

$$N \leqslant |a_N|\rho(|a_N|) + 1.$$

Then from (9"), we get

$$T(r,f) \leqslant O(U(r)^{\lambda} + |a_N|\rho(|a_N|)\mu_N \log r\}, \tag{11"}$$

when r satisfies (10").

For any $\varepsilon > 0$, there is a sequence r_k tending to

∞ such that

$$T(r_k,f) > U(r_k)^{1-\varepsilon}$$

Without loss of generality we may assume that r_k satisfies (10"). In fact, if $|a_N|-\rho_N^* < r_k < |a_N|+\rho_N^*$, then for $r_k' = |a_N|+\rho_N^*$, we have

$$T(r_k',f) > T(r_k,f) > U(r_k)^{1-\varepsilon},$$

$$r_k' - r_k < 2\rho_N^* < \frac{r_k}{\log U(r_k)}.$$

(to see the last inequality, we distinguish two cases: $r_k < |a_N|$ and $r_k \geqslant |a_N|$). Then by the condition (2) in Definition 2, we get

$$T(r_k',f) > U(r_k')^{1-2\varepsilon}.$$

The remaining part of the proof of Theorem 4 can be carried on as in the proof of Theorem 1, provided that here we take

$$\eta_n = \frac{1}{10}\min\{\eta, \frac{|a_n|}{\rho(|a_n|)\log|a_n|}\}$$

instead of the number η in the proof of Theorem 1.

Besides the results on Borel sets of entire functions given in the present paper, we have also obtained some results on Borel sets of meromorphic functions and Nevanlinna sets (corresponding to the deficiency relation $\Sigma\delta(a,f) \leqslant 2$) of meromorphic functions, which will be published elsewhere.

The author is greatly indebted to Professor Chi-tai

Chuang for his kind guidance and is also grateful to Dr Nan-yue Zhang for his help.

REFERENCES

[1] Lehto, O., A generalization of Picard's Theorem, Ark. Mat. 3, 1958, 495-500.

[2] Baker, I.N. and Liverpool, L.S.O., Picard sets of entire functions, Math. Zeit. 126, 1972, 230-238.

[3] Anderson, J.M. and Clunie, J., Picard sets of entire and meromorphic functions, Ann. Acad. Sci. Fenn. ser. A. I. 5, 1980, 27-43.

[4] Langley, J.K., Analogues of Picard sets for entire functions and their derivatives, Contemporary Math., Vol. 25, 1983, 75-88.

[5] Hayman, W.K., Meromorphic Functions, Oxford, 1964.

[6] Yang Lo and Chang Kuan-heo, Recherches sur le nombre des valeurs déficientes et le nombre des directions de Borel des fonctions méromorphes, Sci. Sinica, 18, 1975, 21-37.

[7] Hiong King-lai, Sur les fonctions entières et les fonctions méromorphes d'ordre infini, J. de Math., 14, 1935.

[8] Chuang Chi-tai, Singular directions of meromorphic functions (in Chinese), Beijing, 1982.

GENERALIZATION OF A THEOREM OF EDREI

Chi-tai Chuang and Li-zhi Ma
Department of Mathematics
Peking University
and
Department of Mathematics
Beijing Institute of Technology
Beijing, China

Let $f(z)$ be a meromorphic function of the complex variable $z=re^{i\theta}$. Consider the q radii defined by

$$re^{i\omega_1},\ re^{i\omega_2},\ \ldots.,\ re^{i\omega_q}\quad (r\geqslant 0) \tag{1}$$

where

$$0\leqslant\omega_1<\omega_2<\ldots<\omega_q<2\pi\quad (q\geqslant 1). \tag{2}$$

Following Edrei, we say that the roots of the equation

$$f(z) = a \tag{3}$$

are distributed on the radii (1), if there exist at most a finite number of roots of the equation (3) which do not lie on the radii (1).

With this definition, Edrei proved the following theorem which is well known:

Theorem 1. Let $f(z)$ be meromorphic and such that the roots of the three equations

$$f(z)=0, \quad f(z)=\infty, \quad f^{(\ell)}(z)=1 \quad (\ell>0, \ f^{(0)}\equiv f),$$

be distributed on the radii (1). Denote by $\delta(a,f^{(\ell)})$ the deficiency of the value a, of the function $f^{(\ell)}$, and assume

$$\delta(0,f) + \delta(1,f^{(\ell)}) + \delta(\infty,f) > 0.$$

Then the order ρ, of $f(z)$, is necessarily finite and

$$\rho \leqslant \beta = \sup\{\frac{\pi}{\omega_2-\omega_1}, \frac{\pi}{\omega_3-\omega_2}, \ldots, \frac{\pi}{\omega_{q+1}-\omega_q}\}$$
$$(\omega_{q+1} = 2\pi + \omega_1). \tag{4}$$

The purpose of the present work is to give a generalization of this theorem of Edrei, in replacing $f^{(\ell)}(z)$ by a function $g(z)$ which is a homogeneous differential polynomial of $f(z)$ with constant coefficients. Our result is the following theorem:

Theorem 2. Let $f(z)$ be a meromorphic function and let $g(z)$ be a homogeneous polynomial of $f^{(j)}(z)$ $(j=0,1,2,\ldots,p)$:

$$g(z) = \sum_{n=1}^{k} \varphi_n(z), \quad \varphi_n(z) = C_n \prod_{j=0}^{p} \{f^{(j)}(z)\}^{\lambda_{nj}} \tag{5}$$

where C_n $(n=1,2,\ldots,k)$ are constants, λ_{nj} $(n=1,2,\ldots,k;$ $j=0,1,\ldots,p)$ are non-negative integers such that

$$\sum_{j=0}^{p} \lambda_{nj} = d \ (d>0) \quad (n=1,2,\ldots,k). \tag{6}$$

Assume that the following conditions are satisfied:

(1) The functions $f(z)$ and $g(z)$ are non-constant and have the same order.

(2) The roots of the equations

$$f(z) = 0, \quad f(z) = \infty, \quad g(z) = 1$$

are distributed on the radii (1).

(3) $\delta(0,f)+\delta(\infty,f)+\delta(1,g)>0$.

Then the order ρ of the function $f(z)$ is finite and satisfies the inequality (4).

In the work entitled "On the growth of a meromorphic function and of its differential polynomials" appearing in this book, Chuang obtains certain growth preserving operators $\Omega(f)$ which are homogeneous differential polynomials of f with positive integer coefficients. In the Theorem 2, if the meromorphic function $f(z)$ is transcendental, then the meromorphic function $g(z)=\Omega\{f(z)\}$ is also transcendental and the two functions $f(z)$ and $g(z)$ have the same order. So in Theorem 2, we may take $g(z)=\Omega\{f(z)\}$.

For the proof of Theorem 2, we need the following lemmas:

Lemma 1. Let $f(z)$ be a holomorphic function for $|z|<1$ non-identically equal to zero and let $p\geq 1$ be a positive integer. Then there exist positive constants A_j $(j=1,2,3,4)$ such that for $0<r<R<1$ we have

$$m(r, \frac{f^{(p)}}{f}) < A_1\overset{+}{\log}\, m(R,f) + A_2\log\frac{1}{R-r} + A_3\log\frac{1}{r} + A_4. \tag{7}$$

Proof. For $p=1$, Lemma 1 is known [2], [3]. Assume that Lemma 1 holds for a positive integer p. To show that it also holds for $p+1$, distinguish two cases:

1° $f^{(p)}(z) \equiv 0$. In this case $f^{(p+1)}(z)/f(z) \equiv 0$.

2° $f^{(p)}(z) \not\equiv 0$. In this case we have

$$m(r, \frac{f^{(p+1)}}{f}) \leqslant m(r, \frac{f^{(p+1)}}{f^{(p)}}) + m(r, \frac{f^{(p)}}{f}). \tag{8}$$

By assumption (7) holds. On the other hand, since Lemma 1 is true for $p=1$, there exist positive constants B_j $(j=1,2,3,4)$ such that for $0<r<R<1$ we have

$$m(r, \frac{f^{(p+1)}}{f^{(p)}}) \leqslant B_1 \overset{+}{\log}\, m(\rho, f^{(p)}) + B_2 \log\frac{1}{\rho - r} + B_3 \log\frac{1}{r} + B_4 \tag{9}$$

where $\rho=(r+R)/2$. We have

$$m(\rho, f^{(p)}) \leqslant m(\rho, f) + m(\rho, \frac{f^{(p)}}{f}),$$

$$\overset{+}{\log}\, m(\rho, f^{(p)}) \leqslant \overset{+}{\log}\, m(\rho, f) + \overset{+}{\log}\, m(\rho, \frac{f^{(p)}}{f}) + \log 2$$

$$\leqslant \overset{+}{\log}\, m(R, f) + m(\rho, \frac{f^{(p)}}{f}) + \log 2.$$

Then by (7) we get

$$\overset{+}{\log}\, m(\rho, f^{(p)}) < (A_1+1) \overset{+}{\log}\, m(R, f) + A_2 \log\frac{1}{R-\rho} + A_3 \log\frac{1}{\rho} + A_4 + \log 2. \tag{10}$$

(9) and (10) yield

$$m(r, \frac{f^{(p+1)}}{f^{(p)}}) < B_1' \overset{+}{\log} m(R,f) + B_2' \log\frac{1}{R-r}$$

$$+ B_3' \log\frac{1}{r} + B_4' \qquad (11)$$

where B_j' $(j=1,2,3,4)$ are positive constants. (8), (11) and (7) yield

$$m(r, \frac{f^{(p+1)}}{f}) < (A_1+B_1') \overset{+}{\log} m(R,f) + (A_2+B_2') \log\frac{1}{R-r}$$

$$+ (A_3+B_3') \log\frac{1}{r} + (A_4+B_4')$$

which shows that Lemma 1 also holds for $p+1$.

Lemma 2. Let $U(r)$ be a non-negative and non-decreasing function in an interval $r_o<r<\rho$ $(r_o \geqslant 0)$. Let a and b be two positive numbers such that $b \geqslant 2a$ and $b \geqslant 8a^2$. If for $r_o<r<R<\rho$ we have

$$U(r) < a \overset{+}{\log} U(R) + a\log\frac{R}{R-r} + b,$$

then for $r_o<r<R<\rho$ we have

$$U(r) < 2a\log\frac{R}{R-r} + 2b.$$

This lemma has been used in the above mentioned work of Chuang appearing in this book.

Lemma 3. Let $F(z)$ be a holomorphic function in the circle $|z|<1$ and let $G(z)$ be a homogeneous polynomial of $F^{(j)}(z)$ $(j=0,1,2,...,p)$:

$$G(z) = \sum_{n=1}^{k} \Phi_n(z),$$

$$\Phi_n(z) = a_n(z) \prod_{j=0}^{p} \{F^{(j)}(z)\}^{\lambda_{nj}} \tag{12}$$

where $a_n(z)$ $(n=1,2,\dots,k)$ are holomorphic functions in the circle $|z|<1$ and λ_{nj} $(n=1,2,\dots,k;\ j=0,1,2,\dots,p)$ are non-negative integers such that

$$\sum_{j=0}^{p} \lambda_{nj} = d \quad (d>0) \quad (n=1,2,\dots,k). \tag{13}$$

Assume that the following conditions are satisfied:

(1) The function $F(z)$ has no zero in the circle $|z|<1$.

(2) The function $G(z)$ is non-constant and $G(z)-1$ has no zero in the circle $|z|<1$.

(3) $m(r,a_n) = O(\log\frac{1}{1-r})$ $(r\to1,\ 0<r<1)$ $(n=1,2,\dots,k)$.

Then

$$m(r,F) = O(\log\frac{1}{1-r}),\ m(r,G) = O(\log\frac{1}{1-r}). \tag{14}$$

Proof. We have

$$m(r,\frac{1}{F^d}) \leqslant m(r,\frac{G}{F^d}) + m(r,\frac{1}{G}), \tag{15}$$

$$m(r,\frac{G}{F^d}) \leqslant \sum_{n=1}^{k} m(r,\frac{\Phi_n}{F^d}) + \log k,$$

$$m(r,\frac{\Phi_n}{F^d}) = m(r,a_n \prod_{j=0}^{p} \{\frac{F^{(j)}}{F}\}^{\lambda_{nj}}) \leqslant$$

$$m(r,a_n) + \sum_{j=0}^{p} \lambda_{nj} m(r,\frac{F^{(j)}}{F}),$$

hence

$$m(r,\frac{G}{F^d}) \leqslant \mu(r) + S(r) \tag{16}$$

where

$$\mu(r) = \sum_{n=1}^{k} m(r,a_n) + \log k, \tag{17}$$

$$S(r) = \sum_{j=0}^{p} \Lambda_j m(r,\frac{F^{(j)}}{F}), \quad \Lambda_j = \sum_{n=1}^{k} \lambda_{nj}$$

$$(j=0,1,\ldots,p). \tag{18}$$

By Lemma 1, there exist positive constants B_j $(j=1,2,3,4)$ such that for $0<r<R<1$ we have

$$S(r) < B_1 \overset{+}{\log} m(R,F) + B_2 \log\frac{1}{R-r} + B_3 \log\frac{1}{r} + B_4. \tag{19}$$

On the other hand from the identity

$$\frac{1}{G} = 1 - \frac{G-1}{G'}\frac{G'}{G}$$

we have

$$m(r,\frac{1}{G}) \leqslant m(r,\frac{G-1}{G'}) + m(r,\frac{G'}{G}) + \log 2, \tag{20}$$

$$m(r,\frac{G-1}{G'}) + N(r,\frac{G-1}{G'}) = m(r,\frac{G'}{G-1}) + h$$

where h is a constant. Since

$$N(r,\frac{G-1}{G'}) \geqslant n(0,\frac{G-1}{G'})\log r,$$

we have

$$m(r,\frac{G-1}{G'}) \leqslant m(r,\frac{G'}{G-1}) + \nu\log\frac{1}{r} + h \tag{21}$$

where $\nu = n(0,\frac{G-1}{G'})$. From (20) and (21) we get

$$m(r,\frac{1}{G}) \leqslant m(r,\frac{G'}{G-1}) + m(r,\frac{G'}{G}) + \nu\log\frac{1}{r} + h + \log 2.$$

Then by Lemma 1, we have for $0<r<R<1$,

$$m(r,\frac{1}{G}) < b_1\overset{+}{\log} m(\rho,G) + b_2\log\frac{1}{\rho-r} + b_3\log\frac{1}{r} + b_4 \tag{22}$$

where $\rho=(r+R)/2$ and b_j $(j=1,2,3,4)$ are positive constants. Next we have

$$m(\rho,G) \leqslant m(\rho,F^d) + m(\rho,\frac{G}{F^d})$$

and by (16) and (19),

$$m(\rho,G) \leqslant dm(\rho,F) + \mu(\rho) + S(\rho),$$

$$S(\rho) < B_1\overset{+}{\log} m(R,F) + B_2\log\frac{1}{R-\rho} + B_3\log\frac{1}{\rho} + B_4,$$

hence

$$m(\rho,G) < (d+B_1)m(R,F) + \mu(\rho) + B_2\log\frac{1}{R-\rho}$$

$$+ B_3\log\frac{1}{\rho} + B_4. \tag{23}$$

(22) and (23) yield

$$m(r,\frac{1}{G}) < b_1\overset{+}{\log} m(R,F) + b_1\overset{+}{\log}\mu(R) + b_2'\log\frac{1}{R-r}$$

$$+ b_3'\log\frac{1}{r} + b_4'. \qquad (24)$$

Then from (15), (16), (19) and (24), we get

$$dm(r,\frac{1}{F}) \leqslant m(r,\frac{G}{F^d}) + m(r,\frac{1}{G})$$

$$\leqslant \mu(r) + S(r) + m(r,\frac{1}{G})$$

$$< (B_1+b_1)\overset{+}{\log} m(R,F) + (1+b_1)\mu(R)$$

$$+ (B_2+b_2')\log\frac{1}{R-r} + (B_3+b_3')\log\frac{1}{r} + B_4 + b_4'.$$

Since

$$m(r,F) = m(r,\frac{1}{F}) + h',$$

we have

$$m(r,F) < B_1'\overset{+}{\log} m(R,F)+B_2'\log\frac{1}{R-r} +B_3'\log\frac{1}{r} +B_4'+B_5'\mu(R).$$

Now take a number t such that $1/2<t<1$. Then for $1/2<r<R<t$, we have

$$m(r,F) < a\overset{+}{\log} m(R,F) + a\log\frac{R}{R-r} + b$$

where

$$a = \max(B_1', B_2'), \quad b = (B_2'+B_3')\log 2 + B_4' + B_5'\mu(t).$$

Evidently we may assume that $b \geqslant 2a$ and $b \geqslant 8a^2$. Hence by Lemma 2, for $1/2<r<R<t$, we have

$$m(r,F) < 2a\log\frac{R}{R-r} + 2b.$$

Keeping r fixed and let R tends to t, we get

$$m(r,F) \leqslant 2a\log\frac{t}{t-r} + 2b. \tag{25}$$

(25) holds for $1/2<r<t<1$. In particular, taking $t=(1+r)/2$, we see that for $1/2<r<1$, we have,

$$m(r,F) < 2a\log\frac{1}{1-r} + b' + 2B_5'\mu(\frac{1+r}{2}).$$

Then by (17) and the condition (3) in Lemma 3, we find

$$m(r,F) = O(\log\frac{1}{1-r}). \tag{26}$$

On the other hand, by (23), we have for $0<r<R<1$,

$$m(r,G) < (d+B_1)m(R,F) + \mu(R) + B_2\log\frac{2}{R-r}$$

$$+ B_3\log\frac{1}{r} + B_4.$$

In particular, taking $R=(1+r)/2$ and making use of (26) and the condition (3) in Lemma 3, we get

$$m(r,G) = O(\log\frac{1}{1-r}). \tag{27}$$

The proof of Lemma 3 is now complete.

Now let us come back to the q radii (1). Consider a Sector Δ_k defined by

$$\Delta_k: r > 1, \quad \omega_k < \theta < \omega_{k+1} \tag{28}$$

and set

$$\zeta_k = \frac{\omega_k + \omega_{k+1}}{2} . \tag{29}$$

In the paper [1], Edrei defined a function

$$z = e^{i\zeta_k} \phi_k(w) \tag{30}$$

which maps conformally the circle $|w|<1$ onto the section Δ_k.

Lemma 4. Let $f(z)$ be a meromorphic function in the domain $D: 1<|z|<+\infty$ and $g(z)$ a homogeneous polynomial of $f^{(j)}(z)$ $(j=0,1,...,p)$ with constant coefficients defined by (5) and (6). Assume that $f(z)$ and $g(z)$ are non-constant and such that the roots in D of the equations

$$f(z) = 0, \quad f(z) = \infty, \quad g(z) = 1 \tag{31}$$

all lie on the radii (1). Then if we define

$$F_k(w) = f\{e^{i\zeta_k} \phi_k(w)\}, \quad G_k(w) = g\{e^{i\zeta_k} \phi_k(w)\}$$

$$(k=1,2,...,q), \tag{32}$$

we have for $t \to 1$ $(0<t<1)$,

$$m(t,F_k(w)) = O(\log\frac{1}{1-t}), \tag{33}$$

$$m(t,1/F_k(w)) = O(\log\frac{1}{1-t}), \tag{34}$$

$$m(t, 1/(G_k(w)-1)) = O(\log\frac{1}{1-t}), \tag{35}$$

$$(k=1,2,...,q).$$

Proof. For the sake of simplicity, we drop the subscript k, and write $F(w)=F_k(w)$, $G(w)=G_k(w)$, $\zeta=\zeta_k$, $\phi(w)=\phi_k(w)$. By hypothesis, $F(w)$ and $G(w)$ are holomorphic in the circle $|w|<1$ and such that $F(w)$ and $G(w)-1$ have no zero in the circle $|w|<1$. We have

$$G(w) = \sum_{n=1}^{k} \varphi_n(e^{i\zeta}\phi(w)),$$

$$\varphi_n(e^{i\zeta}\phi(w)) = C_n \prod_{j=0}^{p} \{f^{(j)}(e^{i\zeta}\phi(w))\}^{\lambda_{nj}} .$$

In the paper [1], Edrei has proved that

$$f^{(j)}(e^{i\zeta}\phi(w)) = \sum_{s=1}^{j} \alpha_s(w)F^{(s)}(w), \quad (j=1,2,...)$$

where $\alpha_s(w)$ is a holomorphic function in the circle $|w|<1$ such that

$$m(t,\alpha_s) = O(\log\frac{1}{1-t}).$$

Then it is easy to see that

$$G(w) = H\{F(w),\ F'(w),\ \ldots,\ F^{(p)}(w)\} \tag{36}$$

where $H\{F(w), F'(w), \ldots, F^{(p)}(w)\}$ is a homogeneous polynomial of $F^{(j)}(w)$ $(j=0,1,\ldots, p)$ of degree d, whose coefficients $\beta_i(w)$ are holomorphic functions in the circle $|w|<1$ such that

$$m(t,\beta_i) = O(\log\frac{1}{1-t}). \tag{37}$$

We can therefore apply Lemma 3 to $F(w)$ and $G(w)$ and conclude that we have

$$m(t,F(w)) = O(\log\frac{1}{1-t}),$$

$$m(t,G(w)) = O(\log\frac{1}{1-t}).$$

Next from

$$m(t,\frac{1}{F(w)}) = m(t,F(w)) + h,$$

$$m(t,\frac{1}{G(w)-1}) = m(t,G(w)) + O(1),$$

we also conclude that

$$m(t,\frac{1}{F(w)}) = O(\log\frac{1}{1-t}),$$

$$m(t,\frac{1}{G(w)-1}) = O(\log\frac{1}{1-t}).$$

We also need the following lemma proved in [1].

Lemma 5. Let $h(z)$ be a meromorphic function of

order ρ (not necessarily finite). Assume

(i) that the poles of $h(z)$ either lie on the q radii defined by (1) or else belong to the disk $|z|\leq 1$;

(ii) that the poles of $h(z)$ have a positive deficiency.

Consider the number β defined by

$$\beta = \sup\{\frac{\pi}{\omega_2-\omega_1}, \frac{\pi}{\omega_3-\omega_2}, \ldots, \frac{\pi}{\omega_{q+1}-\omega_q}\}$$

$$(\omega_{q+1} = 2\pi + \omega_1), \tag{38}$$

the numbers ζ_k and the functions ϕ_k defined by (29) and (30).

Then

$$\beta < \rho \tag{39}$$

implies

$$\overline{\lim_{t\to 1}} \frac{m(t, h(e^{i\zeta_k}\phi_k(w)))}{\log\frac{1}{1-t}} = +\infty \quad (0<t<1) \tag{40}$$

for some integer k $(1\leq k\leq q)$.

Now we are in a position to prove Theorem 2. First of all, by the condition (2) in Theorem 2, there is a number $R>0$ such that the roots in the domain $R<|z|<+\infty$ of the equations

$$f(z) = 0, \quad f(z) = \infty, \quad g(z) = 1$$

all lie on the radii (1).

Consider first the case $R=1$. In this case, by Lemma 4, we have (33), (34) and (35) for $k=1,2,\ldots,q$. On the other hand, consider the three functions f, $1/f$, $1/(g-1)$. By the condition (3) in Theorem 2, the poles of at least one of these three functions have a positive deficiency. Denote this function by $h(z)$ whose order ρ is equal to that of $f(z)$, by the condition (1) in Theorem 2. By Lemma 5, if (39) holds, then we would have (40) for some k and hence get a contradiction to (33), (34) and (35) valid for $k=1,2,\ldots,q$. So we must have $\rho \leqslant \beta$.

To prove Theorem 2 in the general case, we introduce the auxiliary function

$$f_1(z) = f(Rz)$$

and the homogeneous polynomial of $f_1^{(j)}(z)$ $(j=0,1,2,\ldots,p)$:

$$g_1(z) = \sum_{n=1}^{k} \psi_n(z), \quad \psi_n(z) = C_n' \prod_{j=0}^{p} \{f_1^{(j)}(z)\}^{\lambda_{nj}}$$

$$C_n' = C_n R^{-\mu_n}, \qquad \mu_n = \sum_{j=0}^{p} j\lambda_{nj}.$$

We have

$$g_1(z) = g(Rz).$$

It is easy to see that the following conditions are satisfied:

(1) ρ being the order of $f(z)$, $f_1(z)$ and $g_1(z)$ are non-constant and have the same order ρ.

(2) The roots in the domain $1<|z|<+\infty$ of the equations

$$f_1(z) = 0, \quad f_1(z) = \infty, \quad g_1(z) = 1$$

all lie on the radii (1).

(3) $\delta(0,f_1) + \delta(\infty,f_1) + \delta(1,g_1) > 0$.

Consequently by the particular just treated, we have $\rho \le \beta$.

REFERENCES

[1] Edrei, A., Meromorphic functions with three radially distributed values (Trans. Am. Math. Soc. 78 (1955), 276-293).

[2] Nevanlinna, R., Le théorème de Picard-Borel et la théorie des fonctions méromorphes, Paris, 1929.

[3] Chuang Chi-tai, Singular directions of meromorphic functions, Beijing, 1982.

THE DISTRIBUTION OF ZEROS OF SOLUTIONS OF A CLASS OF LINEAR DIFFERENTIAL EQUATIONS

Zou Xiu-Lin
Department of Applied Mathematics
Beijing Institute of Technology
Beijing, China

1. INTRODUCTION

On the distribution of zeros of solutions of linear differential equation of order k, S. Bank, G. Frank and I. Laine have recently proved the following results [1]:

Theorem A. Given the linear differential equation

$$w^{(k)} + a_{k-1}(z)w^{(k-1)} + \dots + a_0(z)w = 0 \qquad (1.1)$$

where $k \geq 2$, and the coefficients $a_j (j = 0, 1, \dots, k-1)$ are polynomials satisfying the following conditions: (i). $a_0 \not\equiv$ const; (ii). if $a_j \not\equiv 0$, say $a_j = K_j z^{\alpha_j} + \dots$, then the degree α_j satisfies $\alpha_j \leq (k-j)\alpha_0/k$ for each $j = 1, \dots, k-1$; (iii). all roots of the polynomial $t^k + \sum_j K_j t^j$ (where the sum is over all j for which $a_j \neq 0$ and $\alpha_j = (k-j)\alpha_0/k$) are simple. Then the following hold:

(a). Every solution $f \not\equiv 0$ of (1.1) is an entire function of order of growth $(\alpha_0 + k)/k$.

(b). If a solution $f \not\equiv 0$ of (1.1) has the property that the exponent of convergence of the zero-sequence of f, denoted by $N(f)$, is less than $(\alpha_o + k)/k$, then f has only finitely many zeros.

(c). If $\{f_1, \ldots, f_k\}$ is a fundamental set for (1.1), then $N(f_j) = (\alpha_o + k)/k$ for at least one j in the set $\{1, 2, \ldots, k\}$.

When $(\alpha_o + k)/k$ is a positive integer, the following question was proposed by S. Bank and G. Frank.

How many f_j's exactly in the fundamental set $\{f_1, \ldots, f_k\}$ have the property that $N(f_j) < (\alpha_o + k)/k$?

In paper [2], S. Bank and G. Frank investigated the following special equation of order k

$$w^{(k)} + A(z)w = 0 \tag{1.2}$$

where $A(z)$ is a polynomial of degree n.

They obtained the following.

Theorem B. Let k be an integer greater than 1. Assume that equation (1.2), with $A(z)$ being a polynomial of degree n, possesses a solution $f(z) \not\equiv 0$ such that $N(f) < (n + k)/k$, then if $g(z)$ is any solution of (1.2) which is not a constant multiple of f, we have $N(g) = (n + k)/k$.

This theorem implies that there is at most one f_j in the fundamental set $\{f_1, f_2, \ldots, f_k\}$ of equation (1.2) such that $N(f_j) < (n + k)/k$. Does this conclusion hold for the more general equation (1.1) subject to the conditions (i) - (iii) listed above? In this paper we will give an example to show that the answer is no, and we can prove the following

Theorem. Given the equation

$$w^{(k)} + a_1(z)w' + a_o(z)w = 0 \tag{1.3}$$

where $a_1(z)$ and $a_0(z)$ are polynomials. Suppose that $a_o(z) \not\equiv 0$, and $a_1(z) \not\equiv$ constant, $\alpha_o = \deg(a_0(z))$, $\alpha_1 = \deg(a_1(z))$ and $\alpha_1 < (k-1)\alpha_0/k$. Let $k \geq 3$. If the equation (1.3) possesses two linearly independent solutions $f_1(z)$ and $f_2(z)$ such that $N(f_j) < (\alpha_o + k)/k$ for each $j = 1, 2$, then for any solution of (1.3) $g(z)$ which is neither a constant multiple of f_1, nor a constant multiple of f_2, we have $N(g) = (\alpha_o + k)/k$.

This theorem shows that there exist at most two solutions in the fundamental set of (1.3) having the exponent of convergence of the zero-sequence less than $(\alpha_o + k)/k$, and the example in §4 shows that this is the best possible result.

2. THREE LEMMAS REQUIRED FOR THE PROOF OF THE THEOREM

In this paper the following notations as in [2] were adopted.

(1). If $R(z)$ is a rational function whose Laurent expansion around $z = 0$ has the form

$$R(z) = C_d z^d + C_{d-1} z^{d-1} + \dots, \quad (C_d \neq 0) \tag{2.1}$$

then we will denote d by $\delta(R)$.

(2). $\binom{n}{j}$ represent the binomial coefficient $n!/(j!)((n-j)!)$ where $n \geq j$. As a convention, we put $\binom{n}{j} = 0$ for $n < j$ and $\binom{o}{o} = 0$.

Lemma 1. Let $P(z)$ be a polynomial of degree $m + 1$, where $m \geq 0$. Let $\Delta(z)$ be a rational function with $\delta(\frac{\Delta}{z}) = q$ and assume that $\Delta \neq 0$. Set $h(z) = e^{P(z)}$ and

$f(z) = \Delta(z)\, e^{P(z)}$, then the following are true.

(a). The Laurent expansion of Δ'/Δ around ∞ is of the form

$$\Delta'/\Delta = qz^{-1} + \delta_1 z^{-2} + \dots \tag{2.2}$$

(b). For $n = 0, 1, \dots,$ we have

$$\delta(\Delta^{(n)}/\Delta) \leqslant -n. \tag{2.3}$$

(c). For $n = 0, 1, 2, \dots,$ we have

$$h^{(n)}/h = (P')^n + \binom{n}{2}(P')^{n-2}P'' + V_n , \tag{2.4}$$

where V_n is a polynomial of degree less than $m(n-1) - 1$.

(d). For $n = 0, 1, 2, \dots,$ we have

$$f^{(n)}/f = (P')^n + \binom{n}{2}(P')^{n-2}P'' + n(\Delta'/\Delta)(P')^{n-1} + E_n, \tag{2.5}$$

where E_n is a rational function satisfying $\delta(E_n) < m(n-1) - 1$.

This is a generalization of the lemma in [2] where $\Delta(z)$ is a polynomial. The proofs are similar. We omit it here.

Lemma 2. Let f_1 and f_2 be two linearly independent solutions of equation (1.1), then the equation (1.1) can be transformed into the following form

$$\sum_{i=1}^{k-1}\Big(\sum_{n=i+1}^{k}\Big(\sum_{j=i}^{n-1}\binom{n}{j}(f_1^{(n-j-1)}/f_1)(f^{(j-i+1)}/f)\Big)a_n(z)\Big)v^{(i)} = 0 \tag{2.6}$$

under the change of variable $v_1 = (w/f)'/f'$, where $f = f_2/f_1$ and $a_k(z) = 1$.

Proof. Under the change of variable

$$v_1 = (w/f_1)'$$

the equation (1.1) is tranformed by the method of reduction of order into the equation

$$\sum_{j=1}^{k} \left(\sum_{n=j}^{k} \binom{n}{j} (f_1^{(n-j)}/f_1) a_n \right) v_1^{(j-1)} = 0 \qquad (2.7)$$

and $v_1 = (f_2/f_1)' = f'$ satisfies (2.7). Set $v_1 = f'v$. By the same method we can change the equation (2.7) into (2.6).

Corollary. If f_1 and f_2 are two linearly independent solutions of (1.3), then under the change of variable $v = (w/f_1)'/f'$, the equation (1.3) can be changed into the following equation

$$\sum_{i=1}^{k-1} \sum_{j=i}^{k-1} \binom{k}{j+1}\binom{j}{i}(f_1^{(k-j-1)}/f_1)(f^{(j-i+1)}/f)v^{(i)} = 0 \qquad (2.8)$$

Lemma 3. Suppose that the coefficients $a_0(z)$ and $a_1(z)$ of equation (1.3) satisfy the hypothesis of the theorem. If $f_1(z) = \Delta_1(z)^{P_1(z)}$ and $f_2(z) = \Delta_2(z)^{P_2(z)}$ are two linearly independent solutions of (1.3), where $\Delta_i(z)$ and $P_i(z)$ are polynomials for i = 1, 2. Then degree $(P_i(z)) = m + 1$ (i = 1,2), where $m = \alpha_0/k$ is a positive integer and degree $(P_2(z) - P_1(z)) = m + 1$.

Proof: Corresponding to equation (1.3), (2.7), becomes

$$(a_1(z) + kf_1^{(k-1)}/f_1)V_1 + \sum_{j=1}^{k} \binom{k}{j}(f_1^{(k-j)}/f_1)V_1^{(j-1)} = 0 \qquad (2.9)$$

and $v_1 = (f_2/f_1)' = \Delta_0 e^P$ satisfies (2.9), where $\Delta_0 =$

$(\Delta_2/\Delta_1)' + (\Delta_2/\Delta_1)P'$ is a rational function and $P = P_2 - P_1$. From lemma 1, $\delta(f_1^{(k-j)}/f_1) = m(k-j)$ for $j = 0, 1, 2, \ldots$. By the hypothesis that $\alpha_1 < (k-1)m$, we have $\delta(a_1(z) + kf_1^{(k-1)}/f_1) = m(k-1)$. It now follows from Wiman-Valiron theory [4] that the growth order of $\Delta_o e^{P(z)}$ is $m + 1$. Therefore $P(z) = P_2(z) - P_1(z)$ is a polynomial of degree $m + 1$. The conclusion $\deg(P_i) = m + 1$ can be obtained directly from theorem A(a).

3. PROOF OF THE THEOREM

From the hypothesis of the theorem and theorem A(a), it follows that $m = \alpha_o/k$ is a positive integer. So by theorem A(b) we must have $f_i(z) = \Delta_i(z)^{P_i(z)}$, where $\Delta_i(z)(\not\equiv 0)$ and $P_i(z)$ are polynomials and $\deg P_i(z) = m + 1$, $(i = 1,2)$. Set $f = f_2/f_1$, $\Delta = \Delta_2/\Delta_1$, $P = P_2 - P_1$, then $f = \Delta e^P$ and $\deg P(z) = m + 1$ by lemma 3. If w satisfies (1.3), then $v = (w/f_1)'/f'$ satisfies (2.8) by the corollary of lemma 2.

We now assume that the conclusion of the theorem fails to hold, then (1.3) possesses a solution $g(z)$ which is neither a constant multiple of f_1, nor a constant multiple of f_2, and has the property that $N(g) < (\alpha_o + k)/k$. In view of theorem A(a)(b) we may write $g(z) = H(z)e^{W(z)}$, where $H(z)(\neq 0)$ and $W(z)$ are polynomials and $\deg W(z) = m + 1$, $v = (g/f_1)'/f'$ satisfies (2.8). Set $(g/f_1)'/f' = R\omega$, where $R(z) = ((H/\Delta_1)' + (H/\Delta_1)(W-P_1))/(\Delta'+P'\Delta)$.

$$R(z) = ((H/\Delta_1)' + (H/\Delta_1)(W-P_1))/(\Delta'+P') \tag{3.1}$$

is a rational function. Since $\deg(W-P_1) = m + 1$ by lemma 3, we have

$$\delta(R) = \deg H - \deg\Delta_2, \tag{3.2}$$

$$\omega = e^{Q(z)} = e^{W-P_2} \tag{3.3}$$

and $\deg Q(z) = \deg(W-P_2) = m+1$ by lemma 3. Substituting

$$v^{(i)} = \sum_{t=0}^{i} \binom{i}{t} \omega^{(i-t)} R^{(t)}$$

into (2.8), rearranging terms and dividing the equation by ω, we see that $R(z)$ satisfies the equation

$$\sum_{t=0}^{k-1} \left(\sum_{i=t}^{k-1} \binom{i}{t} G_i(z) \left(\omega^{(i-t)}/\omega\right)\right) R^{(t)} = 0 \tag{3.4}$$

where $G_i = \sum_{j=i}^{k-1} \binom{k}{j}\binom{j}{i}\left(f_1^{(k-j-1)}/f_1\right)\left(f^{(j-i+1)}/f\right)$. We write (3.4) in a short form

$$\sum_{t=0}^{k-1} A_t R^{(t)} = 0 \tag{3.5}$$

where $A_t(z) = \sum_{i=t}^{k-1} \binom{i}{t} G_i\left(\omega^{(i-t)}/\omega\right)$

$$= \sum_{i=t}^{k-1} \sum_{j=i}^{k-1} \binom{i}{t}\binom{k}{j+1}\binom{j}{i}\left(f_1^{(k-j-1)}/f_1\right)\left(f^{(j-i+1)}/f\right)\left(\omega^{(i-t)}/\omega\right)$$

$$= \sum_{j=t}^{k-1} \sum_{i=t}^{j} \binom{i}{t}\binom{k}{j+1}\binom{j}{i}\left(f_1^{(k-j-1)}/f\right)\left(f_1^{(j-i+1)}/f\right)\left(\omega^{(i-t)}/\omega\right). \tag{3.6}$$

It follows from lemma 1 that

$$\delta(A_t) \leqslant (k-t)m. \tag{3.7}$$

Since $g(z)$ is not a constant multiple of f_1, $R(z) \not\equiv 0$. Therefore (3.5) have a rational solution $R(z) \not\equiv 0$. But the following deduction and computation show that it is impossible.

We now compute A_0. From (3.6) we have

$$A_o = \sum_{j=0}^{k-1} \sum_{i=0}^{j} \binom{k}{j}\binom{j}{i}(f_1^{(k-j-1)}/f_1)(f^{(j-i+1)}/f)(\omega^{(i)}/\omega)$$

$$= \sum_{j=1}^{k} \sum_{i=1}^{j} \binom{k}{j}\binom{j-1}{i-1}(f_1^{(k-j)})/f_1)(f^{(j-i+1)}/f)(\omega^{(i-1)}/\omega).$$

Using lemma 1 we write

$$A_0 = F_1 + F_2 + F^* + F_3 \tag{3.8}$$

where

$$F_1 = \sum_{j=1}^{k} \sum_{i=1}^{j} \binom{k}{j}\binom{j-1}{i-1}(P_1')^{k-j}(P')^{j-i+1}(Q')^{i-1} \tag{3.9}$$

$$F_2 = I_1 + I_2 + I_3 \tag{3.10}$$

$$I_1 = \sum_{j=1}^{k} \sum_{i=1}^{j} \binom{k}{j}\binom{j-1}{i-1}\binom{i-1}{2}(P_1')^{k-j}(P')^{j-i+1}(Q')^{i-3}Q''; \tag{3.11}$$

$$I_2 = \sum_{j=1}^{k} \sum_{i=1}^{j} \binom{k}{j}\binom{j-1}{i-1}\binom{j-i+1}{2}(P_1')^{k-j}(P')^{j-i-1}(Q')^{i-1}P''; \tag{3.12}$$

$$I_3 = \sum_{j=1}^{k} \sum_{i=1}^{j} \binom{k}{j}\binom{j-1}{i-1}\binom{k-j}{2}(P_1')^{k-j-2}(P')^{j-i+1}(Q')^{i-1}P_1'', \tag{3.13}$$

$$F^* = I_1^* + I_2^*, \tag{3.14}$$

$$I_1^* = \sum_{j=1}^{k} \sum_{i=1}^{j} \binom{k}{j}\binom{j-1}{i-1}(j-i+1)(\Delta'/\Delta)(P_1')^{k-j}(P')^{j-i}(Q')^{i-1} \tag{3.15}$$

$$I_2^* = \sum_{j=1}^{k} \sum_{i=1}^{j} \binom{k}{j}\binom{j-1}{i-1}(k-j)(\Delta_1'/\Delta_1)(P_1')^{k-j-1}(Q')^{i-1}(P')^{j-i+1} \tag{3.16}$$

F_3 is a rational function satisfying

$$\delta(F_2) < m(k-1) - 1. \tag{3.17}$$

From (3.9) we obtain

$$F_1 = (P') \sum_{j=1}^{k} \sum_{i=0}^{j-1} \binom{k}{j}\binom{j-1}{i}(P_1')^{k-j}(P')^{j-1-i}(Q')^i$$

$$= P' \frac{(P_1' + Q' + P')^k - (P_1')^k}{Q' + P'} = P' \frac{(P_1' + Y')^k - (P_1')^k}{Y'}$$

$$= P' \prod_{i=1}^{K-1} (Y' - b_i P_1') \tag{3.18}$$

where $Y = P + Q = W - P_1$ is a polynomial of degree $m + 1$, b_i $(i = 1, \ldots k-1)$ are non-zero roots of the polynomial

$$\varphi(s) = ((s+1)^k - 1)/s = \sum_{j=1}^{k} \binom{k}{j} s^{j-1}. \tag{3.19}$$

We now divide the proof into two cases.

Case 1°. $Y' \not\equiv b_i P_1'$ for each $i = 1, 2, \ldots, k-1$, then $Y' - b_i P_1'$ is of degree m for all i with the possible exception of one value of i. Hence from (3.18) we have

$$\text{degree} F_1 \geqslant m(k-1) . \tag{3.20}$$

But from (3.13) - (3.17), using lemma 1, we obtain that

$$\delta(F_2 + F^* + F_3) \leqslant m(k-1) - 1 .$$

So in view of (3.8) and (3.20), we have

$$\delta(A_0) \geqslant m(k-1) .$$

It now follows that (3.5) cannot possess a rational solution $R(z) \not\equiv 0$ since we can easily see that

$$\delta(A_t R^{(t)}) < \delta(A_0 R), \quad (t = 1, 2, \ldots, k-1)$$

holds for any rational function $R \not\equiv 0$. This shows that Case 1° is impossible.

Case 2°. There exists an r in $\{1, 2, \ldots, k-1\}$ such that

$$Y' \equiv b_r P_1' \tag{3.21}$$

and thus $F_1 \equiv 0$. Now we compute F_2 and F^*. From (3.11) we have

$$\begin{aligned} I_1 &= \sum_{j=1}^{K} \binom{k}{j}(P')^{k-j} \sum_{i=0}^{j-1} \binom{j-1}{i}\binom{i}{2}(P')^{j-i-1}(Q')^{i-2}(Q'')P' \\ &= \sum_{j=1}^{k} \binom{k}{j}(P')^{k-j}\binom{j-1}{2}(P'+Q')^{j-3}(Q'')P' \\ &= \sum_{j=1}^{k} \binom{k}{j}\binom{j-1}{2} b_r^{j-3} (P_1')^{j-3} Q'' P' \\ &= K_2 (P_1')^{j-3} Q'' P' \end{aligned} \tag{3.23}$$

$$\text{where } K_2 = \sum_{j=1}^{k} \binom{k}{j}\binom{j-1}{2} b_r^{j-3} = \frac{1}{2}\varphi''(b_r)$$

$$= \frac{-k\ (k-1)}{2b_r(b_r+1)^2} - \frac{k}{b_r^2(b_r+1)} . \tag{3.23}$$

From (3.12) we have

$$I_2 = \sum_{j=1}^{k} \binom{k}{j}(P_1')^{k-j} \sum_{i=0}^{j-1} \binom{j-1}{i}\binom{j-i}{2}(P')^{j-i-2}P''(Q')^{i}$$

$$= \sum_{j=1}^{k} \binom{k}{j}(P_1')^{k-j}\left((j-1)(P'+Q')^{j-2} + \binom{j-1}{2}(P'+Q')^{j-3}\right)P''$$

$$= \sum_{j=1}^{k} \binom{k}{j}(j-1)b_r^{j-2}(P_1')^{k-2}P'' + \sum_{j=1}^{k} \binom{k}{j}\binom{j-1}{2}b_r^{j-3}(P_1')^{k-3}P''P'$$

$$= K_1(P_1')^{k-2}P'' + K_2(P_1')^{k-3}P''P' \tag{3.24}$$

where

$$K_1 = \sum_{j=1}^{k} \binom{k}{j}(j-1)b_r^{j-2} = \varphi'(b_r) = \frac{k}{b_r(b_r + 1)} . \tag{3.25}$$

Similarly, from (3.13) we have

$$I_3 = K_3(P_1')^{k-3}P_1''P', \tag{3.26}$$

where

$$K_3 = \sum_{j=1}^{k} \binom{k}{j}\binom{k-j}{2}b_r^{j-1} .$$

Making use of the function

$$\psi(s) = \sum_{j=1}^{k} \binom{k}{j}s^{k-j} = s^{k-1} \varphi(1/s), \tag{3.27}$$

we obtain that $K_3 = \frac{1}{2} b_r^{k-3} \psi''(1/b_r)$. Computing ψ'' and noticing that $\varphi(b_r) = 0$, we obtain that

$$K_3 = (b_r^2 \varphi''(b_r) - (2k-4)b_r\varphi'(b_r))/2$$

$$= - \frac{k(k-1)}{2(b_r + 1)^2} - \frac{k(k-1)}{2(b_r + 1)} . \tag{3.28}$$

From (3.10) and (3.22) - (3.28), we obtain that

$$F_2 = K_1(P_1')^{k-2}P'' + K_2(P_1')^{k-3}(Q''+P'')P' + K_3(P_1')^{k-3}P_1''P'$$

$$= K_1(P_1')^{k-2}P'' + (b_rK_2 + K_3)(P_1')^{k-3}P_1''P'. \qquad (3.29)$$

Denote $b_rK_2 + K_3$ by K_{23}. In view of (3.23) and (3.28) we have

$$K_{23} = -k/b_r + k(3-k)/2(b_r+1) . \qquad (3.30)$$

From (3.15), noticing that $\sum_{j=1}^{k} \binom{k}{j} b_r^{j-1} = 0$, we can obtain

$$I_1^* = K_1(\Delta'/\Delta)(P_1')^{k-2}P'. \qquad (3.31)$$

From (3.16), we can obtain

$$I_2^* = K^*(\Delta_1'/\Delta_1)(P_1')^{k-2}P' \qquad (3.32)$$

where $K^* = \sum_{j=1}^{k} \binom{k}{j}(k-j)b_r^{j-1} = b_r^{k-2}\ \psi'(1/b_r)$. So

$$K^* = -k/(b_r + 1). \qquad (3.33)$$

From (3.14) we can obtain

$$F^* = K_1(\Delta'/\Delta)(P_1')^{k-2}P' + K^*(\Delta_1'/\Delta_1)(P_1')^{k-2}P'. \qquad (3.34)$$

Finally we compute A_1. From (3.6), using lemma 1 and (3.21), we obtain

$$A_1 = K_1(P_1')^{k-2}P' + G \qquad (3.35)$$

where G is a rational function satisfying

$$\delta(G) \leqslant m(k-2) - 1 . \qquad (3.36)$$

Let the Laurent expansion of $R(z) \neq 0$, the rational

solution of (3.5), around ∞ be

$$R(z) = C_d z^d + C_{d-1} z^{d-1} + \dots \quad (C_d \neq 0). \tag{3.37}$$

Suppose that the polynomials P_i', P', which are of degree m, are of the form

$$P_i' = a_i z^m + \dots \quad (a_i \neq 0),\ i = 1,\ 2; \tag{3.38}$$

$$P' = az^m + \dots \quad (a \neq 0)\ . \tag{3.39}$$

Write (3.5) in the following form

$$A_j R' + A_0 R = - \sum_{t=2}^{k} A_t R^{(t)}\ . \tag{3.40}$$

In view of (3.8), (3.29), (3.34) and the fact that $F_1 = 0$, we have

$$A_0 = K_1 (P_1')^{k-2} P'' + K_{23} (P_1')^{K-3} P_1'' P' + K_1 (\Delta'/\Delta)(P_1')^{k-2} P'$$

$$+ K^* (\Delta_1'/\Delta_1)(P_1')^{k-2} P' + F_3 .$$

We may assume that $\deg \Delta_2 \geqslant \deg \Delta_1$, and set

$$\deg \Delta_i = q_i \quad (i = 1,\ 2)\ ,$$

$$\delta(\Delta) = (\deg \Delta_2 - \deg \Delta_1) = q\ ,$$

then $q_i \geqslant 0$ $(i = 1,2)$ and $q \geqslant 0$. In view of lemma 1(a) we have

$$A_0 = (K_1 m \div K_{23} m + K_1 q + K^* q_1) a_1^{k-2} a z^{(k-1)m-1} + F_4$$

$$= K_4\, a_1^{k-2}\, a z^{(k-1)m-1} + F_4 \tag{3.41}$$

where

$$K_4 = K_1 m + K_{23} m + K_1 q + K^* q_1\ , \tag{3.42}$$

and F_4 is a rational function satisfying

$$\delta(F_4) < (k-1)m-1 . \tag{3.43}$$

We assert that $K_4 \neq 0$. If we assume the contrary, then using (3.25), (3.30), (3.33) and (3.42) we would obtain

$$b_r[(m + q_1) + \frac{1}{2}(k-3)m] = q, \tag{3.44}$$

which shows that b_r is real. But the non-zero real root (if any) is only -2. Hence $b_r = -2$. This would imply that the left side of (3.44) is less than zero, while the right side is non-negative. So it is impossible.

From (3.35), (3.38) and (3.39) we have

$$A_1 = K_1 a_1^{k-2} a z^{(k-1)m} + \ldots , \tag{3.45}$$

$$A_1 R' + A_0 R = (K_4 + K_1 d) a_1^{k-2} a C_d z^{(k-1)m+d-1} + F_5 \tag{3.46}$$

where the rational function F_5 satisfies

$$\delta(F_5) < (k-1)m+d-1 .$$

If $K_4 + K_1 d \neq 0$, then $\delta(A_1 R' + A_0 R) = (k-1)m+d-1$. But

$$\delta(A_t R^{(t)}) \leqslant (k-2)m+d-2 \quad \text{for} \quad t \geqslant 2,$$

in view of (3.40), this is impossible. So we must have $K_4 + K_1 d = 0$. This means that

$$b_r(q_1 + \frac{1}{2}(k-1)m) = q + d, \tag{3.47}$$

which as in (3.44) shows that $b_r = -2$ and (3.47) then yields

$$d = -((k-1)m + q_2 + q_1). \tag{3.48}$$

On the other hand, if we set $\deg H = \sigma$, it follows from (3.2)

$$d = \sigma - q_2 . \tag{3.49}$$

It follows from (3.48) and (3.49) that

$$\sigma = -((k-1)m + q_1) < 0 ,$$

since $k > 1$, $m \geqslant 1$ and $q_1 \geqslant 0$. Therefore case 2^0 is also impossible. The proof of the theorem is completed.

4. AN EXAMPLE

The following equation

$$w''' + 6w' + (-8z^3 + 12z)\,w = 0 \tag{4.1}$$

is of the form (1.3) with the coefficients satisfying the conditions of the theorem. It can be easily verified that (4.1) possesses two linearly independent solutions

$$f_1(z) = e^{\lambda_1 z^2}, \quad f_2(z) = e^{\lambda_2 z^2}, \tag{4.2}$$

where λ_1 and λ_2 are $e^{\frac{2\pi}{3} i}$ and $e^{\frac{4\pi}{3} i}$ respectively. $f_1(z)$ and $f_2(z)$ have no zeros.

Finally, the author would like to thank his advisers Prof. He Yu-zhan and Yang Wi-qi for their help.

REFERENCES

1. S. Bank, G. Frank and I. Laine: Über die nullstellen von lösunge linear differentialgeichungen. Math. Z 183, 355-364 (1983).

2. S. Bank and G. Frank: A note on the distribution of zeros of solutions of linear differential equation. Comment. Math. Univ. St. Pauli 33 (1984), 143-151.

3. He Yu-zhan: Nevanlinna theory and its application. The 5th national conference of one complex variable function. Gui Yang, China, 1986.

4. G. Valiron: Lectures on the general theory of integral functions. New York, 1949.

ON BOUNDARY PROPERTIES OF ANALYTIC FUNCTIONS

De-Chang Pu
Department of Mathematics
Yantai University
Yantai, China

In this paper we always denote by D a bounded domain in the complex plane $\mathbb{C}$ and by ∂D the boundary of D.

Definition. Let $\varphi(z)$ be a complex valued function defined in D. A value $\alpha \in \mathbb{C} \cup \infty$ is called a sequential boundary value of $\varphi(z)$ at $\zeta \in \partial D$, if there is a sequence $z_n \in D$ such that $\lim_{n\to\infty} z_n = \zeta$ and $\lim_{n\to\infty} f(z_n) = \alpha$. We denote by $I_\varphi(\zeta)$ the set of all sequential boundary values of $\varphi(\zeta)$ at $\zeta \in \partial D$ and define $I_\varphi(\partial D) = \bigcup_{\zeta\in\partial D} I_\varphi(\zeta)$. If there exists $\beta \in \mathbb{C} \cup \infty$ such that $\beta \bar{\in} I_\varphi(\partial D)$, we write $\varphi(z) \in \mathscr{A}$ in D.

By a theorem of Plessner[1], if ∂D is a rectifiable curve and if $f(z) \in \mathscr{A}$ is analytic in D, then almost everywhere on ∂D, $f(z)$ has angular boundary values.

The purpose of the present paper is to investigate the relationship between the condition $f(z) \in \mathscr{A}$ or the condition $f(z) \bar{\in} \mathscr{A}$ and the set of values taken by the analytic function $f(z)$ in D. Our main results are the following theorems:

Theorem 1. Let $f(z)$ be analytic in D. In order

that $f \in \mathscr{A}$ in D, it is necessary and sufficient that there exist an integer $p \geqslant 0$ and an open set $\Omega \in \mathbb{C}$, such that the number of the zeros of $f(z) - w$ in D (counted with their orders of multiplicity) for each $w \in \Omega$ does not exceed p.

Theorem 2. Let $f(z)$ be analytic in D. In order that $f \in \mathscr{A}$ in D, it is necessary and sufficient that $f(z)$ can be expressed in the form

$$f(z) = \frac{P(z)}{B(z)} + \lambda$$

where $P(z)$ is a polynomial whose zeros all lie in D, $B(z)$ is a bounded analytic function with no zero in D and λ is a constant.

Theorem 3. Let $f(z)$ be analytic in D. In order that $f(z) \bar{\in} \mathscr{A}$ in D, it is necessary and sufficient that for each $w \in \mathbb{C}$, $f(z) - w$ has an infinite number of zeros in D, except perhaps a set of points w of first category with respect to the w-plane.

An interesting consequence of Theorems 2 and 3 is the following.

Corollary. Let $f(z)$ be analytic in $|z| < 1$. If

$$\lim_{r \to 1} \int_0^{2\pi} \log^+ |f(re^{i\theta})| d\theta = \infty,$$

then for each $w \in \mathbb{C}$, $f(z) - w$ has an infinite number of zeros in $|z| < 1$, except perhaps a set of points w of first category with respect to the w-plane.

Proof of Theorem 1.

Suppose that the analytic function $f(z) \in \mathscr{A}$ in D. Since $I_f(\partial D)$ is a closed set, there is a complex number $b \bar{\in} I_f(\partial D)$. Set $F(z) = f(z) - b$, then $0 \bar{\in} I_F(\partial D)$. If $F(z)$ has no zero in D, there is $\varepsilon > 0$ such that

$|F(z)| > \varepsilon$ in D. So $p = 0$ and $\Omega = \{|w - b| < \varepsilon\}$ satisfy the required conditions in Theorem 1. If $F(z)$ has a finite number of zeros a_j $(j = 1,2,\ldots,m)$ in D. Let α_j $(j = 1,2,\ldots,m)$ be respectively their orders of multiplicity and $\alpha = \sum_{j=1}^{m} \alpha_j$. Choose $\delta > 0$ such that the disks $\bar{D}_j : |z - a_j| \leqslant \delta$ $(j = 1,2,\ldots,m)$ are mutually disjoint and within D. Then $F(z) \neq 0$ in $G = D \setminus \bigcup_{j=1}^{m} \bar{D}_j$ and $0 \bar{\in} I_F(\partial G)$. Consequently there is $\varepsilon' > 0$ such that $|F(z)| > \varepsilon'$ in G. For $|w| < \varepsilon'$, $F(z) - w$ has no zero in G and has precisely α_j zeros in D_j, hence $F(z) - w$ has precisely α zeros in D. So $p = \alpha$ and $\Omega = \{|w - b| < \varepsilon'\}$ satisfy the required conditions in Theorem 1.

Conversely suppose that there exist an integer $p \geqslant 0$ and an open set $\Omega \in \mathbb{C}$ such that for each $w \in \Omega$, the number of the zeros of $f(z) - w$ in D does not exceed p. We may assume that p is the smallest of such integers. If $p = 0$, then $w \bar{\in} I_f(\partial D)$ for $w \in \Omega$. If $p > 0$, then there is $w_0 \in \Omega$ such that the number of the zeros in D of $f(z) - w_0$ is p (with due count of order of multiplicity). Let b_j $(j = 1,2,\ldots,n)$ be these zeros and β_j $(j = 1,2,\ldots,n)$ respectively their orders of multiplicity with $\sum_{j=1}^{n} \beta_j = p$. Choose $\delta > 0$ such that the disks $\bar{G}_j : |z - b_j| \leqslant \delta$ $(j=1,2,\ldots,n)$ are mutually disjoint and in D. Set

$$m_j = \min_{z \in \partial G_j} |f(z) - w_0| \qquad (j = 1,2,\ldots,n)$$

and let $0 < \rho < \min(m_1, m_2, \ldots, m_n)$ such that the disk $\Gamma : |w - w_0| < \rho$ is interior to Ω. When $w \in \Gamma$, $f(z) - w$ has precisely p zeros in D, which all belong to $\bigcup_{j=1}^{n} G_j$.

Now we are going to show that $w_0 \bar{\epsilon} I_f(\partial D)$. In fact, if $w_0 \epsilon I_f(\partial D)$, then there is a sequence $z_n \epsilon D$ such that $\lim_{n\to\infty} z_n = \zeta \in \partial D$ and $\lim_{n\to\infty} f(z_n) = w_0$. Hence there is an integer $N > 0$ such that

$$|f(z_N) - w_0| < \rho, \; z_N \bar{\epsilon} \bigcup_{j=1}^{n} \bar{G}_j.$$

Consequently $f(z_N) \epsilon \Gamma$ and $f(z) - f(z_N)$ has a zero $z_N \bar{\epsilon} \bigcup_{j=1}^{n} G_j$. So we get a contradiction.

Proof of Theorem 2.

Assume that $f(z) = (P(z)/B(z)) + \lambda$. Then since all the zeros of $P(z)$ lie in D and $B(z)$ is bounded in D, it is clear that $\lambda \bar{\epsilon} I_f(\partial D)$, hence $f(z) \epsilon \mathscr{A}$ in D.

Conversely assume that $f(z) \epsilon \mathscr{A}$ in D. Then there is a complex number $\lambda \bar{\epsilon} I_f(\partial D)$. $f(z) - \lambda$ has at most a finite number of zeros C_j $(j = 1, 2, \ldots, k)$ in D. Let γ_j $(j = 1, 2, \ldots, k)$ be respectively the orders of multiplicity of these zeros. Then

$$f(z) = F(z) \prod_{j=1}^{k} (z - C_j)^{\gamma_j} + \lambda$$

where $F(z)$ is an analytic function vanishing nowhere in D and $0 \bar{\epsilon} I_f(\partial D)$. Hence there is $m > 0$ such that $|F(z)| > m$ in D. It is then sufficient to take $P(z) = \prod_{j=1}^{k} (z - C_j)^{\gamma_j}$, $B(z) = 1/F(z)$.

Proof of Theorem 3.

Assume that $f(z) - w$ has an infinite number of zeros in D, for each $w \in G$, where G is the complementary set of a set of first category with respect to the w-plane. Then $G \subset I_f(\partial D)$. Next since G is an everywhere dense set in the w-plane and $I_f(\partial D)$ is a closed set, hence $I_f(\partial D) = \mathbb{C} \cup \infty$ and $f(z) \bar{\in} \mathscr{A}$ in D.

Conversely, if $f(z) \bar{\in} \mathscr{A}$ in D, then for any integer $p \geqslant 0$, the set of points w such that the number of the zeros of $f(z) - w$ in D is greater than p, is everywhere dense in the w-plane, by Theorem 1, and it is an open set. Consequently the set E_p of the points w such that the number of the zeros in D of $f(z) - w$ does not exceed p, is nowhere dense in the w-plane. Let $E = \bigcup_{p=o}^{\infty} E_p$. Then E is a set of first category with respect to the w-plane. For $w \bar{\in} E$, the number of the zeros of $f(z) - w$ in D is infinite.

REFERENCES

1. Plessner, A., Uber das Verhalten analytischer Funktion am Rande ihren Definitionsbereich, Journal für die reine und angewante Mathematik 158 (1927).

2. Plivalov, I.I., The Boundary Properties of Analytic Functions, Moscow (1950) (Russian), 31-32.

GENERALIZATION OF THE DE BRANGES THEOREM AND COEFFICIENTS OF THE SYMMETRIC UNIVALENT FUNCTIONS

Wei-qi Yang and Bo-han Liu
Beijing Institute of Technology
Beijing, China

It is well known that the establishment of the de Branges theorem whereby the Bieberbach conjecture was verified is one of the most important mathematical results in recent years. In this paper we shall give a generalized form of the de Branges theorem and use it to study the coefficients of the power series

$$\left(\frac{f(z)}{z}\right)^{\lambda} = \sum_{n=0}^{\infty} D_n(\lambda)z^n, \ (\lambda > 0,\ f \in S).$$

The Grinspan conjecture is verified, that is

$$|D_n(\lambda)| \leqslant d_n(2\lambda), \quad (\lambda \geqslant 1,\ f \in S).$$

For $0 < \lambda < 1$ we give Robertson type estimates and asymptotic estimates. Furthermore, we get some results on the coefficients of p-symmetric univalent functions.

1. GENERALIZATION OF THE DE BRANGES THEOREM

Let S denote the class of functions

$$f(z) = z + \sum_{n=2}^{\infty} a_n z^n \quad (|z| < 1) \tag{1}$$

that are analytic and univalent in the unit disk D. Let

$$\log \frac{f(z)}{z} = 2 \sum_{n=1}^{\infty} \gamma_n z^n \tag{2}$$

where γ_n are called the logarithmic coefficients of f.

For $\lambda > 0$ let

$$\left(\frac{f(z)}{z}\right)^{\lambda} = \sum_{n=0}^{\infty} D_n(\lambda) z^n, \quad D_o(\lambda) = 1. \tag{3}$$

Taking $f(z) = \frac{z}{(1-z)^2}$, the Koebe function, we have

$$D_n(\lambda) = d_n(2\lambda)$$

here the functions $d_n(x)$ are defined by

$$d_o(x) = 1, \; d_n(x) = \frac{x(x+1)\ldots(x+n-1)}{n!}, \; n = 1, 2, \ldots . \tag{4}$$

In 1984, de Branges prove that[1]

de Branges theorem Let $f \in S$ then

$$\sum_{\nu=1}^{n} \sum_{k=1}^{\nu} (k|\gamma_k|^2 - \frac{1}{k}) \leqslant 0, \quad n = 1, 2, \ldots . \tag{5}$$

For each n, equality holds if and only if f is a rotation of the Koebe function.

A generalized form of this theorem can be given as the following.

Theorem 1. Let $f \in S$ and $\{\mu_n\}$ be an arbitrary increasing sequence of positive numbers, then

$$\sum_{\nu=1}^{n} \{\mu_{n-\nu} \sum_{k=1}^{\nu} (k|\gamma_k|^2 - \frac{1}{k})\} \leqslant 0, \quad n = 1, 2, \ldots . \tag{6}$$

For each n, equality holds if and only if f is a rotation of the Koebe function.

Proof. Introduce the notation

$$\sigma_\nu = \sum_{k=1}^{\nu} \left(k|\gamma_k|^2 - \frac{1}{k}\right), \tag{7}$$

the Abel transform gives

$$\sum_{\nu=1}^{n} \mu_{n-\nu}\sigma_\nu = \mu_o \sum_{\nu=1}^{n} \sigma_\nu + \sum_{\nu=1}^{n-1} \left\{(\mu_{n-\nu} - \mu_{n-\nu-1}) \sum_{k=1}^{\nu} \sigma_k\right\}.$$

Thus, (6) and the statement about equality follow from (5).

2. THE VERIFICATION OF THE GRINSPAN CONJECTURE

By virtue of Theorem 1 we verify the Grinspan conjecture[2].

Theorem 2. Let $f \in S$ then, for any $\lambda \geq 1$,

$$|D_n(\lambda)| \leq d_n(2\lambda), \quad n = 1, 2, \ldots \tag{8}$$

and equality occurs if and only if f is a rotation of the Koebe function.

In order to prove Theorem 2, we need two lemmas that are due to Milin [3]. Let

$$w(z) = \sum_{k=1}^{\infty} A_k z^k \tag{9}$$

analytic in D, and

$$\begin{aligned} \phi(z) &= \exp\{w(z)\} \\ &= \sum_{n=0}^{\infty} D_n z^n . \end{aligned} \tag{10}$$

For $x > 0$ we adopt the notation

$$\Delta_n(x) = \frac{1}{x^2} \sum_{k=1}^{n} k|A_k|^2 - \sum_{k=1}^{n} \frac{1}{k}, \quad n = 1, 2, \ldots, \tag{11}$$

$$\theta_n(x) = \frac{1}{d_n(x+1)} \sum_{j=0}^{n} \frac{|D_j|^2}{d_j(x)} \exp\{-\frac{x}{d_n(x+1)} \sum_{\nu=1}^{n} d_{n-\nu}(x)\Delta_\nu(x)\},$$

$$n = 1, 2, \ldots, \tag{12}$$

$$\theta_0(x) = 1.$$

Lemma 1. ([3], p. 33) $\{\theta_n(x)\}$ is a nonincreasing sequence of positive numbers, that is

$$1 \geqslant \theta_1(x) \geqslant \theta_2(x) \geqslant \ldots . \tag{13}$$

The equality $\theta_n(x) = 1$ holds for some $n \geqslant 1$ if and only if

$$A_k = \frac{k}{n} e^{ik\alpha}, \quad k = 1, 2, \ldots, n, \tag{14}$$

where α is a real constant.

Lemma 2. ([3], p. 37) For every $n \geqslant 1$ and any $x > 0$, the coefficients of the expansion (10) satisfy the inequality

$$|D_n|^2 \leqslant \theta_{n-1}(x) d_n^2(x) \exp\{\frac{x}{d_n(x)} \sum_{\nu=1}^{n} d_{n-\nu}(x-1)\Delta_\nu(x)\}. \tag{15}$$

Equality holds for the coefficients (14).

Proof of Theorem 2. Taking

$$w(z) = \log\left(\frac{f(z)}{z}\right)^\lambda, \quad x = 2\lambda,$$

we have

$$A_k = 2\lambda\gamma_k, \ D_k = D_k(\lambda), \ \Delta_k(2\lambda) = \sigma_k.$$

Thus, (15) becomes

$$|D_n(\lambda)|^2 \leqslant d_n^2(2\lambda) \exp\{\frac{2\lambda}{d_n(2\lambda)} \sum_{\nu=1}^{n} d_{n-\nu}(2\lambda - 1)\ \sigma_\nu\}. \tag{16}$$

Notice that $\{d_n(x)\}$ is an increasing sequence of positive numbers of any $x \geqslant 1$. Applying (6) with $\mu_n = d_n(2\lambda-1)$ to (16), the estimates (8) and the assertion about equality follow.

3. ROBERTSON TYPE ESTIMATES

Using the same method, we can obtain Robertson type estimates about $D_n(\lambda)$.

<u>Theorem 3.</u> Let $f \in S$ then

(i) For every $n \geqslant 1$ and any $\lambda \geqslant \frac{1}{2}$

$$\sum_{k=0}^{n} \frac{|D_k(\lambda)|^2}{d_k(2\lambda)} \leqslant d_n(2\lambda + 1). \tag{17}$$

Equality holds if and only if f is a rotation of the Koebe function;

(ii) For any $\lambda > 0$

$$\sum_{k=0}^{n} \frac{|D_k(\lambda)|^2}{d_k(2\lambda)} \leqslant d_n(2\lambda + 1)\ e^{2\lambda\delta} \tag{18}$$

where δ is the Milin constant ([3], p. 53).

<u>Proof.</u> Taking

$$w(z) = \log\left(\frac{f(z)}{z}\right)^\lambda, \ x = 2\lambda,$$

we obtain that

$$\sum_{k=0}^{n} \frac{|D_k(\lambda)|^2}{d_k(2\lambda)} \leqslant d_n(2\lambda + 1) \exp\{\frac{2\lambda}{d_n(2\lambda+1)} \sum_{\nu=1}^{n} d_{n-\nu}(2\lambda)\sigma_\nu\} \tag{19}$$

from (13) and (12).

If $\lambda \geqslant \frac{1}{2}$, then $\{d_n(2\lambda)\}$ is an increasing sequence of positive numbers. The statement (i) follows therefore by applying (6) with $\mu_n = d_n(2\lambda)$ to (19).

If $\lambda > 0$, we can infer (18) from (19) by virtue of the well-known Milin's result

$$\sigma_k < \delta, \ k = 1, 2, \ldots \tag{20}$$

and the identity

$$\sum_{k=0}^{n} d_k(x) = d_n(x + 1). \tag{21}$$

Note that Theorem 2 and Theorem 3 imply the Bieberbach conjecture and the Robertson conjecture respectively.

4. ESTIMATES ON SOME SUBCLASSES OF S

Let $L(x)$ denote the subclass of S with the adding condition

$$|\gamma_n| \leqslant \frac{x}{n}, \quad n = 1, 2, \ldots . \tag{22}$$

As usual, SP, $S^*(\alpha)$ and K denote the spiralike subclass, the α-order starlike subclass and the convex subclass of S respectively.

<u>Theorem 4</u>. If $f \in L(x)(0 < x \leqslant 1)$ then, for any $\lambda > 0$,

$$|D_n(\lambda)| \leqslant d_n(2x), \quad n = 1, 2, \ldots \tag{23}$$

and these inequalities are best possible.

Proof. We start with the identities

$$\sum_{n=0}^{\infty} D_n(\lambda) z^n = \left(\frac{f(z)}{z}\right)^{\lambda}$$

$$= \exp\left\{\lambda \log \frac{f(z)}{z}\right\}$$

$$= \exp\left\{2\lambda \sum_{m=1}^{\infty} \gamma_m z^m\right\}$$

$$= 1 + \sum_{k=1}^{\infty} \frac{1}{k!} \left(2\lambda \sum_{m=1}^{\infty} \gamma_m z^m\right)^k .$$

Comparing the coefficients of $z^n (n = 1, 2, \ldots)$ on both sides, we get

$$D_n(\lambda) = \sum_{k=1}^{n} \frac{(2\lambda)^k}{k!} \sum_{i_1+\ldots i_k=n} \gamma_{i_1} \cdots \gamma_{i_k} . \tag{24}$$

Whence, by taking $f(z) = \dfrac{z}{(1-z)^{2x}}$,

$$d_n(2x\lambda) = \sum_{k=1}^{n} \frac{(2x\lambda)^k}{k!} \sum_{i_1+\ldots+i_k=n} \frac{1}{i_1} \cdots \frac{1}{i_k} . \tag{25}$$

Hence we obtain (23) from (22), (24) and (25).

It is easy to see that $\dfrac{z}{(1-z)^{2x}} \in L(x)$, and then the estimate (23) is sharp. Furthermore, because of the de Branges theorem, the extreme functions for $L(1)$ are merely the Koebe function and its rotations.

Corollary. For any $\lambda > 0$, the following estimates are sharp

(i) If $f \in SP$ then $|D_n(\lambda)| \leqslant d_n(2\lambda)$;

(ii) If $f \in S^*(\alpha)$ $(0 \leqslant \alpha < 1)$ then $|D_n(\lambda)| \leqslant d_n(2(1-\alpha)\lambda)$;

(iii) If $f \in K$ then $|D_n(\lambda)| \leqslant d_n(\lambda)$.

Proof. Consider that

$$\frac{zf'(z)}{f(z)} = 1 + 2 \sum_{n=1}^{\infty} n \gamma_n z^n . \tag{26}$$

If $f \in SP$ then

$$\mathrm{Re}\{e^{-i\theta} \frac{zf'(z)}{f(z)}\} > 0$$

for some $\theta \in (-\frac{\pi}{2}, \frac{\pi}{2})$. Hence

$$\frac{1}{\cos\theta} e^{-i\theta} \frac{zf'(z)}{f(z)} + i\,\mathrm{tg}\theta \in \mathscr{P}$$

where $\mathscr{P}$ denotes the class of functions $p(z)$ analytic in D with

$$p(0) = 1, \ \mathrm{Re}\, p(z) > 0 \quad (|z| < 1).$$

By applying the well-known coefficient estimate on $\mathscr{P}$ (see [4], [5]), it follows that

$$\left|\frac{2e^{-i\theta}}{\cos\theta} n \gamma_n\right| \leqslant 2, \ n = 1, 2, \ldots .$$

Thus, $SP \subset L(1)$. Furthermore, the estimate (i) is sharp because the Koebe function belongs to SP.

If $f \in S^*(\alpha)$ $(0 \leqslant \alpha < 1)$ then

$$\frac{1}{1-\alpha} \frac{zf'(z)}{f(z)} - \frac{\alpha}{1-\alpha} \in \mathscr{P}.$$

Hence

$$\left|\frac{2}{1-\alpha}\, n\, \gamma_n\right| \leq 2, \quad n = 1, 2, \ldots ,$$

that is, $S^*(\alpha) \subset L(1-\alpha)$. Since

$$\frac{z}{(1-z)^{2(1-\alpha)}} \in S^*(\alpha),$$

the estimate (ii) is best possible.

Finally, the estimate (iii) is sharp because

$$\frac{z}{1-z} \in K \subset S^*(\tfrac{1}{2}) .$$

5. ASYMPTOTIC ESTIMATES

Let L denote the subclass of S with the condition

$$\gamma_n = O(\tfrac{1}{n}) \quad (n \to \infty) . \tag{27}$$

Let L_+ denote the subclass of L in which every function f satisfies the condition $\alpha > 0$ where

$$\alpha = \lim_{r \to 1} M(r,f)(1-r)^2$$

called the Hayman number of f.

Theorem 5. If $f \in L$ then, for any $\lambda > 0$,

$$D_n(\lambda) = O(d_n(2\lambda)) = O(n^{2\lambda-1}), \quad n \to \infty. \tag{28}$$

This estimate is best possible.

Proof. Differentiating the identity

$$\sum_{n=0}^{\infty} D_n(\lambda) z^n = \exp\{2\lambda \sum_{m=1}^{\infty} \gamma_m z^m\}$$

and then comparing coefficients we obtain

$$nD_n(\lambda) = 2\lambda \sum_{k=1}^{n} k\ \gamma_k D_{n-k}(\lambda) . \tag{29}$$

$f \in L$ implies

$$k|\gamma_k| \leqslant M, \quad k = 1, 2, \ldots$$

where $M = M(f)$ is a constant only depending upon f.

Since

$$d_n(x) \sim \frac{n^{x-1}}{\Gamma(x)} \quad (n \to \infty) \tag{30}$$

([3], p. 43), hence

$$d_{n-1}(2\lambda+1) \leqslant Kn^{2\lambda}$$

for some constant $K = K(\lambda)$.

Thus, applying Schwarz inequality from (18) to (29) we have

$$\begin{aligned} |D_n(\lambda)| &\leqslant \frac{2\lambda}{n} \Big(\sum_{k=1}^{n} k^2 |\gamma_k|^2 d_{n-k}(2\lambda) \Big)^{\frac{1}{2}} \Big(\sum_{k=0}^{n-1} \frac{|D_k(\lambda)|^2}{d_k(2\lambda)} \Big)^{\frac{1}{2}} \\ &\leqslant \frac{2\lambda}{n} Me^{\lambda\delta}\, d_{n-1}(2\lambda+1) \\ &\leqslant 2\lambda MKe^{\lambda\delta}\, n^{2\lambda-1} \end{aligned} \tag{31}$$

and then (28) follows. The estimate is best possible because the Koebe function belongs to L.

In order to obtain other asymptotic estimate on L_+, we need several lemmas.

Let $w(z)$ and $\phi(z)$ be given by (9) and (10). For $h > 0$, let

$$\phi(z)(1-z)^{-h} = \sum_{n=0}^{\infty} S_n^{(h)} z^n . \tag{32}$$

We shall mention the following conditions on the coefficients of $w(z)$:

(I) $\displaystyle\sum_{k=1}^{\infty} k|A_k|^2 < \infty$;

(II) $\displaystyle\text{Re}\,\{\sum_{k=1}^{n} A_k\} = O(1), \quad n \to \infty;$

(III) $A_k = O(\frac{1}{n})$, $n \to \infty$.

<u>Lemma 3</u>. ([3], p. 44) If the condition (I) holds, then

$$\sum_{k=1}^{n} A_k - w(r) = o(1), \; n \to \infty \tag{33}$$

where r and n satisfy the relation

$$1 - r = \frac{\mu}{n} , \quad m_1 < \mu < m_2 \tag{34}$$

(m_1 and m_2 are positive constants).

<u>Lemma 4</u>. ([3], p. 45) If the conditions (I) and (II) hold then

$$\frac{1}{n} \sum_{k=1}^{n} kD_k = o(1), \; n \to \infty . \tag{35}$$

<u>Lemma 5</u>. [6] If the conditions (I), (II) and (III) hold then, for any $h > 0$,

$$\frac{S_n^{(h)}}{d_n^{(h)}} \sim \sum_{k=0}^{n} D_k , \; n \to \infty . \tag{36}$$

Lemma 6. If the conditions (I), (II) and (III) hold then, for any $h > 0$

$$\frac{S_n^{(h)}}{d_n^{(h)}} \sim \phi(r), \; n \to \infty . \tag{37}$$

where r and n satisfy the relation (34).

Proof. Considering the equality

$$\sum_{k=0}^{n} D_k - \phi(r) = \sum_{k=1}^{n} D_k(1 - r^k) - \sum_{k=n+1}^{\infty} D_k r^k ,$$

we estimate the two terms on the right side.

Since the sequence $\frac{1}{k}(1-r^k)$ decreases monotonically with increasing k, applying the Abel inequality and (35) we deduce

$$\begin{aligned} \left| \sum_{k=1}^{n} D_k(1 - r^k) \right| &= \left| \sum_{k=1}^{n} \frac{1 - r^k}{k} kD_k \right| \\ &\leq (1-r) \max_{1 \leq \nu \leq n} \left| \sum_{k=1}^{\nu} kD_k \right| \\ &= \frac{\mu}{n} \max_{1 \leq \nu \leq n} \left| \sum_{k=1}^{\nu} kD_k \right| \\ &= o(1) \quad (n \to \infty) . \end{aligned}$$

Next, let $N > n$ then

$$\left|\sum_{k=n+1}^{N} D_k r^k\right| = \left|\sum_{k=n+1}^{N} \frac{r^k}{k} kD_k\right|$$

$$= \left|\frac{r^N}{N} \sum_{k=n+1}^{N} kD_k + \sum_{k=n+1}^{N-1} \frac{(k+1)r^k - kr^{k+1}}{k+1} \frac{1}{k} \sum_{\nu=1}^{k} \nu D_\nu\right|$$

$$\leqslant \left(r^{n+1} + \sum_{k=n+2}^{N} \frac{r^k}{k}\right) \max_{n+1 \leqslant k \leqslant N} \left|\frac{1}{k} \sum_{\nu=n+1}^{k} \nu D_\nu\right|$$

$$\leqslant \left(1 + \log \frac{1}{1-r} - \sum_{k=1}^{n} \frac{r^k}{k}\right) 2 \sup_{k \geqslant n} \left|\frac{1}{k} \sum_{\nu=1}^{k} \nu D_\nu\right|$$

$$= o(1) \quad (n \to \infty)$$

where we have applied (35) and the fact

$$0 < 1 + \log \frac{1}{1-r} + \sum_{k=1}^{n} \frac{1-r^k}{k} - \sum_{k=1}^{n} \frac{1}{k} < \log \frac{e}{1-r}$$

$$+ n(1-r) - \log n < \infty .$$

Hence

$$\sum_{k=0}^{n} D_k - \phi(r) = o(1), \quad n \to \infty .$$

Furthermore, the condition (II) implies that $|\phi(r)|$ has a positive lower bound as $r \to 1$, and then

$$\sum_{k=1}^{n} D_k \sim \phi(r) \quad (n \to \infty).$$

(37) follows therefore from (36).

Now the estimate about L_+ can be made.

<u>Theorem 6</u>. If $f \in L_+$ then, for any $\lambda > 0$,

$$\frac{D_n(\lambda)}{d_n(2\lambda)} \sim \alpha^\lambda \exp\{i(\lambda \arg f(re^{i\theta_o}) + (n-\lambda)\theta_o)\},\ n \to \infty \tag{38}$$

where α is the Hayman number of f, and $e^{i\theta_o}$ is the direction such as

$$\lim_{r\to 1} |f(re^{i\theta_o})|(1-r)^2 = \alpha,$$

while r and n satisfy the relation (34).

<u>Proof</u>. Let

$$w(z) = \log\left(\frac{f(z)}{z}(1-z)^2\right)^\lambda$$

then

$$A_k = 2\lambda(\gamma_k - \frac{1}{k}),\ D_n(\lambda) = S_n^{(2\lambda)}.$$

We may assume that $\theta_o = 0$, since otherwise we need only consider the function $e^{-i\theta_o}f(e^{i\theta_o}z)$ instead of f, while we obtain

$$\phi(r) = \left|\frac{f(r)}{r}(1-r)^2\right|^\lambda \exp\{i\lambda \arg f(r)\}$$

$$\sim \alpha^\lambda \exp\{i\lambda \arg f(r)\} \quad (n \to \infty) .$$

According to Lemma 6, it is enough to prove that the coefficients of $w(z)$ satisfy the conditions (I), (II) and (III).

At first, applying the Bazilevic theorem [7] we deduce

$$\sum_{k=1}^{\infty} k|A_k|^2 = 4\lambda^2 \sum_{k=1}^{\infty} k|\gamma_k - \frac{1}{k}|^2$$

$$\leqslant 2\lambda^2 \log\frac{1}{\alpha} < \infty.$$

Next, due to Lemma 3 we have

$$\operatorname{Re}\{\sum_{k=1}^{n} A_k\} - \lambda \log \frac{|f(r)|}{r}(1-r)^2 = o(1), \quad n \to \infty.$$

Hence

$$\lim_{n\to\infty} \operatorname{Re}\{\sum_{k=1}^{n} A_k\} = \lambda \log \alpha,$$

and so the condition (II) holds.

Finally, $f \in L$ implies the condition (III).

6. ESTIMATES OF THE COEFFICIENTS OF SYMMETRIC UNIVALENT FUNCTIONS

Let S_p denote the subclass of S consisting of all the p-symmetric functions. If $f \in S$ then

$$f_p(z) = \sqrt[p]{f(z^p)} = z + \sum_{n=1}^{\infty} a_n^{(p)} z^{np+1} \in S_p . \tag{39}$$

This equation gives a one-to-one correspondence between S and S_p, while

$$a_n^{(p)} = D_n(\tfrac{1}{p}) . \tag{40}$$

Note that

$f_p \in L$ if and only if $f \in L$;

$f_p \in L(x)$ if and only if $f \in L(x)$;

$f_p \in SP$ if and only if $f \in SP$;

$f_p \in S^*(\alpha)$ if and only if $f \in S^*(\alpha)$.

Let L_p, $L_p(x)$, SP_p and $S_p^*(\alpha)$ be the p-symmetric subclasses of L, $L(x)$, SP and $S^*(\alpha)$ respectively. We have the following results.

Theorem 7. Let

$$\phi(z) = z + \sum_{n=1}^{\infty} a_n^{(p)} z^{np+1} \in S_p$$

then the following estimates are sharp:

(i) if $\phi \in L_p(x)$ then $|a_n^{(p)}| \leq d_n(\frac{2x}{p})$, $n = 1,2,\ldots$;

(ii) if $\phi \in SP_p$ then $|a_n^{(p)}| \leq d_n(\frac{2}{p})$, $n = 1,2,\ldots$;

(iii) if $\phi \in S_p^*(\alpha)$ $(0 \leq \alpha < 1)$ then

$$|a_n^{(p)}| \leq d_n(\tfrac{2}{p}(1-\alpha)),\ n = 1,2,\ldots;$$

(iv) if $f \in K$, $\phi = f_p$ defined by (39), then

$$|a_n^{(p)}| \leq d_n(\tfrac{1}{p}),\ n = 1,2,\ldots .$$

Theorem 8. Let

$$\phi(z) = z + \sum_{n=1}^{\infty} a_n^{(p)} z^{np+1} \in S_p$$

then the following asymptotic estimates are best possible:

(i) If $\phi \in L_p$ then

$$a_n^{(p)} = O(d_n(\tfrac{2}{p})) = O(n^{\frac{2}{p}-1}),\ n \to \infty;$$

(ii) If $f \in L_+$, $\phi = f_p$ defined by (39), then

$$|a_n^{(p)}| \sim \alpha^{\frac{1}{p}} d_n(\tfrac{2}{p}),\ n \to \infty$$

where α is the Hayman number of f.

REFERENCES

1. C.H. FitzGerald and Ch. Pommerenke: The de Branges theorem on univalent functions, Trans. Amer. Math. Soc. 290 (1985) no. 2, 683-690.

2. A.Z. Grinspan: Coefficients of powers of univalent functions, Siberian Math. J. 22 (1981) no. 4, 551-554.

3. I.M. Milin: Univalent Functions and Orthonormal Systems, English Transl. Amer. Math. Soc. Providence, R.I. 1977.

4. Ch. Pommerenke: Univalent Functions, Vandenhoeck und Ruprecht Grottingen 1975.

5. P.L. Duren: Univalent Functions, Springer-Verlag, Heidelberg and New York 1983.

6. Hu Ke: The asymptotic behaviour of analytic functions, Chinese annals of Mathematics B, 1983, no. 2, 187-190.

7. I.E. Bazilevic: On a univalence criterion for regular functions and the dispersion of their coefficients, Mat. Sb. 74 (116) (1967), 133-146.

SUPPORT POINTS OF THE CLASS $S_R^*(\alpha,k)$ OF STARLIKE FUNCTIONS

Zhang Yuling and Ma Jinxi
Department of Mathematics
Northwest University
Xian, Shaanxi
The People's Republic of China

ABSTRACT

We determine the set of support points for the class of functions which are starlike of order α, k-fold symmetric, and real on (-1,1).

Let $D = \{z: |z|<1\}$ and A denote the set of all functions analytic in D. Then A is a locally convex linear topology space with respect to the topology given by uniform convergence on compact subsets of D. Suppose that $\mathscr{F}$ is a compact subset of A. A function f is called a support point of $\mathscr{F}$ if $f \in \mathscr{F}$ and there is a continuous linear functional L on A so that

$$\operatorname{Re}L(f) = \operatorname{Max}\{\operatorname{Re}L(g): g \in \mathscr{F}\},$$

and $\operatorname{Re}L$ is nonconstant on $\mathscr{F}$. We shall denote the set of support points of $\mathscr{F}$ by $\operatorname{supp}\mathscr{F}$. A function f is called an extreme point of $\mathscr{F}$ if $f \in \mathscr{F}$ and if $f=tg+(1-t)h$ where $g, h \in \mathscr{F}$ and $0<t<1$ implies that $g=h$. We shall use the notation $E\mathscr{F}$ to denote the set of extreme points of $\mathscr{F}$. And we shall denote the closed convex hull of $\mathscr{F}$ by $H\mathscr{F}$.

Let $0<\alpha<1$ and k be a positive integer. We denote $S_R^*(\alpha,k)$ the class of starlike functions of order α which are real on $(-1,1)$ and have power series developments which are k-fold symmetric. It is known that $S_R^*(\alpha,k)$ is compact. In [1] Hallenbeck discussed the closed convex hull of $S_R^*(\alpha,k)$ and extreme points of the closed convex hull of $S_R^*(\alpha,k)$.

Let $X = \{x: |x|=1, I_m x \geq 0\}$ and P be the set of probability measures on X.

Let

$$k(x,z) = \frac{z}{(1-xz^k)^{\frac{1-\alpha}{k}}(1-\bar{x}z^k)^{\frac{1-\alpha}{k}}}$$

$$B_o = \{k(x,z): x \in X\}.$$

Hallenbeck proved[1]

$$HS_R^*(\alpha,k) = \{\int_X k(x,z)d\mu(x): \mu \in P\},$$

$$EHS_R^*(\alpha,k) = B_o.$$

We prove

$$\text{supp}S_R^*(\alpha,k) = EHS_R^*(\alpha,k) = B_o.$$

<u>Lemma</u>. Let $m \geq 2$. Suppose that $\lambda_j \neq 0$ $(j=1,2,\ldots m)$, $|x_j|=1$, and x_j's are distinct. Then $\sum_{j=1}^{m} \lambda_j k(x_j,z) \notin S_R^*(\alpha,k)$.

Proof. Suppose that $f(z) = \sum_{j=1}^{m} \lambda_j k(x_j,z) \in S_R^*(\alpha,k)$. Then

$$g(z) = z\left(\frac{f(z^{\frac{1}{k}})}{z}\right)^{k\frac{1}{1-\alpha}} \in S^*,$$

where S^* is the usual class of starlike functions. But

we have

$$g(z) = z\left(\sum_{j=1}^{m} \frac{\lambda_j}{(1-x_j z)^{\frac{1-\alpha}{k}}(1-\bar{x}_j z)^{\frac{1-\alpha}{k}}}\right)^{\frac{k}{1-\alpha}}$$

$$= \frac{z}{\prod_{j=1}^{m}(1-x_j z)(1-\bar{x}_j z)}\left(\sum_{j=1}^{m} \lambda_j \prod_{\ell\neq j}[(1-x_\ell z)(1-\bar{x}_\ell z)]^{\frac{1-\alpha}{k}}\right)^{\frac{k}{1-\alpha}}.$$

This indicates that g has poles on the unit circle with a total order of more than 2 and such a function is not univalent in D. This implies that $f\notin S_R^*(\alpha,k)$.

Theorem. $\operatorname{supp} S_R^*(\alpha,k) = EHS_R^*(\alpha,k) = B_0$.

Proof. Let $x_0 = e^{i\alpha_0}$, $x = e^{i\alpha}$ $(\alpha_0, \alpha \in \mathbb{R})$ and $\beta = \frac{1-\alpha}{k}$. Then

$$k(x,z) = z+2\beta\cos\alpha z^{k+1}+(2\beta(\beta+1)\cos^2\alpha-\beta)z^{2k+1}+\ldots$$

$$k(x_0,z) = z+2\beta\cos\alpha_0 z^{k+1}+(2\beta(\beta+1)\cos^2\alpha_0-\beta)z^{2k+1}+\ldots$$

Letting $a=\frac{1}{\beta}\cos\alpha_0$ and $b=-\frac{1}{2\beta(\beta+1)}$, we define a continuous linear functional L for a fixed x_0. $L(g)=aa_{k+1}+ba_{2k+1}$, where $g = \sum_{n=0}^{\infty} a_n z^n \in A$. By simple computation, we get

$$L(k(x,z)) = -\cos^2\alpha + 2\cos\alpha\cos\alpha_0 + \frac{1}{2(\beta+1)},$$

$$L(k(x_0,z)) = \cos^2\alpha_0 + \frac{1}{2(\beta+1)}.$$

It follows that $L(k(x,z)) \leq L(k(x_0,z))$ and equality holds if and only if $x=x_0$. This shows that $k(x_0,z)$ is the unique solution to an extremal problem $\max\{\operatorname{Re}L(f): f\in EHS_R^*(\alpha,k)\}$. Hence $k(x_0,z)\in \operatorname{supp} S_R^*(\alpha,k)$. This proves that $EHS_R^*(\alpha,k)\subset \operatorname{supp} S_R^*(\alpha,k)$. In the following, we shall prove that $\operatorname{supp} S_R^*(\alpha,k) \subset EHS_R^*(\alpha,k)$.

Let $f_o \in \operatorname{supp} S_R^*(\alpha,k)$ and J be the associated continuous linear functional. Then there exists [2, Th.9.1] a finite complex Borel measure μ with compact support E contained in D such that

$$J(g) = \int_E g(z)\,d\mu(z), \qquad g \in A.$$

Suppose that $|x|=1$ and $F(x)=J(k(x,z))$. Then

$$F(x) = \int_E \frac{z}{(1-xz^k)^{\frac{1-\alpha}{k}}\,(1-\bar{x}z^k)^{\frac{1-\alpha}{k}}}\,d\mu(z).$$

Let $G(x) = \frac{1}{2}\,(F(x)+\overline{F(\bar{x}^{-1})})$. Then both F and G are analytic on ∂D and $G(x) = \operatorname{Re}F(x)$ whenever $|x|=1$.

Let $\operatorname{Re}J(f_o)=M$. We claim that there are only a finite number of solutions to $\operatorname{Re}F(x)=M$ with $|x|=1$. Otherwise, since $G(x)=\operatorname{Re}F(x)$ for $|x|=1$, the identity theorem implies that $G(x)\equiv M$ and so $\operatorname{Re}F(x)\equiv M$. Since $EHS_R^*(\alpha,k)=\{k(x,z)\colon |x|=1\}$, it follows that $\operatorname{Re}J(f)=M$ for all f in $HS_R^*(\alpha,k)$, and, in particular $\operatorname{Re}J$ is constant on $S_R^*(\alpha,k)$. This contradicts the hypothesis that $f_o \in \operatorname{supp} S_R^*(\alpha,k)$. Therefore, the set $\{g\colon g\in EHS_R^*(\alpha,k),\ \operatorname{Re}J(g)=M\}$ is finite.

Let $B=\{f\colon f\in HS_R^*(\alpha,k),\ \operatorname{Re}J(f)=M\}$. It follows easily that B is non-empty, compact, convex, and that B is an extremal subset of $HS_R^*(\alpha,k)$. Hence $EB\subset EHS_R^*(\alpha,k)$. Because of what was proved above, there can only be a finite number of functions in EB, say $k(x_1,z)$, $k(x_2,z)\ldots k(x_m,z)$. From Choguet theorem [3, Th.A.4], it follows that

$$B = \{\sum_{j=1}^{m} \lambda_j k(x_j,z)\colon \lambda_j \geq 0,\ \sum_{j=1}^{m} \lambda_j = 1\}.$$

Since $f_o \in B$, we obtain that $f_o(z)=k(x_\ell,z)$ from the above lemma. This proves the theorem.

REFERENCES

[1]. D.J. Hallenbeck, Convex hulls and extreme points of families of starlike and close-to-convex mappings, Pacific J. Math. 57(1975), 167-176.

[2]. P.L. Duren, Univalent Functions, Springer-Verlag, Heidelbergand New York, 1983.

[3]. G. Schober, Univalent functions - Selected topics, Springer-Verlag, Berlin, 1975.

ON THE KIRWAN CONJECTURE FOR TYPICALLY REAL MEROMORPHIC FUNCTIONS

Ma Wancang
Department of Mathematics
Northwest University, Xian
The People's Republic of China

ABSTRACT

Sharp estimates are given for coefficient functionals tb_1-b_n in the class $T\Sigma$. As a consequence, the Kirwan conjecture is proved for functions in Σ with real coefficients.

Let Σ be the family of univalent functions

$$f(z) = z+\sum_{n=o}^{\infty} b_n z^{-n} \tag{1}$$

in $E = \{z;\ |z|>1\}$.

It was conjectured by Kirwan and Schober [1] that

$$\mathrm{Re}(nb_1-b_n) \leqslant n \quad (n\geqslant 2) \tag{2}$$

holds for all functions in the class Σ. Equality always occurs for the function $k(z) = z+1/z$.

In general, the problem is to find the smallest values for t such that

$$\mathrm{Re}(tb_1-b_n) \leqslant t \tag{3}$$

is true in the class Σ. If (3) holds for some t, so does it for all $t' \geq t$ since

$$\mathrm{Re}(t'b_1 - b_n) = \mathrm{Re}(tb_1 - b_n) + (t'-t)\mathrm{Re}\, b_1 \leq t + (t'-t) = t'.$$

For n=2,3, Garabedian and Schiffer [2] proved that (3) is true for $t \geq 2$ and $t \geq 3$ respectively. And t=n is the least value for which the inequality (3) holds when $n=2$ or $n=3$. In addition, Leung and Schober [3] verified that there is an explicit finite number t such that $\mathrm{Re}(tb_1 - b_n) \leq t$.

Let $T\Sigma$ be the class of typically real meromorphic functions $f(z) = z + \sum_{n=o}^{\infty} b_n z^{-n}$ in E, that is, it consists of functions $f(z)$ satisfying the condition

$$(\mathrm{Im} f(z))(\mathrm{Im} z) \geq 0 \tag{4}$$

for all z in E, except for $z=\infty$. It follows that all coefficients of such functions are real. The condition (4) is often written in the equivalent form

$$\mathrm{Im} f(z) = 0 \Leftrightarrow \mathrm{Im} z = 0.$$

Let $T\Sigma'$ denote the class of nonvanishing functions in $T\Sigma$. The subfamily $\Sigma_{\mathbb{R}}$ of Σ consists of those functions $f(z)$ with real coefficients. It is easily seen that $\Sigma_{\mathbb{R}} \subset T\Sigma$ since for functions in $\Sigma_{\mathbb{R}}$ $\mathrm{Im} f(z) = 0$ implies $f(z) = \overline{f(z)} = f(\bar{z})$ and therefore $z = \bar{z}$, that is, $\mathrm{Im} z = 0$.

In this note, we use a representation of functions in $T\Sigma'$ to get sharp upper and lower bounds of $tb_1 - b_n$ for the class $T\Sigma$. As a consequence, we prove the Kirwan conjecture for the family $\Sigma_{\mathbb{R}}$.

Theorem. Let $f(z) = z+\sum_{n=0}^{\infty} b_n z^{-n}$ belong to $T\Sigma$, $n\geq 2$ and $-\infty<t<+\infty$, then we have sharp estimates

$$t+4m(n,t) \leq tb_1-b_n \leq t+4M(n,t), \tag{5}$$

where

$$m(n,t) = \min_{0\leq x\leq \pi} (\sin x \sin nx - t\sin^2 x)$$

and

$$M(n,t) = \max_{0\leq x\leq \pi} (\sin x \sin nx - t\sin^2 x).$$

In particular, we get for $t\geq n$

$$tb_1-b_n \leq t.$$

In some special cases, we can obtain explicit results. For example, $M(4k+1,t) = 1-t$ for $t\leq 0$ and $m(4k-1,t) = -1-t$ for $t\geq 0$, where $k = 1,2,\ldots$.

Proof. At first, we show that $f(z)\in T\Sigma$ omits some real value c'. On the contrary, there is a complex number c ($\mathrm{Im}\,c\neq 0$), $f(z)\neq c$. This implies that $f(z)\neq \bar{c}$, since $f(E)$ is symmetric with respect to the real axis. It follows that $f(z)$ omits some real c' because $\bar{\mathbb{C}}\backslash f(E)$ is connected. This is contrary to our assumption. The function $f(z)-c'$ belongs to $T\Sigma'$. Thus we only need prove the results in $T\Sigma'$.

We know that if $f(z) \in T\Sigma'$, then [4]

$$f(z) = z(1-1/z^2)^2 \int_0^{\pi}(1-2\cos x/z+1/z^2)^{-1}d\mu(x),$$

where $\mu(x)$ is a probability measure on $[0,\pi]$. Thus

$$f(z) = z+2\int_0^{\pi}\cos x\,d\mu(x)+z^{-1}\left(2\int_0^{\pi}\cos 2x\,d\mu(x)-1\right)$$

$$+ 2\sum_{n=2}^{\infty} z^{-n}\int_0^{\pi}(\cos(n+1)x-\cos(n-1)x)\,d\mu(x),$$

$$tb_1-b_n = t\left(2\int_0^{\pi}\cos 2x\,d\mu(x)-1\right)$$

$$- 2\int_0^{\pi}(\cos(n+1)x-\cos(n-1)x)\,d\mu(x)$$

$$= t+4\int_0^{\pi}(\sin x\sin nx-t\sin^2 x)\,d\mu(x).$$

Therefore,

$$t+4m(n,t) \leqslant tb_1-b_n \leqslant t+4M(n,t).$$

The two equalities occur if and only if μ has its support on the subsets

$$\{x;\ \sin x\sin nx-t\sin^2 x = m(n,t),\quad 0\leqslant x\leqslant\pi\}$$

and

$$\{x;\ \sin x\sin nx-t\sin^2 x = M(n,t),\ 0\leqslant x\leqslant\pi\}$$

respectively.

For $t\geqslant n$, we get $M(n,t)=0$ by using $\sin nx/\sin x\leqslant n$. The proof is completed.

Let $\Sigma'_{\mathbb{R}}$ be the class of nonvanishing functions in $\Sigma_{\mathbb{R}}$. The closed convex hull $\overline{co}\Sigma'_{\mathbb{R}}$ of $\Sigma'_{\mathbb{R}}$ is contained in $T\Sigma'$ since $T\Sigma'$ is a closed convex set. So we have the following corollary.

Corollary 1. If $f(z) = z+\sum_{n=0}^{\infty} b_n z^{-n} \in \overline{co}\ \Sigma'_{\mathbb{R}}$, then the estimates (5) hold.

Corollary 2. If $f(z) = z+\sum_{n=o}^{\infty} b_n z^{-n} \in \Sigma_{\mathbb{R}}$, we have sharp inequality

$$tb_1 - b_n \leqslant t \qquad (t \geqslant n).$$

This corollary follows from the Theorem and the fact that $\Sigma_{\mathbb{R}} \subset T\Sigma$.

REFERENCES

1. Kirwan, W.E. and Schober, G.: New inequalities from old ones, Math. Z., 180(1982), 19-40.

2. Garabedian, P.R. and Schiffer, M.: A coefficient inequality for schlicht functions, Ann. of Math., 61(1955), 116-136.

3. Leung, Y.J. and Schober, G.: High order coefficient estimates in the class Σ, Proc. Amer. Math. Soc., 94(1985), 659-664.

4. Goodman, A.W.: Univalent Functions, Volume I. Mariner Publishing Co.: Tampa, Florida, 1983.

CONTRIBUTORS

1. Hai-long Ao — Department of Mathematics, Peking University, Beijing, China

2. I. Noel Baker — Department of Mathematics, Imperial College, London, England

3. Chi-tai Chuang — Department of Mathematics, Peking University, Beijing, China

4. Chong-ji Dai — Department of Mathematics, East China Normal University, Shanghai, China

5. Carl H. FitzGerald — Department of Mathematics, University of California, San Diego, USA

6. David J. Hallenbeck — Department of Mathematical Sciences, University of Delaware, Newark, USA

7. Cheng-qi He — Department of Mathematics, Fudan University, Shanghai, China

8. Yu-zan He — Institute of Mathematics, Chinese Academy of Sciences, Beijing, China

9. Ke Hu — Department of Mathematics, Jiangxi Normal University, Nanchang, China

10. Lu Jin — Department of Mathematics, East China Normal University, Shanghai, China

11. Irwin Kra — Department of Mathematics, State University of New York at Stony Brook, Stony Brook, USA

12. Bo-han Liu — Department of Mathematics, Beijing Institute of Technology, Beijing, China

13. Shu-qin Liu — Department of Mathematics, Northwestern University, Xian, China

14. Jian-ke Lu — Department of Mathematics, Wuhan University, Wuchang, China

15. Jin-xi Ma — Department of Mathematics, Northwestern University, Xian, China

16. Li-zhi Ma — Department of Mathematics, Beijing Institute of Technology, Beijing, China

17. Wan-cang Ma — Department of Mathematics, Northwestern University, Xian, China

18. Ye Mo — Department of Mathematics, Shantung University, Tsinan, China

19. Carl David Minda — Department of Mathematics, University of Cincinnati, Cincinnati, USA

20. Kiyoshi Niino — Faculty of Technology, Kanazawa University, Kanazawa, Japan

21. Cai-heng Ouyang — Institute of Mathematical Sciences, Chinese Academy, Wuhan, China

22. De-chang Pu — Department of Mathematics, Yantai University, Yantai, China

23. Fu-yao Ren — Department of Mathematics, Fudan University, Shanghai, China

24. Guo-dong Song — Department of Mathematics, East China Normal University, Shanghai, China

25. Guo-chun Wen — Department of Mathematics, Peking University (and Yantai University), Beijing (and Yantai), China

26. Chung-chun Yang — Naval Research Laboratory, Washington, D.C., USA

27. Jia-yong Yu — Department of Mathematics, Wuhan University, Wuchang, China

28. Wei-qi Yang — Department of Mathematics, Beijing Institute of Technology, Beijing, China

29. Yu-lin Zhang — Department of Mathematics, Northwestern University, Xian, China

30. Xiu-lin Zou — Department of Mathematics, Beijing Institute of Technology, Beijing, China